Water Science, Policy, and Management

Water Science, Policy, and Management

A Global Challenge

Edited by
Simon J. Dadson, Dustin E. Garrick, Edmund C. Penning-Rowsell,
Jim W. Hall, Rob Hope, and Jocelyne Hughes

WILEY Blackwell

The right of Simon J. Dadson, Dustin E. Garrick, Edmund C. Penning-Rowsell, Jim W. Hall, Rob Hope, and Jocelyne Hughes to be identified as the authors of the editorial material in this work has been asserted in accordance with law.

Registered Office(s)
John Wiley & Sons, Inc., 111 River Street, Hoboken, NJ 07030, USA
John Wiley & Sons Ltd, The Atrium, Southern Gate, Chichester, West Sussex, PO19 8SQ, UK

Editorial Office
9600 Garsington Road, Oxford, OX4 2DQ, UK

For details of our global editorial offices, customer services, and more information about Wiley products visit us at www.wiley.com.

Wiley also publishes its books in a variety of electronic formats and by print-on-demand. Some content that appears in standard print versions of this book may not be available in other formats.

Library of Congress Cataloging-in-Publication Data

Name: Dadson, Simon J., editor.
Title: Water science, policy, and management : a global challenge / edited by
    Simon J. Dadson, University of Oxford [and 5 others].
Description: First edition. | Hoboken, N.J. : John Wiley & Sons, Inc., 2019. |
    Includes bibliographical references and index.
Identifiers: LCCN 2019027165 (print) | LCCN 2019027166 (ebook) | ISBN 9781119520603 (cloth) |
    ISBN 9781119520597 (adobe pdf) | ISBN 9781119520658 (epub)
Subjects: LCSH: Water resources development. | Water-supply–Government policy. |
    Watershed management.
Classification: LCC TC405 .W374 2019 (print) | LCC TC405 (ebook) |
    DDC 628.1–dc23
LC record available at https://lccn.loc.gov/2019027165
LC ebook record available at https://lccn.loc.gov/2019027166

Cover Design: Wiley
Cover Image: © Oleh_Slobodeniuk/Getty Images

Set in 10/12pt Warnock by SPi Global, Pondicherry, India
Printed and bound in Singapore by Markono Print Media Pte Ltd

10 9 8 7 6 5 4 3 2 1

This book is dedicated to the memory of Mike Edmunds whose vision and leadership, with others, created the Water Science, Policy, and Management MSc programme at the University of Oxford.

Contents

List of Contributors

Heather Bond
Flood and Coastal Risk Management
Directorate
Environment Agency
Wallingford, UK

Edoardo Borgomeo
International Water Management
Institute
Colombo, Sri Lanka

David Bradley
School of Geography and the
Environment
University of Oxford
Oxford, UK
and
London School of Hygiene
and Tropical Medicine
London, UK
and
Department of Zoology
University of Oxford
UK

Katrina J. Charles
School of Geography and the
Environment
University of Oxford
Oxford, UK

Alice Chautard
Smith School of Enterprise and the
Environment
University of Oxford
Oxford, UK

Simon J. Dadson
School of Geography and the
Environment
University of Oxford
Oxford, UK

Michaela Dolk
Swiss Re
New York, NY, USA

Nassim El Achi
Global Health Institute
American University of Beirut
Lebanon

Tim Foster
University of Technology Sydney
Sydney, NSW, Australia

Stephen Fragaszy
Water Directorate
New Zealand Ministry for the
Environment
Wellington, New Zealand

National Drought Mitigation Center
School of Natural Resources
University of Nebraska
Lincoln, NE, USA

Dustin E. Garrick
Smith School of Enterprise and the
Environment
University of Oxford
Oxford, UK

David Grey
School of Geography and the
Environment
University of Oxford
Oxford, UK-

Jim W. Hall
Environmental Change Institute
University of Oxford
Oxford, UK

Feyera Hirpa
School of Geography and the
Environment
University of Oxford
Oxford, UK

Rob Hope
School of Geography and the
Environment
and
Smith School of Enterprise and the
Environment
University of Oxford
Oxford, UK

Jocelyne Hughes
School of Geography and the
Environment
University of Oxford
Oxford, UK

Erin Hylton
Concurrent Technologies
Corporation
Washington, DC, USA

David W.M. Johnstone
School of Geography and the
Environment
University of Oxford
Oxford, UK

Clarke Knight
University of California at Berkeley
Berkeley, CA, USA

Johanna Koehler
Smith School of Enterprise and the
Environment
University of Oxford
Oxford, UK

Megan Konar
University of Illinois at
Urbana-Champaign
Urbana-Champaign, IL, USA

Michelle Lanzoni
School of Geography and the
Environment
University of Oxford
Oxford, UK

Rhett Larson
Sandra Day O'Connor College of Law,
Arizona State University,
Arizona, USA

Hannah Leckie
Division of Climate, Biodiversity
and Water
Organisation for Economic Co-operation
and Development
Paris, France

Kelsey Leonard
Department of Political Science
McMaster University
Hamilton, ON, Canada

Homero Paltan Lopez
School of Geography and the
Environment
University of Oxford
Oxford, UK

Rachael McDonnell
Water, Climate Change and Resilience
Strategic Program
International Water Management
Institute
Colombo, Sri Lanka

International Centre for Biosaline
Agriculture
Dubai, UAE

Alex Money
Smith School of Enterprise and the
Environment
University of Oxford
Oxford, UK

Mohammad Mortazavi-Naeini
Land and Water Division
NSW Department of Primary
Industry
Orange, NSW, Australia

Abishek S. Narayan
Aquatic Research
Swiss Federal Institute of Science and
Technology (EAWAG)
Zurich, Switzerland

Saskia Nowicki
School of Geography and the
Environment
University of Oxford
Oxford, UK

Thanti Octavianti
School of Geography and the
Environment
University of Oxford
Oxford, UK

Joanna Pardoe
London School of Economics
London, UK

Jian Peng
School of Geography and the
Environment
University of Oxford
Oxford, UK

Edmund C. Penning-Rowsell
School of Geography and the
Environment
University of Oxford

Oxford, UK
and
Flood Hazard Research Centre
Middlesex University
London, UK

Rebecca Peters
School of Geography and the Environment
University of Oxford
Oxford, UK

Jonathan Rawlins
OneWorld Sustainable Investments
Cape Town, South Africa

Grace Remmington
Cranfield University
Cranfield, UK

Michael Rouse
School of Geography and the
Environment
University of Oxford
Oxford, UK

Richard Rushforth
School of Informatics, Computing, and
Cyber Systems
Northern Arizona University,
Flagstaff, AZ, USA

Julie Self
Government of Alberta
Edmonton, AB, Canada

Ranu Sinha
School of Geography and the Environment
University of Oxford
Oxford, UK

Pauline Smedley
British Geological Survey
Keyworth, Nottingham, UK

Heather M. Smith
Cranfield Water Science Institute
Cranfield University
Cranfield, UK

Kieran Stanley
Institute for Atmospheric and
Environmental Sciences
Goethe University
Frankfurt, Germany

Troy Sternberg
Environmental Change Institute
University of Oxford
Oxford, UK

Abi Stone
University of Manchester
Manchester, UK

Patrick Thomson
School of Geography and the
Environment
and
Smith School of Enterprise and the
Environment
University of Oxford
Oxford, UK

Swathi Veeravalli
Us Army Corp of Engineers
Alexandria, VA, USA

Shuchi Vora
The Nature Conservancy
Delhi, India

Gareth Walker
Insight Data Science
San Francisco, CA, USA

Kevin Wheeler
School of Geography and the
Environment
University of Oxford
Oxford, UK

Paul Whitehead
School of Geography and the
Environment
University of Oxford
Oxford, UK

Foreword

I still recall fragments of some of those conversations that subsequently helped to shape my view of the world.

Over 30 years ago, visiting Egypt as Britain's Development Minister, a young man who worked for an NGO said to me, 'I don't understand why everyone is so obsessed with the politics of oil. In time, they will have far more reason to be worried about the politics of water.'

He was right. And it was, of course, a point made sometime in the millennium before the birth of Christ by Job, who had plenty of other things to worry about. 'But the mountain falls and crumbles away, and the rock is removed from its place; the water washes away the stones; the torrents wash away the soil of the earth; so you destroy the hopes of mortals.'

Hope is not the only casualty. Job could have added that peace and stability are victims too.

A former secretary-general of the UN opined that the next war in north-east Africa would be caused by disputes over access to the waters of the Nile. Equally bleak predictions could be made about the Jordan River, and to the east about the Tigris and Euphrates. In central Asia and the Punjab (which means 'land of five rivers') there are similar concerns.

Populations increase exponentially even in places where there is severe water stress, and larger populations demand more water to help produce their food. Food shortages and price rises have always been sure-fire causes of domestic unrest.

Climate change increases stress in areas that are already suffering economically and politically. Migration from arid to more favoured environments is an inevitable result. We know that economic migration – and the movement of people because of political turbulence – have roiled the prosperous northern democracies to which poor, hungry and thirsty people seek to move. This will get worse unless we act.

Water is such a precious resource which, like others, is not fairly distributed around the globe. The rich get more and invariably pay less for it (unless, I suppose, it is bottled!). So the study of how to do the best job of finding, storing, managing and cleaning water, and paying for it too, deserves more political attention and more research funding. This is true in all parts of the world, not least fast-growing Asia and Africa, where so many cities face imminent water shortages and where so much of the existing supply is polluted.

I hope that the essays in this book will help to stimulate more intense and informed debate about a subject where more international cooperation will be necessary for success. We cannot, to paraphrase Benjamin Franklin, wait to learn these lessons until the wells and aquifers are dry.

The Rt Hon the Lord Patten of Barnes
Chancellor, University of Oxford

Oxford, December 2018.

Acknowledgements

The editors and authors are grateful to Heather Viles for supporting this project, and to Nancy Gladstone, Faith Opio and Ailsa Allen for their careful assistance with the preparation of the manuscript and its figures. Thanks are due to Andrew Harrison at Wiley for taking this project on, and to Antony Sami and Vivek Jagadeesan for seeing the manuscript through to production.

1

Water Science, Policy, and Management

Introduction

Simon J. Dadson[1], Edmund C. Penning-Rowsell[1,2], Dustin E. Garrick[3], Rob Hope[1,3], Jim W. Hall[4], and Jocelyne Hughes[1]

[1] *School of Geography and the Environment, University of Oxford, UK*
[2] *Flood Hazard Research Centre, Middlesex University, London, UK*
[3] *Smith School of Enterprise and the Environment, University of Oxford, UK*
[4] *Environmental Change Institute, University of Oxford, UK*

1.1   Introduction

Understanding the risks and opportunities presented by the changing water cycle, and the intensifying demands and competition for freshwater, is one of the most pressing challenges facing scientists, water managers and policy-makers. In the context of rapid climate, land cover and other environmental changes, the requirement to protect communities against water-related natural hazards, the stewardship of water resources to provide reliable water quantity and quality, and the provision of clean, safely-managed drinking water and improved sanitation to a population predicted to exceed 9 billion, constitute a defining challenge for the twenty-first century. This challenge has inspired the University of Oxford to offer a graduate programme in Water Science Policy and Management since 2004, which to date has seen over 350 students from 57 countries graduate, of which more than half are women. This book is formed from contributions by more than a dozen academics and practitioners who have taught the course, in each case writing in co-authorship with more than two dozen alumni.

This introductory chapter outlines key drivers of change in the water environment and explains how these drivers may evolve into the future, creating new issues and risks requiring interventions. We then outline what we consider to be key challenges facing those responsible for water and its governance and management, globally, in the context of scientific understandings, policy priorities, and management opportunities. We make reference here to the chapters that follow on particular aspects of water science, policy and management, thereby contextualizing those chapters and enabling the reader to see them in a broader context. The final chapter of this book (Chapter 20) gives our vision for the future role of interdisciplinary water education and research in creating greater understanding of the complexities involved and the opportunities for progress.

Water Science, Policy, and Management: A Global Challenge, First Edition. Edited by Simon J. Dadson, Dustin E. Garrick, Edmund C. Penning-Rowsell, Jim W. Hall, Rob Hope, and Jocelyne Hughes.
© 2020 John Wiley & Sons Ltd. Published 2020 by John Wiley & Sons Ltd.

1.2 Drivers of Change: Environment, Politics, Economics

The water sector is strongly impacted by the recent, rapid, and widespread effect of economic growth, often exacerbated by weak governance and inequality. Alongside these human drivers, environmental change, including climate change and variability and changes in land cover and land management, exerts impacts which are often felt most acutely in societies least able to adapt. The opportunities and challenges presented by growing and moving populations, in the context of changing water availability, threaten the sustainable, equitable and efficient use of water resources for economic development.

As demonstrated in Chapter 2, the evidence is overwhelmingly in support of anthropogenic global warming, and it is notable that climate science has unequivocally demonstrated that observed historical climate change is due to anthropogenic emission of fossil carbon (Box 1.1). In the presence of overwhelming evidence, the debate has now shifted towards understanding the regional and local consequences of warming, and their impacts on hydroclimatic variability and extremes. Revealing the regional picture adds additional uncertainty and raises the crucial question of how much additional evidence must we wait for before we act, either in mitigation of future change, or in order to adapt to what may constitute a 'new normal' range of climatic variability? There are

---

Box 1.1 The Paris Agreement

The *Accord de Paris* is an agreement within the United Nations Framework Convention on Climate Change (UNFCCC 2015), dealing with greenhouse-gas emissions mitigation, adaptation, and finance, starting in the year 2020. The agreement was negotiated by representatives of 196 countries at the 21st Conference of the Parties of the UNFCCC in Le Bourget, France, and adopted by consensus on 12 December 2015. The Agreement's long-term goal is to keep the increase in global average temperature to well below 2°C above pre-industrial levels, and to limit the increase to 1.5°C, to substantially reduce the risks and effects of climate change. Under the Agreement, each country must determine, plan, and regularly report on the contribution that it undertakes to mitigate global warming. No mechanism forces a country to set a specific target by a specific date, but each target should go beyond previously set targets. In addition to reporting information on mitigation, adaptation and support, the Agreement requires that the information submitted by each country undergoes international technical expert review.

The consequences of failing to meet the Paris commitments for flooding and water resources are potentially serious, although there is considerable uncertainty in current projections (see Chapter 2). Even with 1.5°C warming, significant increases in rainfall and therefore flood risk are likely, particularly in flood-prone south-east Asia. The outlook for water resources is also strongly dependent on the Paris Accord, with projections of exacerbated water scarcity in already drought-prone areas, should the 1.5°C commitment not be met. Nonetheless, much uncertainty remains, not least because the pathways to 1.5°C involve changes not only to greenhouse gas concentrations but also to atmospheric aerosols and land use.

UNFCCC. (2015). *Paris Agreement*. Available at: https://unfccc.int/files/meetings/paris_nov_2015/application/pdf/paris_agreement_english_.pdf.

many tools at our disposal to answer such questions. Indeed, the challenge to policy-makers and their advisors, and to practitioners in the field of water management, is to extract the salient information on which to base decisions from the plethora of data currently available on the subject (see Chapters 2 and 14).

Whilst global attention has quite properly focused on climate change, widespread policy-driven changes in land cover and management also rank amongst the most striking perturbations to the natural environment that impinge on the water sector. Land cover changes may occur by direct policy intervention; they may also occur as land managers respond individually to market forces and the regulatory environment. Together these changes can also impact land use (tree planting, agricultural practices) by affecting what it is economic to do in the rural environment. The impact, for example, of nitrate on long-lived groundwater quality is of particular note (Chapters 3 and 4), as is the impact of regulatory practices on water quality as evidenced by the EU Water Framework Directive, which is credited with driving a significant, but small, improvement in aquatic biodiversity (Chapter 5). Policies and economic incentives exert a powerful control over land management and agriculture, with impacts that are often felt more immediately and with greater certainty than climatic variability or change but which also act as threat multipliers or stressors of freshwater ecosystems when combined with climate change (e.g. algal blooms, Chapter 4; invasive species proliferation, Chapter 5).

Demographic drivers of change include the growth of global population centres in Asia and Africa, including 'mega-cities' with populations greater than 10 million. Nonetheless, the reality of population growth, urbanization, and the growth of agriculture to support a growing affluent population in the developing world will have profound consequences for water consumption (Chapter 8) and for water quality. As such it is vital to consider not only the physical and natural consequences but also the potential political responses in the light of projected growth of urban populations (Chapters 12 and 18).

Water plays a crucial role in many sectors of the economy and is frequently analysed as a factor of production or as a public economic good. Connections with the energy and agricultural sectors are often highlighted, not least because agriculture consumes by far the most water of any economic sector, and reliable water supplies are needed for energy production. These linkages serve both to amplify the sensitivities of the water sector to global change, and to mandate broad consideration of water-related impacts on other economic sectors in policy development and the consequent enactment of management decisions, particularly in relation to water allocation and reallocation in a rapidly changing world (Chapter 8).

The global importance of water in industrialized and developing economies is also recognized via the Sustainable Development Goals (SDGs), which explicitly mandate universal and equitable drinking water supplies and improved sanitation services, sustainable water withdrawals and protection of ecosystems (Box 1.2). Compared with the earlier Millennium Development Goals, the SDGs bring a stronger and broader framing of sustainable management of water resources to meet human and environmental needs. The role of water for development is partly to alleviate acute poverty and to protect vulnerable populations from water-related risks, especially due to extreme hydrological variability and disease associated with poor sanitation. But there is also a role for water systems to remove the time and effort burden associated with more costly, labour-intensive means of water service provision, which are borne largely by women in rural Africa and Asia, so that individuals and societies are free to devote more attention

---

Box 1.2 The Sustainable Development Goals

The SDGs are a collection of 17 global goals set by the United Nations General Assembly in 2015 (United Nations 2015). The goals are broad and interdependent, yet each has a separate list of targets to achieve. The SDGs cover social and economic development issues including poverty, hunger, health, education, global warming, gender equality, water, sanitation, energy, urbanization, environment and social justice. The Sustainable Development Goal No. 6 for water and sanitation has eight targets and 11 indicators that are being used to monitor progress towards the targets. Most are to be achieved by the year 2030. The first three targets relate to drinking water supply and sanitation. Worldwide, 6 out of 10 people lack safely managed sanitation services, and 3 out of 10 lack safely managed water services. Targets 4 to 6 relate to water-use efficiency, integrated water resources management and transboundary cooperation, and the protection and restoration of freshwater ecosystems, acknowledging that water-related ecosystems underpin many other SDGs. The final two targets are to do with the implementation of SDG 6; international cooperation and capacity building, and stakeholder participation, are both essential if the SDG 6 targets are to be achieved.

United Nations (2015). Transforming our world: the 2030 Agenda for Sustainable Development; Resolution adopted by the General Assembly on 25 September 2015. New York, United Nations.

United Nations (2018). Sustainable Development Goal 6, Synthesis Report 2018 on Water and Sanitation, New York, United Nations.

---

to other activities. Whether the SDGs lead to lasting change or whether they present an impossible-to-fulfil dream is the subject of discussion in Chapter 17.

1.3   Responses to Change: Technology, Information, Equity

Much has been done already by scientists, policy-makers and water managers to respond to the challenges set out above. Their complexity is overwhelming, and it is essential that we learn from those past experiences in order to refine our understanding of potential responses, which must draw not only on the development of new technology, but on social, political and economic innovation. As Grey highlights in Chapter 19, it is insufficient to consider water infrastructure in isolation. Investment in infrastructure can succeed only if accompanied by investment in information systems and in building strong institutions for managing the resulting systems. Moreover, as with any large investment, it is necessary to consider not just economic efficiency in the comparison of benefits and costs, but also to consider how benefits accrue amongst groups different from those who bear the social, political, and economic costs. A broader view of hydrological infrastructure is that it is ultimately for mitigating the effects of hydrological variability (Chapter 19). This view sees an increasingly important role for distributed storage, and acknowledges the links between water, energy and food, any of which can be stored or traded as a substitute for lacking a reliable supply of the other (Chapter 8).

Wastewater treatment technology has also developed rapidly over the past several decades. As noted in Chapter 16, technology capable of treating water more cheaply brings human and environmental benefits in developing countries. It also generates valuable waste streams, which can dramatically alter the investment case for wastewater treatment in many settings. The funding case for wastewater treatment is strengthened if collecting and treating wastewater becomes an economic opportunity rather than a cost. Indeed, the challenge moves rapidly from being one of technological innovation to one of governance in which the need is to finance deployment of such infrastructure (Chapters 12 and 15) and to operate and maintain it efficiently (Chapter 17) and with equitable access (Chapter 18).

The challenge to strengthen the funding case for investment in the water sector is taken up in Chapter 9 in conjunction with another emerging technology – smart infrastructure and environmental informatics. Monitoring technology built into new infrastructure in order to operate, manage and maintain it effectively yet further enhances its useable benefit. The point that management of the process is as vital as initial capital investment is pressed home in Chapters 16 and 17 in relation to regulation of water supply and treatment technology. New in-situ environmental sensors also promise to revolutionize water management, including novel biosensors based on designs inspired by nature to trace and prevent emerging organic pollutants, and the use of eDNA to trace and monitor endangered or invasive species (Chapters 5 and 6). Likewise, our understanding of the changing water cycle in the Earth system will be transformed by the results of hydrological Earth observation missions like NASA's GPM, SMAP and GRACE, the EU Copernicus Sentinel programme, and micro-satellite technology (see Chapter 7).

The promise of new data and monitoring technologies (see Chapter 7) is particularly compelling if harnessed alongside properly-maintained existing observation networks, together with new computational methods (Chapter 14). Yet, if the coming decades are characterized by new data streams, the challenge will be to ensure that these 'big' datasets yield information that can most helpfully inform the decisions needed both for real-time management and long-range planning (see Chapter 7). As Hall and Penning-Rowsell argue in Chapters 11 and 14, it will not be sufficient simply to collect data: it must be integrated within operational systems and its efficient use governed equitably. 'Big data' are expensive to collect and require skills and training to use and interpret – it is necessary to ensure that their benefits accrue equitably and do not lead to unfair information asymmetries, nor become a source through which well-resourced groups can crowd out the poor (Chapters 18 and 19).

Alongside the emerging technical innovations that stand to propel growth and development in the water sector, there is an acute and growing need for social, political, legal and governance structures to tackle the world's biggest challenges (Chapter 10). A wider, more general debate concerns the balance between equity and efficiency in water infrastructure investment and allocation decisions (Chapter 8). This debate emerges no more acutely than in discussions of whether there ought to be a human right to water, as expounded by Larson et al. in Chapter 10. It is vital that governance and legal frameworks take account of these developments, both theoretically and in practice. The theme of national vs. private infrastructure is explored by Smith and Walker in Chapter 12, along with the comparison of centralized and decentralized approaches to water supply (Chapters 8 and 9).

The search for solutions has been marked with silver bullets that ignore or underestimate the politics of water. Water policy involves a sequence of decisions in response to evolving challenges and values, yet policy processes rarely follow a linear path. Responding to crises, such as floods and droughts, has produced fragmented policy and legal frameworks, and often perverse incentives that can increase vulnerability to future risks and concentrate their impact on the most marginal. Solutions have proven partial and provisional, where our past efforts lead to lock-in and path dependency, constraining our capacity to establish new policies, laws and incentives to tackle dynamic water risks and values.

1.4 Science, Policy and Management

Many of the drivers of change identified above can be partly addressed by technology: improvements in monitoring (Chapter 7), water-efficient technologies for agriculture, industry and wastewater treatment (Chapters 13 and 16), and smart infrastructure (Chapter 9) capable of being used in support of allocation decisions. But technology is not the only adaptive mechanism available for human societies to deploy. Without an understanding of the political and societal context within which water challenges emerge, it is impossible to appreciate how the intertwined trajectories of historical contingency have led us to where we are, and yet more difficult to construct an acceptable future course in any given system. As reviewed in this book, our political, legal and economic institutions shape the interactions between policy and technological changes.

Box 1.3 The Sendai Framework for Disaster Risk Reduction

The Sendai Framework was adopted by UN Member States on 18 March 2015 at the Third UN World Conference on Disaster Risk Reduction in Sendai City, Japan. Seven Global Targets were set, including: (i) substantially to reduce global disaster mortality by 2030, aiming to lower average per 100 000 global mortality rate in the decade 2020–2030 compared with the period 2005–2015; (ii) to reduce direct disaster economic losses in relation to global gross domestic product (GDP) by 2030; and (iii) substantially to increase the availability of and access to multi-hazard early warning systems and disaster risk information and assessments to the people by 2030. Four priorities are encapsulated in the Agreement:

Priority 1. Understanding disaster risk.
Priority 2. Strengthening disaster risk governance to manage disaster risk.
Priority 3. Investing in disaster risk reduction for resilience.
Priority 4. Enhancing disaster preparedness for effective response and to 'Build Back Better' in recovery, rehabilitation and reconstruction.

These priorities are especially relevant to water-related natural hazards, which account for 50% of all weather-related disasters and 90% of major disasters since 1990 (including floods, droughts, storm surges and tidal waves; United Nations International Strategy for Disaster Reduction (UNISDR 2018).

United Nations International Strategy for Disaster Reduction (UNISDR). (2015). Sendai Framework for Disaster Risk Reduction. Geneva: UNISDR.

If understanding the natural environment and its relations with the political systems that have shaped humanity's encounters with water is challenging, the difficulty is nothing compared with that facing those whose decisions enact the policies made in response to political, financial, and other constraints. Sound governance and implementation of such strategies requires technical systems of control (planning, targets, incentives, and oversight; covered in Chapter 17), and the instruments for planning and informing decisions proposed or taken (e.g. Chapter 14). The relation between science, policy and management is, of course, complex rather than linear. In practice it may be quite legitimate to find linkages across and between each component. Policy, for example, exerts a reciprocal influence on science – both directly in that policy-makers support and commission evidence-based programmes, and indirectly in that policy often determines tacitly what is considered an appropriate scientific question worthy of study, particularly where limited funding is available for research. Less controversially, scientific evidence is a crucial ingredient in managing water systems.

The chapters contained within the book unpack these challenges both individually and in combination, from a range of perspectives. The range of disciplinary expertise required to solve water challenges extends vastly beyond the water sector alone. The chapters contained within this book draw on ideas in climate science, water chemistry, health and medical sciences, biological and environmental sciences, engineering and information technology. But important though these technical subjects are, the study of economics, finance, politics, history, philosophy, management, governance and administration are necessary in order to reach durable solutions which account for water's role in societies, economies and ecosystems. Taken together, they inspire a vision for interdisciplinary water research and education to solve some of society's most pressing threats in the modern era.

Part I

Water Science

The politics and economics of the global water challenge are influenced by the quality and scope of scientific information available to support decisions. For example, improved climate predictions can benefit long-term decision-making by improving our capacity to manage climate and hydrological risks. This section tackles the major scientific challenges facing the water sector today. The branches of hydrology upon which water managers might draw include climatology and meteorology (Chapter 2), soil physics, hydraulics, and groundwater science (Chapter 3). But there is a broader sense in which the challenge of bringing water security to a growing population draws on science outside the traditional hydrological remit. Coverage in these chapters of the relations between water and health draws on findings in microbiology, epidemiology and public health, and the relations between water quality (Chapter 4) and aquatic habitats and biodiversity (Chapter 5) are probed in detail for their functional value within the ecosystem and the wider catchment, as well as for their intrinsic benefits. Chapter 6 surveys the emerging new monitoring technologies being developed to support decision-making, and reviews the rapid growth of large-scale modelling of the water cycle.

Whilst there is an important role for scientific information in policy-making, attention must also be paid to the receptivity of those charged with making long-lived, long-term investment and allocation decisions, as well as those responsible for day-to-day management of water risks and resources. The production of appropriate scientific information is only one entry point to the dialogue between scientists and decision-makers. Institutional change and enhanced technical capacity are also vital to promote resilient adaptation policies. Institutional transformation within governments, the private sector, NGOs and community participation entails a change in the culture around which those responsible for making decisions take up scientific information.

Without capacity to use new scientific information and to understand its relevance and limitations such approaches are unlikely to realise their potential. Enhanced technical capacity to use climate information requires internal and external actors to consider the most useful form in which decision-makers might request climate information, and to integrate new and uncertain sources of information into decision-making. One way to develop this capacity is through the exchange of personnel, and the long-term effects of policies such as improved education and knowledge exchange with research

Water Science, Policy, and Management: A Global Challenge, First Edition. Edited by Simon J. Dadson, Dustin E. Garrick, Edmund C. Penning-Rowsell, Jim W. Hall, Rob Hope, and Jocelyne Hughes.
© 2020 John Wiley & Sons Ltd. Published 2020 by John Wiley & Sons Ltd.

organizations. The availability of appropriate climate information to inform decisions is key when relevant aspects of future climate and hydrology are conditioned by (i) the temporal extent of decision-making processes, (ii) the spatially variegated nature of impacts (which may span large regions or be focused on critical locations), and (iii) the potential effects of extremes compared with long-term variability. Future challenges will draw upon many strands of scientific evidence, and will undoubtedly present new demands for the scientific community. The contributions in this section do not provide all the answers; they outline the chart on which those who take up the challenge might plot the course ahead.

2

Hydroclimatic Extremes and Climate Change

Simon J. Dadson[1], Homero Paltan Lopez[1], Jian Peng[1], and Shuchi Vora[2]

[1] *School of Geography and the Environment, University of Oxford, UK*
[2] *The Nature Conservancy, India*

2.1   Introduction

Understanding the risks and opportunities presented by the changing water cycle is one of the most pressing challenges facing scientists, water managers and policymakers in the twenty-first century (World Economic Forum 2017). The Paris Agreement challenges scientists to improve our understanding of the water cycle's role in Earth systems, to quantify the risks and impacts of anthropogenic warming and environmental change on hydrological extremes, and to provide more precise estimates of the availability of water resources worldwide. The Sendai Framework for Disaster Risk Reduction reinforces the importance of understanding hydro-meteorological hazards to support risk-based management and to provide timely and accurate predictions. Arising from the 2030 agenda, the Sustainable Development Goals (SDGs) emphasize the need to remove the constraints that water insecurity places on economic growth (Hall et al. 2014; Sadoff et al. 2015), and highlight the transformative benefits arising in both developed and developing economies from well-managed water resources and reduced vulnerability to extremes (World Bank 2017). Moreover, the resilience of key economic and infrastructure assets to current and future water resources availability and extremes demands robust, defensible, physically consistent projections to support long-lived investment decisions (National Infrastructure Commission 2017). This chapter begins with an overview of some recent advances in our understanding of key aspects of the climate system, including a discussion of the principal modes of atmospheric variability that influence precipitation, such as the impact of El Niño–Southern Oscillation (ENSO) on climate and weather in the tropical Pacific and beyond, and the South Asian Monsoon and controls on its variability. The role of anthropogenic processes in climate change is also reviewed, with particular reference to its effects on the water cycle both at a global and regional level. Future changes in hydroclimatic risks manifest themselves largely through changes in extremes. The chapter concludes with a consideration of how climate variability and change can best be understood and incorporated into policy- and decision-making processes.

Water Science, Policy, and Management: A Global Challenge, First Edition. Edited by Simon J. Dadson, Dustin E. Garrick, Edmund C. Penning-Rowsell, Jim W. Hall, Rob Hope, and Jocelyne Hughes.
© 2020 John Wiley & Sons Ltd. Published 2020 by John Wiley & Sons Ltd.

2.2   Key Concepts in Climate Science

2.2.1   The Water Cycle in the Earth System

Water plays a fundamental role in the Earth's climate system. Energy evaporated in warm regions is stored as latent heat and released when water vapour condenses to form cloud, thus transferring energy from Equator to poles. The textbooks by Barry and Chorley (2009), O'Hare et al. (2004) and McIlveen (2010) provide a suitable introduction to the drivers of hydroclimatic variability that are relevant to water resources managers and policy-makers.

2.2.2   Radiative Energy Transfer in the Atmosphere

The Earth's energy budget is dominated by electromagnetic radiation from the sun. Short-wave radiation passes easily through the troposphere and reaches the surface, where a fraction, α, known as the planetary albedo, is reflected back to space. The planetary albedo is approximately 0.3, although this figure varies with surface type, time of year, angle of incidence and wavelength (Figure 2.1). The remaining energy heats the surface, causing it to emit electromagnetic radiation of its own, according to Stefan-Boltzmann's law which states that any object with temperature greater than absolute zero ($-273.15°C$) will emit radiation, with total radiated energy in proportion to the fourth power of the object's temperature. Planck's law further states that the range of wavelengths emitted will depend upon temperature. Hotter objects emit shorter wavelength radiation than cooler ones. For the Earth, whose average surface temperature is approximately 14°C, the typical wavelength emitted is in the infrared part of the

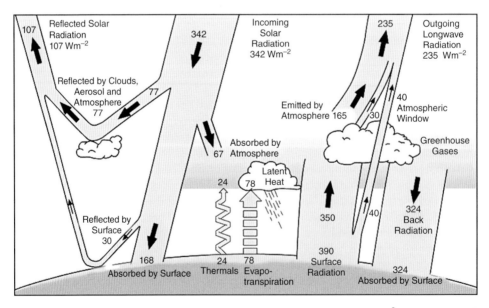

Figure 2.1 Earth's radiation budget averaged at the planetary scale. Figures in W/m². *Source:* Kiehl and Trenberth (1997) redrafted by Le Treut et al. (2007). Reproduced with permission of the American Meteorological Society.

spectrum (visible to certain reptiles and those wearing 'night-vision' goggles, but not to the human eye). Infrared radiation is absorbed by the atmosphere – most notably by water vapour and similarly-sized molecules like carbon dioxide and methane. This absorption is the basis of the natural greenhouse effect, by which the equilibrium temperature of the surface and the atmosphere is significantly elevated above its equilibrium value in the absence of those trace gases. Anthropogenic emissions of additional quantities of radiatively active trace gases to the atmosphere have resulted in additional warming (see Section 2.4).

2.2.3 Convection and Atmospheric Stability

The fact that the troposphere is effectively transparent to short-wave radiation, whilst absorbing long-wave radiation, means that the atmosphere is heated from below. Air temperature therefore decreases with elevation, at a rate of approximately 6.5°C/km. When air at the surface is caused to rise, either by being heated relative to its surroundings, or forced upwards over a mountain range or a cooler, denser air mass, it rises and continues to rise until it is no longer buoyant. As rising air moves into the lower pressure regions aloft, it expands and cools adiabatically – that is, without gaining or losing heat directly across its boundary. When an air parcel has cooled to the point that water vapour held within it can no longer remain as a vapour, liquid water droplets form by condensation. These then coalesce and coagulate, either in their liquid state or, if the temperature is low enough, as colliding crystals of ice, to form hydrometeors. When water droplets or ice crystals have agglomerated to sufficient size, gravity causes them to fall as precipitation. From the point of view of hydrological hazards such as surface water and fluvial flooding, extreme precipitation can accumulate on a range of timescales, from highly intense, convective cloudbursts that can cause short-term surface water flooding, to long-duration rainfall maxima which have long-lived impacts (see Figure 2.2).

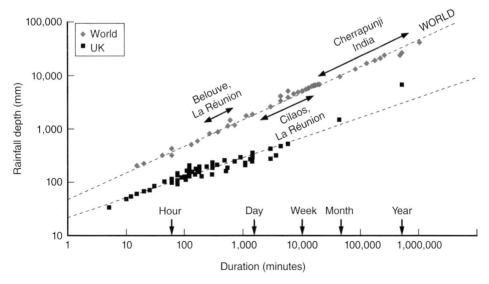

Figure 2.2 Maximum recorded point rainfalls of different durations across the world and the UK. *Source:* Robinson and Ward (2017). Reproduced with permission of IWA Publishing.

2.2.4 The General Circulation

The Earth's surface radiation balance varies with latitude. At the Equator and in the tropics, the net radiation budget is positive; at the mid-latitudes and the poles it is negative. Therefore, without the general circulation, tropical regions would get hotter and polar regions would become colder. The latitudinal transfer of energy in the climate system is accomplished through the general circulation, which begins through vigorous uplift of warm moist air at the Inter-tropical Convergence Zone (ITCZ) in the equatorial region. This air then cools adiabatically and forms cloud, which provides precipitation to sustain the widespread regions of equatorial forest in the Amazon and Congo. Having risen through the troposphere the air cools by radiation of its heat to space and moves poleward to approximately 30° in each hemisphere where it begins to descend slowly and over a wide area, warming and drying through adiabatic compression, leading to the formation of sub-tropical high-pressure zones with significant aridity (e.g. the Sahara, Kalahari, Atacama, and deserts of the American SW and Western Australia). The returning air flow at the surface produces a reliable easterly wind – the trade wind – completing the cycle of the Hadley cell. The ITCZ moves northward and southward in the boreal summer and winter respectively, affecting the passage of rains across the tropical regions it traverses.

Energy transfer in the mid-latitudes is through indirect eddy transfer. Winds acceler-ate aloft at the northern and southern boundaries of the Hadley cell, influenced by the Coriolis force which affects fast-moving flows on a rotating planet. The upper westerly winds 8–10 km above the surface were discovered as aircraft increasingly found their passage was helped or hindered by the jet stream. The structure and pattern of the jet streams is now known: they develop increasingly complex meandering patterns called Rossby waves which develop over time and then decay, only to emerge again in several days' time. The accelerations and decelerations involved in the passage of Rossby waves impact the convergence and divergence of air flows at the surface, and therefore control the passage of mid-latitude weather systems.

The majority of the energy is transported through sensible heat transfer in the atmos-phere (60%). The latent heat flux (i.e. the transfer of energy through the evaporation of water in one region and its condensation in another) contributes 15% and the oceans account for approximately 25% of the energy transferred in the general circulation.

2.3 Hydroclimatic Variability and Extremes

2.3.1 Modes of Hydroclimatic Variability

Many hydroclimatic extremes result from natural climate variability. In this section the most important modes of variability will be discussed, including ENSO, the South Asian monsoon, the North Atlantic Oscillation (NAO) and the Pacific Decadal Oscillation (PDO) (Figure 2.3).

2.3.2 El Niño–Southern Oscillation (ENSO)

The ENSO is the dominant climate mode that influences global climate extremes. It occurs naturally over the tropical Pacific, alternating between anomalously warm (El Niño) and cold (La Niña) phases. The two phases are related to the Walker Circulation,

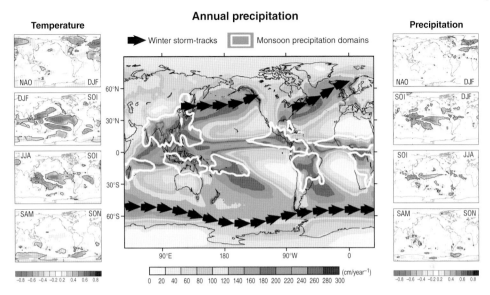

Figure 2.3 Global distribution of average annual rainfall from 1979 to 2010 with monsoon precipitation domain and winter storm-tracks shown with white contours and black arrows respectively. The left and right columns show the seasonal correlation maps of North Atlantic Oscillation (NAO), Southern Oscillation Index (SOI, the atmospheric component of El Niño – Southern Oscillation [ENSO]) and Southern Annular Mode (SAM) mode indexes compared with monthly temperature (precipitation) anomalies in boreal winter (December–February [DJF]), austral winter (June–August [JJA]) and austral spring (September–November [SON]). *Source:* Box 14.1, Figure 1 from Christensen et al. (2013). Reproduced with permission of IPCC. (*See color representation of this figure in color plate section*).

which is caused by the pressure gradient force associated with the regional high-pressure system over the eastern Pacific Ocean and a low-pressure system over Indonesia. El Niño results from a weaker-than-normal Walker Circulation. It is characterized by an east–west atmospheric pressure gradient, slackened Trade Winds, and a tilt in the Pacific Ocean thermocline. By contrast, La Niña is associated with an especially strong Walker Circulation, and generally induces the opposite features (Neelin et al. 1998). These warm and cool phases induce changes in the ocean–atmosphere systems within and outside the tropical Pacific, which can increase the occurrence of extreme events around the globe. Strong El Niño events with exceptionally warm sea surface temperature (SST) anomalies were seen during 1972–1973, 1982–1983, 1997–1998, and 2015–2016, whilst La Niña events, characterized by cold SST anomalies, occurred during 1973–1974, 1975–1976, 1988–1989, 1998–1999, 1999–2000, 2007–2008, and 2010–2011, which translated to significant droughts, floods, human casualties and economic losses (McPhaden et al. 2006).

The IPCC has concluded that ENSO will remain the dominant mode of interannual climate variability with global impacts in the twenty-first century, implying that the extreme events at regional scales induced by ENSO variability will intensify. The most recent climate models show some improvements in simulation of ENSO amplitude and display similar interdecadal modulations to that observed by instrumental records. Improvements in ENSO simulation will help us to understand the complex nonlinear process that determine ENSO extremes, which will lead to more accurate predictions of ENSO. However, there are still many challenges: for example, it is not clear how other modes of climate variability may influence ENSO extremes, due to the complex ENSO teleconnections via atmosphere and ocean; nor is there agreement on whether decadal variability in the amplitude and spatial pattern of ENSO is due to natural variability or anthropogenic effects (Christensen et al. 2013).

2.3.3 South Asian Monsoon

Monsoons are the most important seasonal mode of climate variability within the tropics, and are responsible for conveying atmospheric moisture and energy at the global scale. The monsoon systems have impacts on precipitation over all tropical continents: Asia, Australia, the Americas, and Africa. All monsoons are characterized by life-cycle phases of onset, maturity and decay. They normally feature abundant rainfall during summer and dry conditions during winter. The strength and timing of monsoons are controlled by many factors such as land–sea temperature contrast, atmospheric moisture content, atmospheric aerosol loading, and land cover and use. The IPCC has concluded that monsoon rainfall will become intense and influence large areas in the future due to an increase in atmospheric moisture content (Christensen et al. 2013). However, the strength and variability of regional monsoon systems are complex and more uncertain.

Amongst all the regional monsoon systems, the South Asian monsoon is the strongest and affects about half of the population on the planet. It is often the cause of floods, landslides, droughts, and other hazardous extreme weather events. The South Asian monsoon is complex, due to the unique geographical characteristics of the Indian subcontinent, and associated atmospheric, oceanic, and geophysical factors (see the classic work of Hahn and Manabe 1975). The monsoon circulation is driven by differential heating and cooling between land and sea; by the strength of atmospheric circulation and moisture content of air determining the duration and amount of monsoon rains; and by topography and ENSO. For example, the Tibetan Plateau modulates the strength of the South Asian monsoon system via its variation of snow cover and surface heating. Our capability to simulate and predict the South Asian monsoon has increased substantially in the latest generation of global circulation models (GCMs), although much progress is still needed. It is expected that an increase in both mean and extreme monsoon precipitation will occur in the future: the monsoon circulation will weaken but its interannual variability is projected to increase. However, quantitative simulation of monsoon dynamics – and therefore accurate prediction of the extent and intensity of the South Asian monsoon – remains limited by sensitivity to model resolution, uncertainties in ENSO variability projections, reorientation of aerosol effects, and uncertainties in the Madden–Julian Oscillation (MJO) simulation.

2.3.4 North Atlantic Oscillation (NAO)

The NAO, a dominant mode of atmospheric variability over Europe and the North Atlantic, is known to impact many climatological, hydrological, biological and ecological variables, with its most direct effects being in Europe (Hurrell 1995). Together with the low-pressure system over Iceland and the permanent high-pressure system over the Azores, the NAO determines the direction and strength of westerly winds in Europe. A larger than normal pressure difference, referred to as a positive NAO, leads to wetter and warmer conditions over northwest Europe and drier and cooler conditions across southern Europe. By contrast, a negative NAO is caused by weaker than normal pressure and leads to the opposite conditions.

Over the past decades, the NAO pattern has shifted from a persistent negative phase in the 1960s to a positive phase between the 1960s and 1990s. This tendency switched to a negative state during 1989–2011 but has since changed to a positive phase again. Whilst the possible causal mechanisms underlying NAO variability remains a matter for research – they include Arctic amplification of climate change and changes in the Sun's energy output – the state of the NAO system exerts a strong control on weather in Europe.

Seasonal NAO predictions currently offer between one and four months' lead time for Europe, but there remain large uncertainties in the long-term NAO predictions given by global climate models. Whilst climate models are in general able to simulate the gross features of the NAO, they typically underestimate its long-term variability, leading to biased predictions of regional climate.

2.3.5   Other Modes of Variability

Besides the dominant modes of climatic variability introduced above, three well-known modes, the Atlantic Multi-decadal Oscillation (AMO), the Pacific Decadal Oscillation (PDO) and Madden-Julian Oscillation (MJO) are also of note to water managers. The AMO has been identified as a mode of natural variability occurring in the North Atlantic Ocean with a quasi-periodicity of about 70 years. It is based upon the average anomalies of 150 years worth of instrumental SSTs recorded throughout the North Atlantic basin and is associated with changes in the thermohaline circulation. The AMO exerts important impacts on regional climates observed via instrumental and paleoclimatic records and reproduced in climate models. These connections include summer climate in North America (where warm-phase AMO has been linked to twentieth-century droughts), Atlantic hurricane frequency, and an influence on Sahel rainfall and the West African monsoon. Climate model simulations exhibit long-lived multi-decadal variability of AMO, suggesting that no fundamental changes in AMO variability will be seen in the twenty-first century (Ting et al. 2011).

The PDO is a mode of Pacific climate variability. It has similarities to ENSO but with a much longer timescale of 20–30 years, whilst ENSO cycles typically persist for 6–18 months. Similarly to ENSO, two extreme phases of PDO – positive and negative – have been defined by SST anomalies in the northeast and tropical Pacific Ocean. The PDO is also linked with ENSO, with more La Niña activity during the negative PDO phase and more El Niño activity during positive PDO. Drought patterns and frequencies in the United States are thought to be influenced by PDO phases. The South Asian monsoon is also affected, with increased rainfall and decreased temperatures seen over India during negative PDO phases. The PDO pattern has been considered as the result of internal climate variability and does not have a long-term trend, and the IPCC has reported low confidence in projections of future changes in PDO.

The MJO is the dominant mode of tropical intra-seasonal (30–90 days) variability. It is a large-scale coupling between atmospheric circulation and tropical deep convection and propagates slowly eastward along the Equator to the Pacific Ocean (Zhang 2005). The MJO impacts intra-seasonal monsoon fluctuations, modulates tropical cyclone activity, and has teleconnections with the extra-tropics. Inter-annual variability of extreme precipitation events is directly related to the MJO in many areas. Currently, simulation of the MJO using climate models is challenging, with large uncertainties in climate models limiting accurate projection of MJO variability in the future.

Taken together, these modes of variability, which span a range of time and space scales, help water managers and policy-makers to contextualize intra-seasonal to inter-decadal variability and to understand the physical mechanisms responsible.

2.4 Climate Change and Hydrology

2.4.1 Understanding the Link Between Climate Change and Hydroclimatic Extremes

There is unequivocal evidence that global temperature has increased in response to anthropogenic emissions of CO_2; but the link between global warming and local changes in precipitation has proved harder to establish. Whilst a key challenge in meteorology is to disentangle the signal of climate change from the noise of climate variability, the quantification of each component is paramount. Three specific strands of research have been explored: (i) the identification and extraction of trends and cycles in observed records of meteorological and hydrological extremes; (ii) the attribution of such trends to historical climate change and variability; (iii) evaluation of the projected effects of climate change on future extremes. Whilst there is a general agreement based on fundamental physical principles that a warmer atmosphere should exhibit a greater frequency of more intense hydroclimatic extremes (Bates et al. 2008; Trenberth 2011), complexities associated with energy transfer in the troposphere mean that this simple relation cannot be expected to hold in reality (Allen and Ingram 2002).

2.4.2 Climate Models and Climate Projections

Further activity to quantify the likely changes in hydro-meteorological extremes driven by global climate change has centred on the provision of projections of the relevant meteorological components of the water balance into the future. It is widely recognized that all climate models contain uncertainties of various kinds, including uncertainties in (i) initial conditions, (ii) the dynamic and physical formulation including parameter choices (model uncertainty), and (iii) the scenarios of economic activity upon which they are based (scenario uncertainty). At the timescales relevant to flood risk assessment, the latter two uncertainties dominate; at the shorter timescales of long-range weather forecasting and seasonal prediction, the uncertainty associated with lack of knowledge of the initial state of the atmosphere and ocean (initial condition uncertainty) is dominant, although well-established procedures to quantify this uncertainty, such as ensemble forecasting, are currently in use as a matter of routine within national meteorological services (Hawkins and Sutton 2009; Slingo and Palmer 2011). As a result of these uncertainties, globally-orchestrated climate modelling activities including the Coupled Model Intercomparison Project (CMIP3 and CMIP5, Taylor et al. 2012), upon which the IPCC assessments (Figures 2.4 and 2.5) are based, have been designed to permit quantification of model structural uncertainty and to establish a common framework for the use of agreed scenarios of future emissions of radiatively active gases (IPCC 2000; van Vuuren et al. 2011).

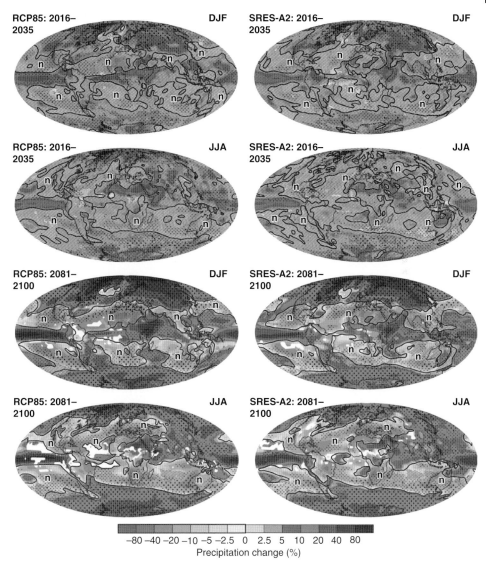

Figure 2.4 Spatial patterns of projected change in precipitation, relative to a 1986–2005 computed baseline, for the two seasons December–February (DJF) and June–August (JJA). Results from the CMIP3 experiments are shown on the left; results from the later CMIP5 experiments are on the right. Stippling indicates areas of model agreement, while hatching shows areas where projected change is within the range of natural variability. Areas with negative changes are labelled 'n' to distinguish them in monochrome. *Source:* Knutti and Sedláček (2013). Reproduced with permission of SpringerNature. (*See color representation of this figure in color plate section*).

2.4.3 Downscaling and Uncertainty

The horizontal grid spacing used by most global climate models is widely regarded as too coarse to inform local-scale adaptation decisions (Maraun et al. 2010), despite recent developments that have seen global models now run at resolutions as high as 25 km. Downscaling to resolutions which are thought to be more relevant to decision-makers is typically either statistical (i.e. based on empirical relations between predictor

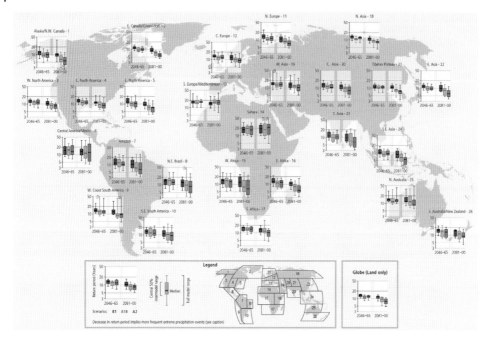

Figure 2.5 Change in return period for a 1-in-20 year daily precipitation event (with baseline 1981–2000). Decreased return periods imply more frequent extremes. Box plots give regional projections for 2046–2065 and 2081–2100, for scenarios B1, A1B, and A2. Results are based on 14 CMIP3 Global Circulation Models (GCMs). *Source:* Figure SPM.4B in IPCC (2012). Reproduced with permission of IPCC.

variables which are available in a global model and variables of ultimate interest) or dynamical (in which a higher resolution regional model is used to simulate meteorological properties at a higher resolution within a limited-area subdomain of the global model). Weather generators can be used to add temporal detail, but these have remained relatively simple in their formulation, only recently offering the prospect of spatial coherence (Wilks 2012). All of these additional modelling steps introduce additional uncertainty. An important scientific challenge that emerges as projections move to finer spatial resolution is to improve the representation of locally-extreme convective rainfall (Kendon et al. 2014) and to improve the representation of blocking anticyclones in winter, which determine the tracks of mid-latitude depressions responsible for winter rainstorms (Slingo and Palmer 2011).

2.5 Managing Hydroclimatic Extremes

2.5.1 Quantifying Risk and Uncertainty

A range of tools is available to evaluate and manage the impact of climate variability and extremes. In some cases, it is the hydrometeorological extreme that is the object of ultimate interest (e.g. flash flooding from surface water in urban areas), but in the vast majority of cases the impact of extremes is felt by a complex social and economic system of people, assets and infrastructure which find themselves in harm's way.

Box 2.1 – Examining the Hydrological Implications of the Paris Agreement

Whilst the Paris Agreement of 2015 aims to limit global warming to 'well below 2°C and to pursue efforts to limit the temperature increase to 1.5°C above pre-industrial levels', little is known about the hydrological implications of these targets. As such, Paltan et al. (2018) utilized the outputs of climate models to run a range of hydrological models and project shifts in global runoff and river flow characteristics that may result from meeting these climate thresholds. These values were then used to estimate changes in global flow statistics and hazards when compared with historical conditions. The study found that at 1.5°C, shifts in the duration of rainy seasons would lead to a notable increase in the occurrence of high flows mostly in catchments in South and East Asia. Yet, at a 2.0°C warmer world, additional catchments in South America, central Africa, central-western Europe, the Mississippi river area, central Asia, and various Siberian catchments, would also experience intensification of high flows. The quantification and mapping of the characteristics of this hazard permits further assessments of social, economic, energy, or infrastructure systems exposed and vulnerable to hydroclimatic shifts. Combining modelling techniques with other relevant information within a specific risk-management method would in turn support the development of adaptation strategies, at the local or catchment scale, to the targets agreed to in Paris.

A risk-based framework for quantifying hydroclimatic extremes gives rigour to policy and management decisions because the damage caused can be assessed in relation to the likelihood of the event, and the costs of providing a given level of protection can be assessed against the benefits across the range of plausible future event sizes (Hall and Borgomeo 2013). Risk has three components: *hazard* is the phenomenon with the potential to do harm, *exposure* refers to the assets or people subject to the hazard, and *vulnerability* characterizes the susceptibility to loss when exposed.

Risk itself is a measure of the probability of an event, and is often based on historical information, under the assumption that the catalogue of events in the past are independent of each other and that each event is drawn from the same distribution of past events. The probability of a hazard is often described using Annual Exceedance Probability (AEP), which is the chance of an event of a given magnitude occurring in any given year. For example, a flood with an AEP of 0.05 has a 5% probability of occurring in any given year. This probability can be expressed – sometimes unhelpfully – as a return period, which is defined as 1/AEP, in this case a 1-in-20-year flood. In order to avoid the erroneous interpretation of regularity, the use of return periods to communicate probability is not recommended (Volpi et al. 2015).

A major concern in the face of climate and other environmental changes is that these external shifts in climate and land cover do not simply alter the magnitude of an individual event – they alter the distribution itself. In other words, the distribution of extremes is non-stationary (Milly et al. 2008, 2015). Under non-stationary conditions the assumption that the past offers an unbiased assessment of future hydroclimatic risk is no longer valid.

Statistical techniques for dealing with non-stationarity are emerging in the hydrological literature, although in many hydrological systems, persistence – that is, the tendency of the system to remain preferentially in one of several states – can lead the analyst to conclude that the system is non-stationary when in fact there is simply insufficient data to characterize the full range of historical variability (Vogel et al. 2011).

2.5.2 Planning for Extremes in Flood Risk and Water Resources Management

Quantifying and managing this combined risk requires a complex process chain which may begin with the study of hydrometeorological variability, extremes, or both, but rarely ends there, instead requiring a series of models each to feed its output into the next in order to calculate the response of large-scale processes into local-scale human or natural systems (Wilby and Dessai 2010) (Figure 2.6). Each step in the end-to-end modelling chain introduces uncertainty into the final calculation. For example, in studies of the impacts of climate change on hydrology, the projected response of the climate system is obtained from GCMs with outputs that include several variables of hydrological interest, such as precipitation, temperature, or energy fluxes. Yet, the coarse spatial resolution of GCM projections (typically ~100–150 km, although in specific high-resolution configurations now reaching 25 km horizontal resolution) is often poorly suited to the assessment of hydrological implications at the catchment scale, so it is usually transformed to a finer resolution using dynamic or statistical downscaling methods (Maurer and Hidalgo 2008).

Dynamic downscaling uses regional climate models (RCMs) to produce high-resolution simulations (~12–25 km, although in some recent experiments down to ~1 km) for a specific region. RCM simulations are constrained at their boundaries by GCM output, but otherwise free to produce physically consistent hydrometeorological data within their limited-area boundaries. By contrast, statistical downscaling offers efficient computation and simple data manipulation (Wilby and Dawson 2012). This method establishes relationships between output of large-scale GCMs or RCMs and local climatic variables in order to predict future states of such variables (Maurer and Hidalgo 2008). Downscaled meteorological forcing is used to drive hydrological, hydrodynamic or other physical models, which generate future responses in hydrological variables including soil moisture, groundwater levels, snowpack properties, river flow, or inundation levels and extents. Finally, impacts models are added to evaluate the impacts on human, economic and natural systems.

Within this framework, several studies have been able to infer impacts of climate disturbances in global or regional economies, societies and infrastructure (e.g. Alfieri et al. 2016; Ward et al. 2013; Winsemius et al. 2018). At the local scale, by applying transfer functions, several analysts have estimated runoff sensitivities and subsequent impacts of climate change in sectors such as irrigation, water supply and hydropower (e.g. Ficklin et al. 2013; Hay et al. 2000; Vano et al. 2010).

2.5.3 Comparing Top-down with Bottom-up Approaches

In each step of the process chain outlined above, the uncertainty introduced by each model subcomponent becomes compounded in the next. At the end of a long process chain the accumulated uncertainty in the system response can be so large that the end result does little to constrain the options that might reliably be chosen by those responsible for implementing any response. Moreover, uncertainties in demographic trends,

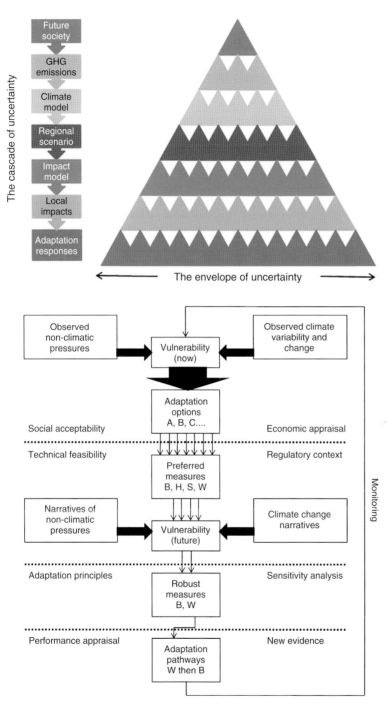

Figure 2.6 The cascade of uncertainty in projections of the impacts of climate change on water resources systems and its impact on water resources decision-making. (a) The cascade of uncertainty. (b) Approaches to decision scaling to overcome uncertainty. *Source:* Wilby and Dessai (2010). Reproduced with permission of John Wiley and Sons.

land use projections, technological transformations, or economic growth hamper the selection of a most likely scenario upon which decisions can be based.

Alternatives to the top-down approach, which use GCM data to predict hydroclimatic risks to design suitable policy and management responses, include a range of methods that turn the question on its head and ask: what determines the resilience of the system being managed and how might future change move the performance of that system out of acceptable tolerances (Brown et al. 2012; Hallegatte 2009; Herman et al. 2015; Ludwig et al. 2014; Wilby and Dessai 2010)? This alternative approach requires the analyst to evaluate the present characteristics of the managed, or planned, system in order to determine metrics and thresholds that would define its acceptable state. The performance of these metrics is then tested against a range of possible climatic and non-climatic scenarios.

Several bottom-up methods exist, such as decision scaling (Brown et al. 2012, 2015), the scenario-neutral approach for managing flood risks (Prudhomme et al. 2010), or information-gap decision theory (Hine and Hall 2010). In the decision-scaling method, for example, plausible climate scenarios are given by climate models, and the level of complexity of data and models required scales up according to the types of risks encountered. Next, a variety of management or policy actions are tested against the range of plausible futures to examine the response of performance metrics. This permits the identification of robust solutions that would perform well independently of any particular realised scenario (Groves et al. 2008).

Box 2.2 – Climate Resilience in the Masai Mara – a Scenario-Neutral Analysis

The Mara river nurtures some of the last wild landscapes of the Masai-Mara and Serengeti reserves in Kenya and Tanzania. Tourism from wildlife ecosystems is an important source of revenue for the countries, but as the countries chart a course towards water and energy security, dams as water storage structures and for producing hydropower are planned on the rivers (McClain 2013). Water security is important in the basin as it performs the difficult balancing act of managing growth and increasing demands with the ecological requirements of the river. The Mara and its tributaries are highly variable rivers, with added uncertainties of insufficient resources, data and infrastructure for water resources management.

There is considerable uncertainty in the way that East African rainfall responds to global climate change, including lack of observational data, previous research and climate complexity. Although most models show an increase in rainfall, a recent increase in droughts has been observed in East Africa (Rowell et al. 2015).

Vora (2018) used WEAP21 (Seiber and Purkey 2015) to model the climate, hydrology and demands in the Mara basin with a scenario-independent approach (Ray and Brown 2015) to test the climate sensitivity of policy responses such as dam building, inter-basin transfers, improvement in compliance with agricultural water use permits, and introduction of piped water supply. The results showed that natural variability and increases in demand are just as likely to impact the management of reserve flows in the Mara river basin as climate change. Thus, policy responses designed for water allocations in the Mara river basin should be tested for robustness to the risks posed by increased demand and natural variability of the system alongside a focus on climate adaptation.

The popularity of this family of approaches can be seen in studies that explore the sensitivity of water supply systems under climate scenarios and other uncertain stressors in areas such as the Great Lakes of North America, the Saskatchewan River Basin in Canada, Lima, Peru, and the Ashokan Reservoir in New York (Hassanzadeh et al. 2016; Kalra et al. 2015; Moody and Brown 2012). Similarly, bottom-up approaches have been used to assess potential climate threats to planned infrastructure investments in areas such the Upper and Middle Niger Basin and in Cameroon (Ghile et al. 2014; Grijsen and Patel 2014). This approach has also been applied in the Netherlands to assess the reliability of current water management strategies with sea-level rise and climate change (Kwadijk et al. 2010).

2.6 Conclusion

This chapter has considered the role of hydroclimatic variability and extremes in determining water availability and hydrological hazards. Our emerging understanding of climatic drivers of variability is revealing new modes of behaviour in the climate system, which help improve predictions across a range of timescales, from short-term forecasts through seasonal predictions to climate projections long into the future. A pressing challenge remains to translate improvements in predicting climate and weather-related phenomena to better predictions of water availability and extremes. In many ways this challenge can be addressed only by reconciling the spatial and temporal scales at which predictions are available to support decisions, with those scales on which decisions are made. As expected, much of the scientific literature in hydrometeorology and climate science concerns and often concludes with reference to uncertainty. Yet the challenge for policy-makers and water managers is to make sound decisions in the face of uncertainty.

References

Alfieri, L., Feyen, L., Salamon, P. et al. (2016). Modelling the socio-economic impact of river floods in Europe. *Natural Hazards and Earth System Sciences* 16: 1401–1411.

Allen, M.R. and Ingram, W.J. (2002). Constraints on future changes in climate and the hydrological cycle. *Nature* 419: 224–232.

Barry, R.G. and Chorley, R.J. (2009). *Atmosphere, Weather and Climate*. Abingdon: Routledge.

Bates, B.C., Kundzewicz, Z.W., Wu, S., and Palutikof, J.P. (2008). *Climate Change and Water*. Geneva: IPCC Secretariat.

Brown, C., Ghile, Y., Laverty, M., and Li, K. (2012). Decision scaling: linking bottom-up vulnerability analysis with climate projections in the water sector. *Water Resources Research* 48 (9) https://doi.org/10.1029/2011WR011212.

Brown, C.M., Lund, J.R., Cai, X. et al. (2015). The future of water resources systems analysis: toward a scientific framework for sustainable water management. *Water Resources Research* 51 (8): 6110–6124.

Christensen, J.H., Krishna Kumar, K., Aldrian, E. et al. (2013). Climate phenomena and their relevance for future regional climate change. Box 14.1, Figure 1. In: *Climate Change 2013: The Physical Science Basis* (eds. T.F. Stocker et al.). Cambridge, UK and New York, NY, USA: Cambridge University Press.

Ficklin, D.L., Stewart, I.T., and Maurer, E.P. (2013). Climate change impacts on streamflow and subbasin-scale hydrology in the Upper Colorado River Basin. *PLoS One* 8: e71297.

Ghile, Y., Taner, M., Brown, C. et al. (2014). Bottom-up climate risk assessment of infrastructure investment in the Niger River Basin. *Climatic Change* 122 (1): 97–110.

Grijsen, J. and Patel, H. (2014). *Understanding the impact of climate change on hydropower: the case of Cameroon – climate risk assessment for hydropower generation in Cameroon.* Report No. 87913. Washington, DC: World Bank Group.

Groves, D.G., Yates, D., and Tebaldi, C. (2008). Developing and applying uncertain global climate change projections for regional water management planning. *Water Resources Research* 44 (12) https://doi.org/10.1029/2008WR006964.

Hahn, D.G. and Manabe, S. (1975). The role of mountains in the South Asian monsoon circulation. *Journal of the Atmospheric Sciences* 32: 1515–1541.

Hall, J. and Borgomeo, E. (2013). Risk-based principles for defining and managing water security. *Philosophical Transactions of the Royal Society A: Mathematical, Physical and Engineering Sciences* 371 (2002): 20120407.

Hall, J.W., Grey, D., Garrick, D. et al. (2014). Coping with the curse of freshwater variability. *Science* 346: 429–430.

Hallegatte, S. (2009). Strategies to adapt to an uncertain climate change. *Global Environmental Change* 19: 240–247.

Hassanzadeh, E., Elshorbagy, A., Wheater, H. et al. (2016). Integrating supply uncertainties from stochastic modeling into integrated water resource management: case study of the Saskatchewan River Basin. *Journal of Water Resources Planning and Management* 142 (2): 5015006.

Hawkins, E. and Sutton, R. (2009). The potential to narrow uncertainty in regional climate predictions. *Bulletin of the American Meteorological Society* 90: 1095–1107.

Hay, L.E., Wilby, R.L., and Leavesley, G.H. (2000). A comparison of delta change and downscaled GCM scenarios for three mountainous basins in the United States. *Journal of the American Water Resources Association* 36: 387–397.

Herman, J., Reed, P., Zeff, H., and Characklis, G. (2015). How should robustness be defined for water systems planning under change? *Journal of Water Resources Planning and Management* 141: 4015012.

Hine, D. and Hall, J.W. (2010). Information gap analysis of flood model uncertainties and regional frequency analysis. *Water Resources Research* 46: W01514. https://doi.org/10.1029/2008WR007620.

Hurrell, J.W. (1995). Decadal trends in the North Atlantic Oscillation: regional temperatures and precipitation. *Science* 269: 676–679.

IPCC (2000). *Special Report on Emissions Scenarios.* Cambridge, UK: Cambridge University Press.

IPCC (2012). *Managing the Risks of Extreme Events and Disasters to Advance Climate Change Adaptation* (eds. C.B. Field, V. Barros, T.F. Stocker, et al.). Cambridge, UK, and New York, NY, USA: Cambridge University Press.

Kalra, N., Groves, D.G., Bonzanigo, L. et al. (2015). *Robust decision-making in the water sector: a strategy for implementing Lima's long-term water resources master plan*, Policy research working paper No. WPS 7439. Washington, DC: World Bank Group.

Kendon, E.J., Roberts, N.M., Fowler, H.J. et al. (2014). Heavier summer downpours with climate change revealed by weather forecast resolution model. *Nature Climate Change* 4: 570–576.

Kiehl, J.T. and Trenberth, K.E. (1997). Earth's annual global mean energy budget. *Bulletin of the American Meteorological Society* 78: 197–208.

Knutti, R. and Sedláček, J. (2013). Robustness and uncertainties in the new CMIP5 climate model projections. *Nature Climate Change* 3: 369.

Kwadijk, J.C.J., Haasnoot, M., Mulder, J.P.M. et al. (2010). Using adaptation tipping points to prepare for climate change and sea level rise: a case study in the Netherlands. *Wiley Interdisciplinary Reviews: Climate Change* 1: 729–740.

Le Treut, H., Somerville, R., Cubasch, U. et al. (2007). Historical overview of climate change. In: *Climate Change: The Physical Science Basis. Contribution of Working Group I to the Fourth Assessment Report of the Intergovernmental Panel on Climate Change* (eds. S. Solomon et al.), 94–127. Cambridge, UK and New York, USA: Cambridge University Press.

Ludwig, F., Van Slobbe, E., and Cofino, W. (2014). Climate change adaptation and Integrated Water Resource Management in the water sector. *Journal of Hydrology* 518: 235–242.

Maraun, D., Wetterhall, F., Ireson, A.M. et al. (2010). Precipitation downscaling under climate change: recent developments to bridge the gap between dynamical models and the end user. *Reviews of Geophysics* 48 (3) https://doi.org/10.1029/2009RG000314.

Maurer, E.P. and Hidalgo, H.G. (2008). Utility of daily vs. monthly large-scale climate data: an intercomparison of two statistical downscaling methods. *Hydrology and Earth System Sciences* 12: 551–563.

McClain, M.E. (2013). Balancing water resources development and environmental sustainability in Africa: a review of recent research findings and applications. *AMBIO* 42: 549–565.

McIlveen, R. (2010). *Fundamentals of Weather and Climate*. Oxford: Oxford University Press.

McPhaden, M.J., Zebiak, S.E., and Glantz, M.H. (2006). ENSO as an integrating concept in earth science. *Science* 314 (5806): 1740–1745.

Milly, P.C.D., Betancourt, J., Falkenmark, M. et al. (2008). Stationarity is dead: whither water management? *Science* 319 (5863): 573–574.

Milly, P.C., Betancourt, J., Falkenmark, M. et al. (2015). On critiques of "stationarity is dead: whither water management?". *Water Resources Research* 51 (9): 7785–7789.

Moody, P. and Brown, C. (2012). Modeling stakeholder-defined climate risk on the Upper Great Lakes. *Water Resources Research* 48 (10) https://doi.org/10.1029/2012WR012497.

National Infrastructure Commission (2017). *The Impact of the Environment and Climate Change on Future Infrastructure Supply and Demand*. London: National Infrastructure Commission.

Neelin, J.D., Battisti, D.S., and Hirst, A.C. (1998). ENSO theory. *Journal of Geophysical Research: Oceans* 103: 14261–14290.

O'Hare, G., Sweeney, J., and Wilby, R. (2004). *Weather, Climate and Climate Change: Human Perspectives*. Abingdon: Routledge.

Paltan, H., Allen, M., Haustein, K. et al. (2018). Global implications of 1.5°C and 2°C warmer worlds on extreme river flows. *Environmental Research Letters* 13: 094003. https://doi.org/10.1088/1748-9326/aad985.

Prudhomme, C., Wilby, R.L., Crooks, S. et al. (2010). Scenario-neutral approach to climate change impact studies: application to flood risk. *Journal of Hydrology* 390 (3–4): 198–209.

Ray, P.A. and Brown, C.M. (2015). *Confronting Climate Uncertainty in Water Resources Planning and Project Design – A Decision Tree Framework*. Washington, D.C.: The World Bank.

Robinson, M. and Ward, R.C. (2017). *Hydrology: Principles and Processes*. London: IWA Publishing.

Rowell, D.P., Booth, B.B., Nicholson, S.E., and Good, P. (2015). Reconciling past and future rainfall trends over East Africa. *Journal of Climate* 28: 9768–9788.

Sadoff, C.W., Hall, J.W., Grey, D. et al. (2015). *Securing Water, Sustaining Growth*. Report of the GWP/OECD Task Force on Water Security and Sustainable Growth. Oxford: University of Oxford.

Seiber, J. and Purkey, D. (2015). *WEAP – Water Evaluation and Planning System User Guide for WEAP 2015*. Somerville: Stockholm Environment Institute.

Slingo, J. and Palmer, T. (2011). Uncertainty in weather and climate prediction. *Philosophical Transactions of the Royal Society A: Mathematical, Physical and Engineering Sciences* 369 (1956): 4751–4767.

Taylor, K.E., Stouffer, R.J., and Meehl, G.A. (2012). An overview of CMIP5 and the experiment design. *Bulletin of the American Meteorological Society* 93 (4): 485–498.

Ting, M., Kushnir, Y., Seager, R., and Li, C. (2011). Robust features of Atlantic multidecadal variability and its climate impacts. *Geophysical Research Letters* 38 https://doi.org/10.1029/2011GL048712.

Trenberth, K.E. (2011). Changes in precipitation with climate change. *Climate Research* 47: 123–138.

Van Vuuren, D.P., Edmonds, J., Kainuma, M. et al. (2011). The representative concentration pathways: an overview. *Climatic Change* 109: 5–31.

Vano, J.A., Scott, M.J., Voisin, N. et al. (2010). Climate change impacts on water management and irrigated agriculture in the Yakima River Basin, Washington, USA. *Climatic Change* 102: 287–317.

Vogel, R.M., Yaindl, C., and Walter, M. (2011). Nonstationarity: flood magnification and recurrence reduction factors in the United States. *Journal of the American Water Resources Association* 47: 464–474.

Volpi, E., Fiori, A., Grimaldi, S. et al. (2015). One hundred years of return period: strengths and limitations. *Water Resources Research* 51 (10): 8570–8585.

Vora, S. (2018). Managing the Mara: Towards a risk-based framework for water management in the Mara River, Kenya. MSc thesis. University of Oxford.

Ward, P., Jongman, B., Weiland, F. et al. (2013). Assessing flood risk at the global scale: model setup, results, and sensitivity. *Environmental Research Letters* 8: 44019–44019.

Wilby, R.L. and Dawson, C.W. (2012). The Statistical DownScaling Model: insights from one decade of application. *International Journal of Climatology* 33 (7): 1707–1719.

Wilby, R.L. and Dessai, S. (2010). Robust adaptation to climate change. *Weather* 65 (7): 180–185.

Wilks, D.S. (2012). Stochastic weather generators for climate-change downscaling, part II: multivariable and spatially coherent multisite downscaling. *Wiley Interdisciplinary Reviews: Climate Change* 3 (3): 267–278.

Winsemius, H.C., Jongman, B., Veldkamp, T.I.E. et al. (2018). Disaster risk, climate change, and poverty: assessing the global exposure of poor people to floods and droughts. *Environment and Development Economics* 23 (3): 328–348.

World Bank (2017). *Turbulent Waters: Pursuing Water Security in Fragile Contexts*. World Bank Group: Washington, DC.

World Economic Forum (2017). *Global Risks 2017: A Global Risk Network Report*. Geneva: World Economic Forum.

Zhang, C. (2005). Madden-Julian oscillation. *Reviews of Geophysics* 43 (2) https://doi.org/10.1029/2004RG000158.

3

Groundwater Resources

Past, Present, and Future

Abi Stone[1], Michelle Lanzoni[2], and Pauline Smedley[3]

[1] *University of Manchester, UK*
[2] *School of Geography and the Environment, University of Oxford, UK*
[3] *British Geological Survey, Keyworth, Nottingham, UK*

3.1 Introduction to Groundwater Science

Current scientific research and understanding brings groundwater into sharp focus, relieving pressure on the well-known adage of 'out of sight, out of mind'. This increasing clarity on groundwater as a resource includes: (i) estimates of how much groundwater exists globally (Döll et al. 2012; Margat and van der Gun 2013); (ii) recognition that the spatial distribution of groundwater is often more evenly distributed than that of surface water (Margat and van der Gun 2013; Morris et al. 2003); (iii) appreciation of how groundwater quality relates both to natural water–rock interactions (Edmunds and Shand 2009), and to the addition of pollutants (Morris et al. 2003); and (iv) the framing of groundwater as situated within the physical spheres of the Earth, and 'entrenched in a web of interdependencies' (van der Gun 2012, p. 3), which ensures groundwater science is not considered in isolation from policy and management. Whilst groundwater is estimated to constitute 97% of liquid freshwater on Earth, the total physical amount of groundwater is not the same as abstractable groundwater, nor as renewable groundwater, nor as that considered high enough quality for use. Not all groundwater is part of the active hydrogeological cycle – less than 6% of it is estimated to have been emplaced within the last 50 years (Gleeson et al. 2016). When it comes to the concept of exploitable groundwater resources, a sound scientific understanding is necessary but not sufficient for assessing what water can, and will, be used, because use is dependent on the strategies employed within water resources management (Margat and van der Gun 2013). Currently only ~20% of the world's water needs are met from groundwater, of which 70% goes to irrigation, 21% to domestic uses (including drinking water) and 9% to industry or mining.

Understanding groundwater is intricately linked to understanding geology – rocks are the static foundation underpinning the flow pathways and fluxes of groundwater, and also the stores (which move slowly). It is important to consider the position and role of groundwater within the hydrological cycle, the Earth's spheres, and economic,

Water Science, Policy, and Management: A Global Challenge, First Edition. Edited by Simon J. Dadson, Dustin E. Garrick, Edmund C. Penning-Rowsell, Jim W. Hall, Rob Hope, and Jocelyne Hughes.
© 2020 John Wiley & Sons Ltd. Published 2020 by John Wiley & Sons Ltd.

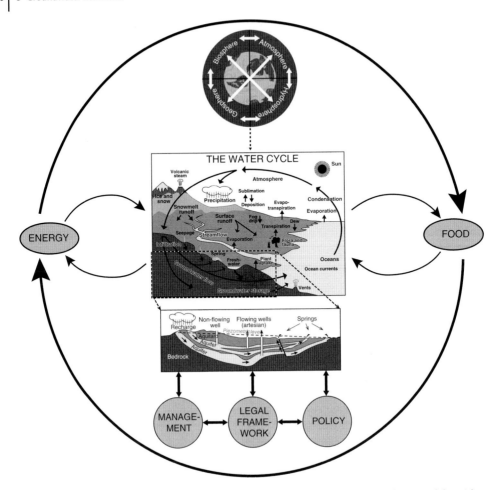

Figure 3.1 Conceptual diagram of groundwater within the water cycle, the Earth's spheres and the wider 'web of interdependencies'. *Source:* Abi Stone. (*See color representation of this figure in color plate section*).

political and social contexts (Figure 3.1). This allows us to appreciate that groundwater is not just a sluggish component of the hydrological cycle situated within the lithosphere, but is interconnected with the atmosphere (water inputs) and cryosphere (store of frozen freshwater and an input during melt). Within the lithosphere there is a zone of transport (the unsaturated zone, USZ) and a zone of groundwater storage and flow, and both zones are a medium for water–rock interactions. Groundwater also has clear interdependencies with the biosphere.

Scientific insights have a wide and important role in addressing the key challenges of groundwater over-abstraction, deterioration in water quality, and adaptation to climate change. Scientific developments include technical innovations, such as the use of managed artificial recharge. Hydrogeological science does not stand in isolation, but is closely linked with understanding hydroclimate science and its variability (Chapter 2), freshwater ecology (Chapter 5), monitoring and modelling of hydrological processes (Chapter 7), and with the social science and humanities approaches taken within groundwater policy (Chapters 8–13) and management (Chapters 14–19).

This chapter aims to provide a (necessarily) selective and partial overview of groundwater science with some insight into where the science connects to policy and management. Whilst it is desirable that the use and protection of groundwater resources should be based on sound scientific understanding, it is pragmatic to recognise that 'national projections for (groundwater) … abstraction are almost never specified by (the) source of water' (Margat and van der Gun 2013, p. 129), and are driven by demand. Not all countries have sustainable development policies towards groundwater use, although use is also influenced by the economic costs and risks associated with over-abstraction, including salinization and resource 'mining'. This chapter focuses on groundwater quantity and quality and explores the link between groundwater and climatic change, over long timescales, in the modern day, and into the future. A brief overview of two other key issues will be given – urban groundwater, and managed artificial recharge. Readers will find that groundwater re-emerges throughout the book, and you are invited to look for, and integrate, the linkages between hydrogeology and the facets of policy and management in later chapters.

3.2 Quantities of Groundwater: Storage, Recharge, and Abstraction

3.2.1 What Do We Know?

The basis of understanding groundwater is to know the geology of a region and how the water interacts with the different types of rock, physically and chemically: different rock types have different abilities to store and to transmit water, based on their composition, and related properties of porosity, permeability and solubility. There are a variety of ways to subdivide rock types into hydrogeological environments (Table 3.1 presents two approaches). Furthermore, the configuration of rocks controls whether an aquifer (water-bearing, permeable rock) is unconfined (in direct contact with the atmosphere, including through open pore spaces of soil cover) or confined (covered by a layer of rock). When confined groundwater is being held at positive pressure by surrounding rock layers, it is known as artesian (pertaining to Artois, France, where monks first sunk wells into such an aquifer in the twelfth century); the pressure causes water to rise up within any well drilled into it, and when it reaches the surface it is described as flowing. A formalization of the laws governing groundwater flow relies on the work of a French municipal engineer, Henri Philiber Gaspart Darcy, in the 1850s, and the equations of Pierre-Simon Marquis de Laplace, in the 1770s (see Hiscock and Bense 2014, for a thorough treatment of physical hydrogeology).

To make an estimate of groundwater volume, information about aquifer composition and its regional structure (thickness, lateral extent) is combined with measurements of the depth of the groundwater table and its thickness (from the base of the aquifer unit to the groundwater table). We must also know how fast and in what direction groundwater is flowing, because aquifers (like rivers) are transport channels, not just reservoirs. Global-scale estimates of the total volume of groundwater vary from 8 to 10 million km^3 (van der Gun 2012), to 22.6 million km^3 (Gleeson et al. 2016). A meaningful insight into the heterogeneity of this resource requires a spatial division into hydrogeological classes (e.g. WHYMAP 2008) (Figure 3.2a), or aquifer systems (similar to approaches used for surface water river basins) (Margat and van der Gun 2013) (Figure 3.2b).

Table 3.1 A comparison of two categorization approaches for principal hydrogeological environments.

	Morris et al. (2003)		MacDonald et al. (2012)	
Type	Hydrogeological environment	Lithology	Aquifer flow/storage type	Effective porosity
Unconsolidated	Alluvial and coastal plain	G, Sa, Si, C	Unconsolidated	25–30%
	Intermontane colluvial and volcanic	P, G, Sa, C & interbedded LT		
	Glacial and minor alluvial	B, P, G, Sa, Si, Sa-C		
	Loessic	Si, fine Sa, C		
Consolidated	Sedimentary	Sandstone		
			inter-granular	25–30%
		Limestone	fractured	3–15%
			both	10–30%
	Recent coastal calcareous	Limestone and calcareous sand		
	Extensive volcanic terrain	LT and ash intercalations	Volcanic	1–10%
	Weathered basement complex	Crystalline rocks	Basement	1–10%

Source: Adapted from Morris et al. (2003) and MacDonald et al. (2012).
B = boulders, P = pebbles, G = gravel, Sa = sand, Si = silt, C = clay, LT = lava and tuffs.

Groundwater flows are in flux, not static, and aquifers are open systems. Groundwater quantity, therefore, needs to be considered in relation to inputs (recharge) and outputs (both natural outflows and abstraction). There is also a need: (i) to consider whether groundwater is part of the active hydrological cycle, in order to establish whether active recharge to groundwater is occurring; (ii) to consider what proportion of losses is countered by recharge; and (iii) to understand whether groundwater is renewable. Quantifying recharge is one of two approaches to calculating total groundwater flux through an aquifer (called the upflow side approach); the other (called the downflow side approach) is based on analysing data for groundwater discharge, including baseflow and estimates of the less visible outputs to oceans, lakes, neighbouring aquifers, as well as human abstraction (Margat and van der Gun 2013).

There is an array of approaches to assess recharge, which can either be field-based or remote, such as using satellite data or modelling approaches. The field-based approaches include measuring tracers within the USZ (or vadose zone) and below the groundwater table, and installing physical equipment in the USZ, such as lysimeters, which measure changes to moisture content within a soil/USZ (Allen et al. 1991). In semi-arid and arid environments USZ moisture contents are low, and it is difficult to measure small (and heterogeneous) quantities of precipitation and changes in sediment moisture, which makes physical methods and numerical modelling challenging. Environmental tracers

(a)

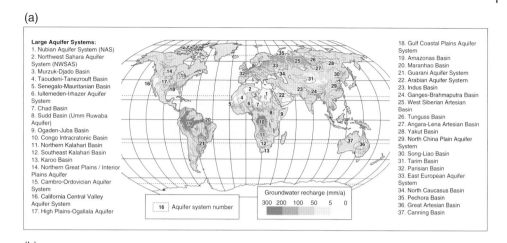

Large Aquifer Systems:
1. Nubian Aquifer System (NAS)
2. Northwest Sahara Aquifer System (NWSAS)
3. Murzuk-Djado Basin
4. Taoudeni-Tanezrouft Basin
5. Senegalo-Mauritanian Basin
6. Iullemeden-Irhazer Aquifer System
7. Chad Basin
8. Sudd Basin (Umm Ruwaba Aquifer)
9. Ogaden-Juba Basin
10. Congo Intracratonic Basin
11. Northern Kalahari Basin
12. Southeast Kalahari Basin
13. Karoo Basin
14. Northern Great Plains / Interior Plains Aquifer
15. Cambro-Ordovician Aquifer System
16. California Central Valley Aquifer System
17. High Plains-Ogallala Aquifer

18. Gulf Coastal Plains Aquifer System
19. Amazonas Basin
20. Maranhao Basin
21. Guarani Aquifer System
22. Arabian Aquifer System
23. Indus Basin
24. Ganges-Brahmaputra Basin
25. West Siberian Artesian Basin
26. Tunguss Basin
27. Angara-Lena Artesian Basin
28. Yakut Basin
29. North China Plain Aquifer System
30. Song-Liao Basin
31. Tarim Basin
32. Parisian Basin
33. East European Aquifer System
34. North Caucasus Basin
35. Pechora Basin
36. Great Artesian Basin
37. Canning Basin

16 Aquifer system number

Groundwater recharge (mm/a)
300 200 100 50 5 0

(b)

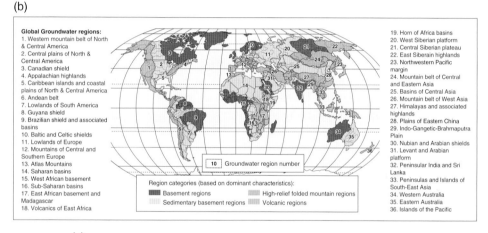

Global Groundwater regions:
1. Western mountain belt of North & Central America
2. Central plains of North & Central America
3. Canadian shield
4. Appalachian highlands
5. Caribbean islands and coastal plains of North & Central America
6. Andean belt
7. Lowlands of South America
8. Guyana shield
9. Brazilian shield and associated basins
10. Baltic and Celtic shields
11. Lowlands of Europe
12. Mountains of Central and Southern Europe
13. Atlas Mountains
14. Saharan basins
15. West African basement
16. Sub-Saharan basins
17. East African basement and Madagascar
18. Volcanics of East Africa

19. Horn of Africa basins
20. West Siberian platform
21. Central Siberian plateau
22. East Siberian highlands
23. Northwestern Pacific margin
24. Mountain belt of Central and Eastern Asia
25. Basins of Central Asia
26. Mountain belt of West Asia
27. Himalayas and associated highlands
28. Plains of Eastern China
29. Indo-Gangetic-Brahmaputra Plain
30. Nubian and Arabian shields
31. Levant and Arabian platform
32. Peninsular India and Sri Lanka
33. Peninsulas and Islands of South-East Asia
34. Western Australia
35. Eastern Australia
36. Islands of the Pacific

10 Groundwater region number

Region categories (based on dominant characteristics):
☐ Basement regions   High-relief folded mountain regions
☐ Sedimentary basement regions   Volcanic regions

(c)

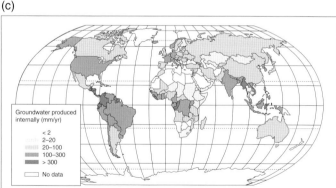

Groundwater produced internally (mm/yr)
< 2
2–20
20–100
100–300
> 300
No data

Figure 3.2 Global assessments of groundwater volumes, including (a) WHYMAP (2008) depicting the outline of the 37 'mega aquifers' and grey shading for recharge rates in the major groundwater basins. *Source:* see WHYMAP (2008) for a full version of the figure, depicting in three colours additional information about the recharge rate in two other groundwater types (areas with complex hydrogeological structure and areas with local and shallow aquifers) (reproduced in modified form with permission from WHYMAP); (b) the 36 global GW regions (IGRAC 2009), which are different from the 37 'mega aquifers'. *Source:* Margat and van der Gun (2013). Reproduced with permission from IGRAC; and (c) using country borders. *Source:* Authors using AQUASTAT data, FAO (2016). (*See color representation of this figure in color plate section*).

(entering meteorically) and applied tracers (added artificially) that behave conservatively are extremely useful for tracking the amount of moisture moving to recharge groundwater. These can be applied in both arid and humid environments, and include chloride, sulphate, tritium, oxygen isotopes and chlorine-36 (see Scanlon et al. 2002; Box 3.1).

It may seem surprising that approaches to collecting groundwater abstraction data are similarly diverse. However, given a lack of full regulation of groundwater boreholes and wells, scientific approaches that monitor abstraction volume are useful. Those approaches based on data are: (i) integrating the change in groundwater head (liquid pressure above a geodetic datum) across an aquifer using point measurements multiplied by the storage coefficient (volume of water released per unit decline in hydraulic head) (Konikow and Kendy 2005); (ii) inference from establishing the loss of water pressure, revealed by mapping the resulting land-surface subsidence (e.g. Ortiz-Zamora and Ortega-Guerrero 2010); and (iii) inferring changes from geophysical gravity measurements using the GRACE (Gravity Recovery and Climate Experiment) satellite (e.g. Castelazzi et al. 2016). Modelling approaches are also used, such as macro-scale hydrological models coupled with water demand models to simulate withdrawal and consumptive use (e.g. Wada et al. 2014).

The best global- and national-scale compilation of groundwater abstraction data can be found in Appendix 5 of Margat and van der Gun (2013). Estimates of global use for the three main usage categories, irrigation, domestic and industrial, are 688, 209 and 85 km^3/year respectively, and national-scale data are depicted in Figure 3.3a–c. Figure 3.3d–g depicts the share of freshwater supply that is contributed from groundwater (globally this is 25% for irrigation, 45% for domestic and 15% for industrial uses), to highlight where groundwater is the principal source (Margat and van der Gun 2013).

There are three groundwater abstraction policy and management points to highlight. First, a 'silent revolution' of increased abstraction in arid and semi-arid regions (Llamas and Martínez-Santos 2005) has taken place, enabled by a large reduction in the cost of groundwater abstraction, owing to technological advances in well-drilling and submersible pumps. A second, that of addressing the problem of over-abstraction, is not only a problem of data scarcity (addressed by global state-of-play studies such as Seibert et al. (2010) for irrigation), but also data interpretation challenges (Konikow and Kendy 2005), effective communication of that data, and appropriate management, policy and governance structures in order to implement change(s) to use. Changes to policy and management require political will to balance priorities within water use and water conservation. Thirdly, the impacts of (over)abstraction go beyond a reduction in the groundwater resource, involving serious challenges relating to ground subsidence, such as the well-known example of Mexico City (Ortiz-Zamora and Ortega-Guerrero 2010).

3.2.2 Future Outlook on Measuring Groundwater Quantity

In celebrating the wealth of information about groundwater, it is important to recognize limitations to the available scientific data, in terms of gaps, uncertainties and inconsistencies. This relates, in part, to a non-uniformity of adopted definitions of variables (such as abstraction and recharge) and concepts (such as non-renewable and over-exploitation) across the world. Here we draw attention to five key areas of focus for the future of the science of groundwater quantity, recognizing that this is not an exhaustive list.

(a)

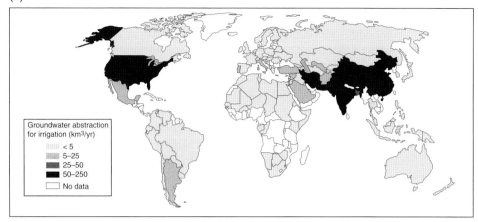

(b)

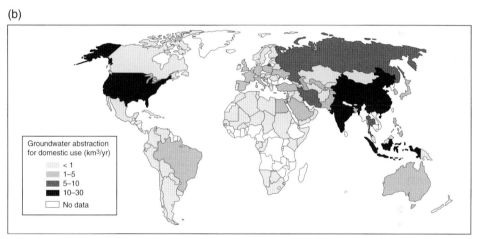

(c)

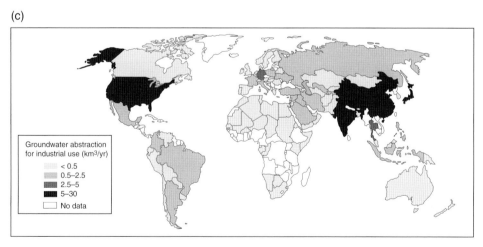

Figure 3.3 Estimates of current groundwater withdrawal as rates and as share of total freshwater withdrawal by country: withdrawal rates for (a) irrigation, (b) domestic supply, (c) industrial use; and share of groundwater in total freshwater in each country, (d) overall, and for (e) irrigation, (f) domestic supply and (g) industry. *Source:* Adapted from Margat and van der Gun © (2013). Reproduced by permission of Taylor and Francis Books UK. (*See color representation of this figure in color plate section*).

(d)

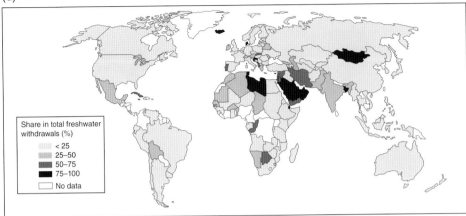

(e)

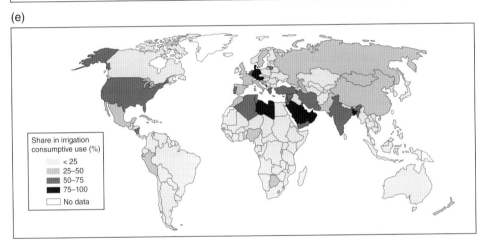

(f)

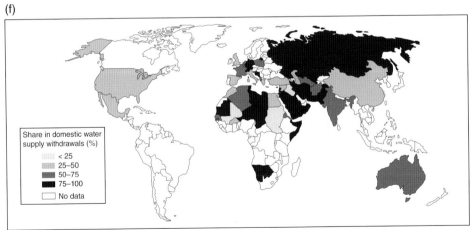

Figure 3.3 (*Cont'd*)

(g)

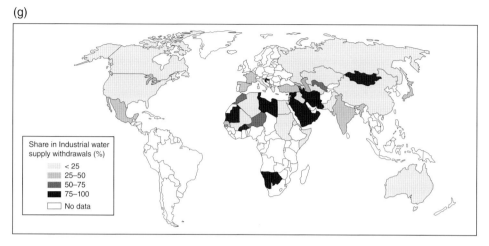

Figure 3.3 (*Cont'd*)

Firstly, there is a continuing need to quantify and map groundwater stores and fluxes at the subregional scale, which is the scale at which people (not just national populations) access and use water. Effective strategies for the use and management of groundwater resources need to be based on scientifically well-founded approaches. The focus on 37 mega aquifers by the World-wide Hydrogeological Mapping and Assessment Programme (WHYMAP) (Figure 3.2a), or 36 global groundwater regions by the International Groundwater Resources Assessment Centre (IGRAC) (Figure 3.2b), means that aquifers between these areas run the risk of remaining data-poor.

Secondly, routine use of GRACE satellites to measure groundwater mass changes, coupled with InSAR (Interferometry of Synthetic Aperture Radar) techniques to quantify land-surface subsidence (due to changes in fluid pressure) will continue to be a powerful approach for remotely detecting the depletion and dynamics of groundwater (e.g. Castelazzi et al. 2016). A third area of focus is the use of smart handpumps, which have the potential to turn handpumps into a real-time groundwater monitoring network in rural areas where groundwater data are often otherwise scarce (e.g. Chapter 7, this volume; Colchester et al. 2017).

The continued incorporation of numerical groundwater modelling is a fourth key area of focus. Since its development in 1988, the US Geological Survey (USGS) software MODFLOW is considered the global industrial standard for groundwater modelling (Zhou and Li 2011), due to the comprehensive coverage of hydrogeological processes and the flexibility offered in terms of model grid geometry (equations are solved on a finite difference grid), which only requires information about cell properties and their connections, and nothing about cell shape or positioning. Anticipated developments in modelling include: (i) continued increase in computing power, to simulate groundwater systems either at higher resolutions or at a larger spatial scale; (ii) collection of higher-quality static data (geological, hydrogeological and geophysical) to build more realistic model frameworks; and (iii) collection of more dynamic data (groundwater table, recharge and discharge) to improve model calibration. The fifth key area – getting to

---

Box 3.1 Groundwater Recharge in the Butana Region of Sudan

Michelle Lanzoni

As a general rule, 200 mm/year rainfall for diffuse recharge represents a 'cusp of sustainability'. In these areas, focused recharge is a substantial component of groundwater renewal. Across North Africa and the Middle East, wadis (ephemeral channels), formed following intermittent rain, provide focused recharge. The Nubian sandstone aquifer system (NSAS), underlying parts of Libya, Egypt, Chad and Sudan, is the largest transboundary aquifer in the world, composed of a number of interconnected aquifer units, and assessing how these are recharged is critical in understanding groundwater resources in this arid region.

In 1982, a four-year study in northern Sudan used conservative tracers and environmental isotopes to determine the rate and extent of recharge to the shallow, alluvial aquifer and to the underlying Nubian sandstone unit. Edmunds et al. (1992) showed whilst the majority of the NSAS groundwater is paleowater formed during a wetter climate, Nile River water does replenish, to some extent, both aquifers locally. Three decades later, in a subset of the original study area away from the Nile, Lanzoni et al. (2018) found that focused wadi recharge does locally reach the Nubian sandstone unit. Figure 3.4 shows groundwater isotopes plotted using the Global Meteoric Water Line (GMWL) for reference. Isotope signatures from boreholes drilled in the Nubian sandstone less than 300 m from the wadi suggest focused wadi recharge, whereas those more than 5 km away retain a paleowater signal. A further discussion of North African paleowater is found in Section 3.4.1 and Edmunds (2009).

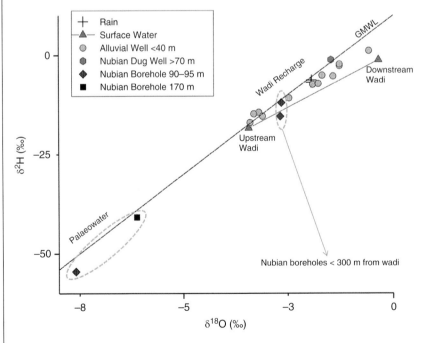

Figure 3.4 Isotopic composition of alluvial and Nubian sandstone groundwater within the wadi catchment in the Butana region of Sudan. *Source:* Adapted from Lanzoni et al. (2018). Reproduced with permission from Elsevier.

grips with predicted changes to groundwater storage and recharge under future climate change – is considered in Section 3.4.

3.2.3 Improving Scientific Knowledge of Groundwater Volumes and Fluxes

There are a number of potential constraints to future progress in quantifying the volume and fluxes of groundwater. For example, there is a 'methodological gap' within remote-sensing technologies (GRACE and InSAR) in so much as they lack the regional- or local-scale resolution to adequately match the typical scale required and used for water management (Castelazzi et al. 2016). Within regional-scale numerical modelling, the primary challenges, or bottlenecks, tend to be a lack of: (i) sufficient 'static' data on aquifer properties to produce a reasonable spatial resolution for the hydrogeological framework of the model, remembering that understanding the geological setting is fundamental to understanding groundwater; and (ii) sufficient dynamic data, including time-series, to be used as calibration datasets for the model (Zhou and Li 2011). Approaches that couple groundwater models with land surface models (and atmospheric climatic models) will also prove very valuable for assessing groundwater volumes and fluxes (see Chapter 7).

One of the biggest constraints to improving our scientific knowledge of groundwater quantity is a pragmatic, rather than a technical one. There is a lack of resources (not just technological infrastructure, but political and institutional capacity) to install and develop groundwater monitoring networks. There is a need for further installation, and consistent recording, of hydrometeorological data and borehole monitoring technology. Approaches using automatic data recording and mobile technologies, such as the use of smart handpumps (Chapter 7), offer promising solutions to this challenge.

Communicating the right message about groundwater science is just as vital as the scientific understanding itself, as highlighted by Edmunds' (2012) perspective piece on MacDonald et al.'s (2012) 'Quantitative maps of groundwater resources in Africa', where he argued that headline figures (in this case, groundwater storage being 100 times that of annually renewable surface water for Africa) must be communicated with important caveats, so that findings are not 'seized upon to justify unsustainable groundwater exploitation' (Edmunds 2012, p. 1), or used as justification to cut funding for water provision for those in need. This message about the importance of building effective management as well as scientific understanding is also well-captured by Giordano (2009, p. 172) – 'whatever deficiencies we still have in basic data and scientific understanding, it is the fundamental understanding of how to determine and implement location-appropriate frameworks for groundwater management in which we appear to be most deficient'.

3.3 Groundwater Quality

3.3.1 The Composition of Groundwater: Natural Baselines and Pollution

The conceptualization of groundwater as an open system is equally useful for thinking about groundwater quality. Aquifers represent 'gigantic open system(s)', with rainwater, soils and rock strata involved within these 'unevenly equilibrated chemical reactor(s)' (Edmunds and Shand 2008, p. 4). The major distinction to make within groundwater quality is between natural baseline geochemistry, where solutes are acquired during processes and reactions in the hydrogeological cycle (Edmunds and Shand 2008), and pollution, which is 'contamination caused or induced by human activities and typically measured by reference to predetermined permissible or recommended maximum limits' (Morris et al. 2003, p. 33). There are five broad approaches to defining the baseline (Table 3.2) and these can be used in combination. The interpretation of variations in baseline composition must be specific to the groundwater body in question. Accurate interpretations require knowledge of: (i) aquifer parameters, because these have a control over geochemical processes, and (ii) groundwater age(s), because processes are time-dependent and because the impact of pollution on fossil resources (those not currently part of the active hydrogeological cycle) will be greater than for renewable resources with definable turnover times (Edmunds and Shand 2008). Groundwater age can be assessed using inert tracers, stable isotopes or noble gases (as described in relation to recharge in Section 3.2.1).

Groundwater composition carries the fingerprint of atmospheric inputs, which get modified via evaporative enrichment, by processes in the soil, and by water–rock interactions at the soil–bedrock interface (these can be the most intense within the geochemical cycle), as well as water–rock interactions occurring along the flow pathway. An overview of 25 European aquifers is provided in Edmunds and Shand (2008), and these underpinned the EU-funded BaSeLiNe project that aimed to develop a standardized set of definitions and guidelines to be used within the European Union Water Framework Directive. This work clearly illustrated that within a groundwater body there are both geochemical gradients, which are sequential changes in chemical properties with time (and distance) from the recharge source, and geochemical boundaries, which are relatively abrupt regions of change in water quality that relate to changes in solution limits, mineral saturation, redox boundaries and exchangeable cation boundaries.

In terms of natural baseline chemistry, 99% of the solute content of groundwater is represented by nine major chemical elements or compounds (Ca, Na*, Mg*, K, HCO_3, Cl*, SO_4*, NO_3*, Si), and those starred are considered particularly undesirable for human and animal health at concentrations above guideline values (see Box 3.2 for NO_3 and Selinus et al. (2013) for more detail on medical geology). Two trace elements cause particular concern – arsenic (Box 3.3) and fluoride (Table 3.3). It is useful to reiterate that determining groundwater quality requires measurements of (i) the parameters representing quality (~50 inorganic properties and organic carbon) compared to a baseline, and (ii) properties or composition that allows an assessment of water age (using inert tracers, stable isotopes, radiocarbon, or noble gases) to establish the antiquity of the water quality issue and also to make an assessment of how rapidly changes to the quality parameters might occur in the future.

Anthropogenic activities can reduce groundwater quality by: directly adding pollutants, for example from waste disposal and chemical spills at the land surface (Morris et al. 2003); induced quality changes from changes to the groundwater regime (Margat and van der Gun 2013, p. 190); and by subsurface injection of waste water. Human-induced movement of groundwater can induce large-scale movement of water of different qualities, for example, over-abstraction of groundwater near the coast leads to saline intrusion into the aquifer (see Section 3.4). The return flows from abstracted groundwater (and surface water) can also cause problems, for example, waterlogging and salinization caused by irrigation in semi-arid and arid regions such as southern Australia and the Indus Valley of Pakistan (Morris et al. 2003). Table 3.4 provides a summary and classification of activities that can produce contaminants.

Table 3.2 Overview of approaches used to determine the baseline chemistry of groundwater.

Approach	Description	Advantages	Disadvantages	Example
Historical data	Provides data on groundwater composition for points in time, and for trends through time.	Valuable records.	Long-term datasets are rare, range of solutes recorded often small. Solutes measured and detection limits may vary through time.	Illinois State Water Survey and Environmental Protection Agency, 1890 onwards (major ions, total dissolved solids) (Kelly and Wilson 2008).
Down-gradient profiles	Data from profiles along the groundwater flow gradient reveal 2D pattern of groundwater quality evolution through time (and across space).		Limited use in unconfined aquifers. Care in interpreting depth and reach of abstraction between boreholes.	East Midlands Triassic sandstone flow path data on chloride, inert gases and stable isotopes (Bath et al. 1979).
Extrapolation from adjacent areas	Provides first-order approach.	Informative in lack of area-specific data.	No substitute for examining each water body. Each area has unique geological and hydrogeological properties.	Interpolation between the ~50 000 borehole data points for map of nitrate concentrations across southern Africa (Tredoux et al. 2001).
Geochemical modelling	An independent means of testing hypotheses of groundwater evolution.	Can back up empirical data.	Requires good knowledge of hydrodynamic situation, end-member compositions and aquifer minerology.	pH and Eh (oxidation-reduction potential) baselines at Olkiluoto site, Finland (Pitkänen et al. 2004).
Statistical methods	Provide summary of median and ranges of data for a water body. Distinguish anomalies and outliers.	Useful (only) alongside hydrogeochemical studies.	Cannot identify cause of outliers (natural or anthropogenic).	Trace elements within major aquifers in the United States (Lee and Helsel 2005).

Source: Adapted from Edmunds and Shand (2009). Reproduced with permission of John Wiley and Sons.

Table 3.3 Overview of two key natural trace elements in groundwater that cause particular hazards for health, and some approaches to treatment.

Trace element	WHO guideline value	Significance for health	Controls on occurrence	Approaches to treatment
Arsenic (As)	10 μg/l (p)	Toxic and carcinogenic, inorganic form (arsenite or arsenate) usually present	Geologically recent aquifers, strongly reducing conditions	Most promising are coagulation, co-precipitation, and adsorption
Fluoride (F)	1.5 mg/l	There is a narrow desirable range (<500 μg/l can result in dental caries, >2000 μg/l can cause fluorosis when general health status and nutrition is poor)	Granitic or volcanic aquifer geologies (F-bearing), facilitated by slow circulation (so semi-arid and arid)	Gypsum and lime/alum mix encourages precipitation. Ion-exchange resins for filtration
Cadmium (Cd)	3 μg/l	Probably carcinogenic, accumulates in the kidneys	Associated with zinc-ores and coal seams. Low levels of natural leaching from rocks, exacerbated by mining	Coagulation, or precipitation using a variety of antioxidants
Chromium (Cr)	50 μg/l	Probably carcinogenic	Ultra-mafic rocks, Cr-bearing minerals, in particular chromite found near convergent plate margins. The toxic Cr(IV) predominates under oxidizing conditions	Coagulation, or chelation with ethylenediamine-tetraacetic acid
Lead (Pb)	10 μg/l (p)	Neurodevelopmental effects, cardiovascular disease, impaired renal function, hypertension, impaired fertility	Occurs in a variety of igneous, metamorphic, and sedimentary rocks. Solubility increased under acidic conditions. No poisoning-level conditions reported from natural settings	Chelation with ethylenediamine-tetraacetic acid

Sources: adapted from Foster et al. (2006) and WHO (2011).
(p) denotes provisional.

Table 3.4 Summary of activities capable of producing contaminants for groundwater and characteristics of these as a pollution load.

| Activity | Character of pollution load | | | |
	Types	Location	Category	Soil zone bypassed
Agricultural				
Agrochemicals	n, o	r, u	D	N
Irrigation	s, n, o	r	D	N
Sludge/slurry	n, o, s	r	D	N
Unlined effluent storage	p, n, o	r	P	Y
Land discharge of effluent	n, s, o, p	r	P–D	Y
Stream discharge of effluent	o, n, p	r	P–L	Y
Industrial				
Effluent lagoons	o, h, s	u	P	Y
Tank/pipe leakage	o, h	u	O	U
Accidental spillages	o, h	u, r	P	N
Land discharge of effluent	o, h, s	r	P–D	N
Stream discharge of effluent	o, h, s	r	P–L	Y
Landfill disposal residues/waste	o, h, s	u, r	P	Y
Well disposal of effluent	o, h, s	r	P	Y
Aerial fallout	a	u, r	D	N
Mining				
Mine drainage discharge	s, h, a	r, u	P–L	Y
Sludge lagoons	h, a, s	r, u	P	Y
Solid mine tailings	h, a, s	r, u	P	Y
Oilfield brine disposal	s	r	P	Y
Hydraulic disturbance	s	r, u	D	N/A
Urban				
Unsewered sanitation	p, n, o	u, r	P–D	Y
Land discharge of sewage	n, s, o, p	u, r	P–D	N
Stream discharge of sewage	n, o, p	u, r	P–L	Y
Sewage oxidation lagoons	o, p, n	u, r	P	Y
Sewer leakage	o, p, n	u, r	P–L	Y
Landfill/solid waste disposal	o, s, n, h	u, r	P	Y
Highway drainage soakaways	s, o, h	u, r	P–L	Y
Wellhead contamination	p, n	u, r	P	Y
Resource management				
Saline intrusion	s	u, r	D–L	N/A
Recovering water levels	s, o, a	u	D	N/A

(Continued)

Table 3.4 (Continued)

Activity	Character of pollution load			
	Types	Location	Category	Soil zone bypassed
Endocrine-disrupting substances				
Natural				
Plant phytoestrogens	o	r, u	D	N
Oeostrone, oestradiol for animals (incl. humans)	o	u, r	P–D	N
Man-made				
Polychlorinated organic compounds (incineration and electrical equipment)	o	u, r	P	N
Organochlorine pesticides	o	r	D	N
Organotine (anti-fouling agent)	o	u, r	P	N
Alkylphenols (detergents, paints, cosmetics)	o	u, r	P	N
Phthalates (plasticizers)	o	u, r	P	N
Bi-phenolic compounds (plastics/resins)	o	u, r	P	N
Synthetic steroids (contraceptives)	o	u, r	P	N

Source: Adapted from Morris et al. (2003).
u = urban, r = rural; D = diffuse, P = point, L = linear; p = faecal pathogens, n = nutrients, o = organic micropollutants, h = heavy minerals, s = salinity, a = acidification; Y = yes, N = no, N/A = not applicable.

The soil zone, and entire USZ, is vitally important with respect to pollution, as this is where the majority of attenuation is possible, due to the presence of oxygen and biological activity that can remove, transform or retard pollutants (in the saturated zone this occurs much more slowly) (see Chapter 5). Groundwater pollution risk is the product of the nature of the contaminant load (magnitude, duration and composition) and the natural aquifer pollution vulnerability (which is determined by characteristics of the aquifer) (Figure 3.5) (see Morris et al. (2003) for a comprehensive treatment). Good scientific information about the aquifer characteristics and surface hydraulic conditions is essential for producing maps of aquifer pollution vulnerability and/or pollution risk, and these are best created on a site-by-site basis, along with good knowledge of the contaminant load. Generalized maps are more difficult to make and characterization by broad hydrogeological setting, or aquifer type, (such as those in Table 3.1), can only provide an initial rough guide to vulnerability or attenuation potential. An overlay of maps of aquifer vulnerability and of pollution threats is needed to guide efforts to prevent pollution (Margat and van der Gun 2013).

Management of groundwater quality is arguably most vital for drinking-water, and an excellent treatment of approaches to the management and protection of drinking-water sources and pollution source management is given by Schmoll et al. (2006). With regards to public health, 17% of the global population lack access to 'improved' water supply, whilst

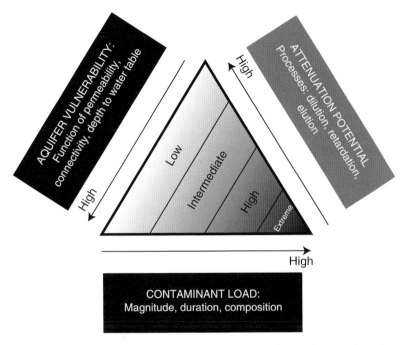

Figure 3.5 Schematic overview of pollution risk in groundwater. *Source:* Adapted from Morris et al. (2003). (*See color representation of this figure in color plate section*).

40% lack access to 'improved' excreta disposal (see WHO website; Schmoll et al. 2006); diarrhoeal diseases are largely derived from these poor conditions, and account for ~1.8 million deaths per annum (~4% of global disease). It is challenging to determine the proportion of water-related disease attributable to groundwater, and groundwater has advantages over surface water in generally exhibiting good microbial quality, although contamination can occur during withdrawal, transport and storage (Schmoll et al. 2006; Chapter 6 this volume).

3.3.2 Future Outlook on Groundwater Water Quality: Key Constraints and Approaches to Addressing Them

There is a continuing need to improve and consistently update temporal trends within our databases on groundwater quality, through monitoring, effective mapping, and storage of high-quality scientific data. Groundwater quality monitoring needs to be undertaken to identify poor water quality for drinking supply, and within agricultural food chains, and only then can strategies to minimize negative impacts be designed and implemented. It is anticipated there will be continued growth in research activity into endocrine (hormone) disrupting pollutants, which have been added to Table 3.4 as a separate category, even though their presence stems from agricultural, industrial and urban activities. The effects of these substances on wildlife and human populations are still poorly understood (Morris et al. 2003), and future progress will require collaborative research across hydrogeological and medical sciences.

There are recent developments in managed aquifer recharge (MAR) through the re-use of treated wastewater, a practice with implications for groundwater quality. In some

---

Box 3.2 Nitrate in a Multilayer Dryland Aquifer and UK Chalk Aquifer

Abi Stone

Nitrate in groundwater is a widespread groundwater quality and pollution issue because of its toxicity to humans and animals (indirectly, where reduction of nitrate, post-ingestion, leads to oxygen starvation) and links to increased gastric cancer, in addition to reported detrimental effects for crops and freshwater ecology. Nitrate can enter groundwater: (i) naturally, via leaching of nitrate-rich geological sediments, or through infiltration of nitrate produced by nitrate-fixing vegetation (e.g. legumes) and fauna (e.g. termites) and (ii) as a source of pollution from pit latrines, animal excreta from pastoral farming, and over-application of fertilizer for arable farming. The USZ is a crucial place to help attenuate nitrate before it reaches the groundwater table, because the exposure to oxygen facilitates denitrification of nitrate (NO_3^-) to N_2, unlike within the anoxic environment below the water table. Two different sources of nitrate and hydrogeological settings are outlined below.

In the chalk aquifer regions of England, agricultural nitrate application peaked (~70 kg N/ ha) in the mid-1970s and 1980s, and is moving downwards through the USZ at a rate of 1–2 m/year (Geake and Foster 1989). The thickness of the USZ (tens of metres here) means that these peaks have been described as the nitrate 'time bomb' in terms of the delay of this pollution entering the groundwater resource. Wang et al. (2012) have modelled the nitrate travel time across the range of hydrogeological settings in Great Britain, and for 60% of the chalk aquifers the peak has not yet reached the groundwater table, and is predicted to do so in 50–60 years from the date of the publication of their study.

High concentrations of nitrate (>100 mg/l as NO_3-N, WHO guideline ~11 mg/l) are common across dryland southern Africa, including the multilayered, transboundary Stampriet Aquifer Basin (SAB) (Tredoux et al. 2001). Whilst there is low-density sheep grazing in parts of the SAB, nitrate concentrations in the sand dune (non-fertilized land) USZ away from animal enclosures suggest there is a patchy, yet pervasive natural source of nitrate (Stone and Edmunds 2014), and ongoing stable isotope analysis will aid source identification. The presence of nitrate in the confined Auob aquifer layer (radiocarbon ages of 30 000–40 000 years) also indicates a natural origin (Heaton et al. 1983). The implication for the SAB is that nitrate pollution (from farming, pit latrines, or other anthropogenic sources) will be particularly problematic in addition to an already high baseline concentration. This study confirms that detailed site-specific studies are vital in improving our understanding of nitrate in groundwater, and suggests that pulses of naturally produced nitrate in the USZ of dryland settings migrate towards the groundwater table.

---

cases MAR is used to control saline intrusion in near-coastal boreholes. A pioneering example of wastewater for irrigation, whilst protecting the groundwater resources that it will recharge, is that of the Mezquital Valley, Mexico (Caucci and Hettiarachchi 2017). Scaling up for a wider use of this approach will require a great deal of work by authorities and managers to manage the perceptions and preconceptions of water users around re-using wastewater (see Chapter 6).

A vicious cycle of a lack of resources, lack of knowledge, and lack of planning can exacerbate groundwater quality degradation (Morris et al. 2003). However, even when the monitoring infrastructure is in place to provide a good understanding, the major

---

Box 3.3 Arsenic in Groundwater

Pauline Smedley

Understanding of the scale and causes of chronic health problems related to arsenic in drinking water has advanced in the last couple of decades, driven mainly by the discovery of arsenic problems in populations of the Bengal Basin, India and Bangladesh (Smedley 2008). Today, some 140 million people globally are considered to be exposed to arsenic in drinking water.

The WHO guideline value for arsenic in drinking water stands provisionally at $10\,\mu g/l$, although many developing countries use the pre-1993 guideline value of $50\,\mu g/l$. Long-term use of drinking and cooking water containing concentrations much above $50\,\mu g/l$ can lead to skin disorders including melanosis, keratosis, and skin cancer, other internal cancers, diabetes, and a range of other health problems.

Arsenic is more prevalent in groundwater because of natural water–rock interactions in aquifers, and occurrences of high concentrations have been found in a number of countries across the world, albeit with a majority in south and east Asia. Groundwater from young sediments in large alluvial and deltaic plains is especially vulnerable, with occurrences noted in the Indus, Red River, Yellow River, Mekong, Irrawaddy, Terai plains and Bengal Basin. Others have been found in groundwater from sediments containing volcanic material (e.g. the Pampean Plain of Argentina), geothermal waters, and groundwater in metal mineralized areas (e.g. the goldfields of Ghana and Burkina Faso).

The mechanisms of arsenic release to groundwater are complex, but pH and redox conditions play a key part, as do concentrations of other solutes. Iron oxides are the most likely source of arsenic in the alluvial and deltaic aquifers of south and east Asia, and high concentrations are typically found in anoxic groundwater within young (around 10 000 years old) sediments. One feature of the affected aquifers is the large short-range variability in groundwater arsenic concentrations, with high-arsenic boreholes often in close proximity to low-arsenic boreholes, with concentrations being unpredictable on a local scale. Ensuring quality compliance inevitably requires a detailed borehole testing and monitoring programme.

Mitigation of the arsenic problems in affected areas ranges from locating alternative arsenic-safe sources through to arsenic removal at domestic to municipal scale. In Bangladesh, short-range variability in arsenic concentrations has led to borehole switching as a common mitigation measure, at least in the short term. Other arsenic-safe alternatives include use of groundwater from shallow hand-dug wells, surface water via pond-sand filters, deep groundwater from pre-Holocene aquifers, rainwater harvesting or piped supplies (either from surface water or deep groundwater).

Mitigation of arsenic problems in affected areas requires a multidisciplinary approach involving water monitoring and management, health surveillance and provision, and education. As microbiological safety, exposure to other potentially harmful solutes and security of supply are also important, mitigation requires a strategic approach to avoid substituting one hazard with another.

challenge for science, policy and management is establishing remediation strategies. Options include technical solutions for treating poor-quality water, and also assessments of the viability of alternative supplies (either groundwater or surface) (Foster et al. 2006).

3.4 Groundwater and Climate Change

There is a growing body of evidence about the impacts of climate change on groundwater resources; see for example Crosbie et al. (2013), Jiménez Cisneros et al. (2014) and Ranjan et al. (2006). This section aims to highlight the influence of climate change on recharge and output zones, such as the freshwater–saline water boundary, and the ways that changes to groundwater can influence climate, emphasizing that this connection has a long history.

3.4.1 Long-term Climatic Influences on Groundwater

The volume of groundwater that is considered to be renewable (still part of the active hydrological cycle) is lower than the volume of water that is physically abstractable. The analysis of groundwater age (using dissolved radiocarbon or noble gases) demonstrates that groundwater in many aquifers around the world, particularly those that are deep and confined, is thousands to many hundreds of thousands of years old, with the oldest pocket of groundwater between 1.1 and 2.6 billion years old, in northern Ontario, Canada (Li et al. 2016). This demonstrates very clearly that much groundwater is a 'fossil' resource, being mined in the same way that hydrocarbons are. Less than 6% of groundwater in the top 2 km of the continental crust is <50 years old (Gleeson et al. 2016). The climatic conditions responsible for groundwater emplacement in the past may have been quite different from conditions experienced today, and these chemical and isotopic signatures within deep groundwater, and the USZ, can themselves be used to reconstruct the nature of these past climatic conditions (Edmunds 2009; Stone and Edmunds 2016).

Good examples of groundwater recharge under a different climatic regime than present come from arid North Africa. Here, the combination of groundwater stable isotopic signatures and radiocarbon contents reveals hiatuses in recharge through time and also that the moisture source responsible for emplacing particular packages of groundwater has changed through time (Edmunds 2009). This is best expressed in the northeast of the region, with a lack of recharge at the end of the Late Pleistocene, which is consistent with other paleoclimatic evidence for aridity in this region at this time. The more negative $\delta^{18}O$ values for the older waters indicate an abundance of an Atlantic moisture source, which moved west to east, depositing the lightest isotopes last.

3.4.2 Current and Future Influences of Climate Change on Groundwater

It is challenging to attribute the proportion of changes in groundwater levels to climate change, because of the overprint of resource abstraction and the influence that anthropogenic land use changes can have on recharge rates (Jiménez Cisneros et al. 2014). However, the IPCC 5th Assessment Report highlights regional-based studies using modelling approaches that attribute both decreases in spring discharge in Kashmir and declining groundwater levels in karstic aquifers in Spain to observed reductions in precipitation, and increases in evapotranspiration. Climate-change induced sea-level rise is well accepted, as is the impact it has on the intrusion of saline water across the freshwater–saline water interface in coastal aquifers, leading to a loss (degradation) of the resource (Margat and van der Gun 2013). The Ghyben-Herzberg relation captures the magnitude of the impact of sea-level rise: a 1 m decrease of the distance between groundwater table and seawater in an unconfined aquifer is accompanied by a 40 m decrease in the thickness of the freshwater layer. This has significant repercussions for the sustainability of ground-water abstraction in these regions, as abstraction can exacerbate saline intrusion by caus-ing a cone of groundwater depression around the borehole, and the influence of abstraction may exceed that of sea-level rise in the twenty-first century (Jiménez Cisneros et al. 2014).

Future predictions, based on the Hadley Centre climate model, forced with high and low emission scenarios, show that coastal groundwater resources will be further lost for all regions except northern Africa (Ranjan et al. 2006). Portmann et al. (2013) used an ensemble of 5 CMIP5 global climate models alongside the WaterGAP hydrological model to demonstrate where significant changes to groundwater recharge are expected. This a powerful message, suggesting that lower, rather than higher, emission scenarios will: (i) reduce the percentage of the projected global population suffering from a sig-nificant (>10%) loss of groundwater recharge from 38% to 24% and (ii) increase the percentage of population spared from significant increases in groundwater recharge (flooding and salinity issues) from 29% to 47%.

Only a few studies exploring projections in groundwater recharge consider the feed-backs from changes in vegetation response to temperature, precipitation changes and increases in CO_2, or incorporate the influence of climate-driven changes in land-cover and land-use on recharge (Jiménez Cisneros et al. 2014). There are also reciprocal influ-ences of changes in groundwater flux to the global climate. Groundwater depletion has increased rapidly since the 1990s, with an estimated global rate for the year 2000 of $204 \pm 30 \, \text{km}^3$/year, and this contributes to a sea-level rise of 0.57 ± 0.09 mm/year (Wada et al. 2014). Past abstraction data and future predictions are used by Wada et al. (2014) to show this is a change from 0.035 ± 0.009 mm/year in 1900, which may rise to 0.82 ± 0.13 mm/year by 2050.

3.5 Continuing Challenges for Groundwater Science

This short section highlights three areas of increasing interest and focus for groundwa-ter science. One area is the mapping and quantification of low-salinity (fresh and brack-ish) groundwater found within the rock of continental shelves, which have been termed vast meteoric groundwater reserves (VMGRs) (Post et al. 2013). So far these have been recognized by direct observation in the Beaufort-Mackenzie Basin, Canada, Florida, Nantucket, and New Jersey in the USA, Suriname, the Niger Delta, Bredasdorp basin, Jakata, East China Sea, and Perth Basin and Gippsland in Australia, and other potential regions have been identified based on either porewater composition or extrapolating spatial patterns of onshore paleogroundwater (see Post et al. 2013). There is a clear incentive for future research into these as freshwater resources, whilst recognizing that these represent another non-renewable portion of the global resource. They were emplaced during the Last Glacial Maximum (~23–19 ka) during a low-stand of sea-level in which the shelf areas were exposed and covered with river systems and freshwater lakes, and infiltration of meteoric water and glacial meltwater occurred (Post et al. 2013).

A second key issue is the challenge of groundwater provision and quality in the urban setting, given the continued acceleration of urbanization and associated peri-urban growth (Morris et al. 2003). The urban subsurface is under competing demands as an aquifer, a transfer zone (and repository) for disposing of wastewater and solid waste, and to house engineering infrastructure, such as pipes, wires, tunnels and foundations. All three groups of activities contribute to effects on the groundwater system, which include: (i) aquifer over-abstraction; (ii) problems from reductions in abstraction, such as flooding of subsurface infrastructure, as seen in de-industrialized central London, since 1967 (the GARDIT [General Aquifer Research Development and Investigation Team] strategy has re-commissioned pumping in the centre of the Thames basin, and redirected some of this water into artificial recharge elsewhere in the basin), excess infiltration (bringing in aggressive chemicals and raising the water table), reduced infiltration from more impermeable surfaces, and excessive contaminant loads from wastewater disposal and other sources, reducing groundwater potability and clogging wells. Morris et al. (2003) provide a useful overview of the proposed general model for the stages of evolution in urban water infrastructure following the initial development of the urban settlement. In China, a legacy of groundwater over-abstraction and urban flooding influenced a policy change in 2015 towards the 'sponge city initiative' that aims to enhance the infiltration capacity and replenish groundwater. This is part of an ambitious Chinese goal for more sustainable water use, in which 80% of urban areas should absorb and re-use ≥70% of rainwater by 2020, through the use of green infrastructure and permeable surfaces.

MAR is the third area of pressing interest, and shows promise in the urban setting (for example the GARDIT strategy in London, above, and the Mezquital Valley in Mexico in Section 3.3.2) as well as in aquifers experiencing declining yields, in aquifers at risk of coastal saline intrusion (see Section 3.3.1) and in aquifers prone to land subsidence (see Section 3.2.1). MAR is the intentional storage of water in aquifers. In addition, the biological and chemical filtration capacity of the USZ and deep groundwater can be utilized to improve water quality (see Section 3.3.1). MAR schemes have been applied at small and large scales at locations from the Coimbatore District of Tamil Nadu, India, through to Windhoek in Namibia (see van der Gun 2012). MAR's success as a groundwater harvesting and reuse tool relies on excellent scientific understanding of the physical and chemical hydrogeological setting of the aquifer, to understand the directions and rates of groundwater flow and to reduce negative effects of possible chemical disequilibrium induced in the aquifer system through the injection of water.

3.6 Concluding Points

Groundwater is a vital part of the hydrological cycle, accounting for 97% of liquid freshwater. Scientific understanding of groundwater volumes, fluxes and quality continues to improve. The collection of data at regional, or aquifer-specific, scales has been key to demonstrating that groundwater is not a homogeneous resource, just as the world's reliance upon it is not. Only a few regions are entirely dependent on groundwater, such as Saudi Arabia; however, the challenge remains keeping checks on increasing utilization of groundwater so that it falls within levels supported by recharge of the resource, and this requires good data on groundwater recharge, volume and fluxes.

Given groundwater is used at the local scale, the devil is in the detail when it comes to providing an adequate scientific understanding of its physical and chemical behaviour in order to firmly inform policy and management strategies. Capturing this detail requires more funding and infrastructure for field measurements and monitoring, as well as continued developments in remote sensing and numerical modelling technologies. As stressed, it is crucial that scientific data goes beyond providing estimates of groundwater yields and provides clearly-communicated information about: (i) the extent to which the source is a renewable, or a fossil, resource, and (ii) up-to-date information about groundwater quality (both natural baselines and anthropogenic pollution). Scientific understanding does not sit alone, but should be situated in the 'web of interdependencies'. This means it must be communicated effectively, even when it calls on the precautionary principle, in order to protect non-renewable groundwater sources from irreversible over-exploitation or reduction in quality.

Three key future priorities all involve a demonstrable role for: scientific and technological innovation to understand the potential of groundwater resources within continental shelves, to maintain groundwater supply and quality in urban environments, and to facilitate MAR. Scientific research alongside nuanced policy and management decisions will allow the global community to face the challenges of changing quantities and qualities of water across the globe, from coastal to inland regions, humid to arid environments, and rural to urban settings.

References

Allen, R.G., Howell, T.A., Pruitt, W.O. et al. (eds.) (1991). Lysimeters for evapotranspiration and environmental measurements. *Proceedings of the International Symposium on Lysimetry*, Honolulu, Hawaii (23–25 July 1991). New York: ASCE.

Bath, A.H., Edmunds, W.M., and Andrews, J.N. (1979). Palaeoclimatic trends deduced from the hydrochemistry of a Triassic sandstone aquifer. *Proceedings of the International Symposium on Isotope Hydrology,* Vol. II. IAEA-SM-228/27. Vienna: IAEA.

Castelazzi, P., Martel, R., Galloway, D.L. et al. (2016). Assessing groundwater depletion and dynamics using GRACE and InSAR: potential and limitations. *Groundwater* 54 (6): 768–780.

Caucci, S. and Hettiarachchi, H. (2017). Wastewater irrigation in the Mezquital Valley, Mexico: solving a century-old problem with the Nexus Approach. *Proceedings of the International Capacity Development Workshop on Sustainable Management Options for Wastewater and Sludge*, Mexico (15–17 March 2017). Dresden: United Nations University Institute for Integrated Management of Material Fluxes and of Resources (UNU-FLORES).

Colchester, F.E., Marais, H.G., Thomson, P. et al. (2017). Accidental infrastructure for groundwater monitoring in Africa. *Environmental Modelling and Software* 91: 241–250.

Crosbie, R.S., Pickett, T., Mpelasoka, F.S. et al. (2013). An assessment of the climate change impacts on groundwater recharge at a continental scale using a probabilistic approach with an ensemble of GCMs. *Climatic Change* 117 (1–2): 41–53.

Döll, P., Hoffmann-Dobrev, H., Portmann, F.T. et al. (2012). Impact of water withdrawals from groundwater and surface water on continental water storage variations. *Journal of Geodynamics* 59–60: 143–156. https://doi.org/10.1016/j.jog.2011.05.001.

Edmunds, W.M. (2009). Palaeoclimate and groundwater evolution in Africa – implications for adaption and management. *Hydrological Sciences Journal* 54 (4): 781–792.

Edmunds, W.M. (2012). Limits to the availability of groundwater in Africa. *Environmental Research Letters* 7: 021003. https://doi.org/10.1088/1748-9326/7/2/021003.

Edmunds, W.M. and Shand, P. (2008). Natural Groundwater Quality. Oxford: Blackwell.

Edmunds, W.M., Darling, W.G., Kinniburgh, D.G. et al. (1992). Sources of recharge and Abu Delaig, Sudan. *Journal of Hydrology* 131 (1–4): 1–24.

FAO. 2016. *AQUASTAT Main Database, Food and Agriculture Organization of the United Nations* (FAO). Accessed October 2018. Available online at: http://www.fao.org/nr/water/aquastat/data/query/index.html?lang=en.

Foster, S., Kemper, K., Tuinhof, A. et al. (2006). Natural Groundwater Quality Hazards: Avoiding Problems and Formulating Mitigation Strategies (English), GW Mate Briefing Note Series No. 14. Washington, DC: World Bank.

Geake, A.K. and Foster, S.S.D. (1989). Sequential isotope and solute profiling in the unsaturated zone of British Chalk. *Hydrological Sciences Journal* 34 (1): 79–95.

Giordano, M. (2009). Global groundwater? Issues and solutions. *Annual Review of Environmental Resources* 34: 153–178.

Gleeson, T., Befus, K.M., Jasechko, S. et al. (2016). The global volume and distribution of modern groundwater. *Nature Geoscience* 9: 161–167. https://doi.org/10.1038/ngeo2590.

Heaton, T.H.E., Talma, A.S., and Vogel, J.C. (1983). Origin and history of nitrate in confined groundwater in the western Kalahari. *Journal of Hydrology* 62: 243–262.

Hiscock, K.M. and Bense, V.F. (2014). Hydrogeology: Principles and Practice, 2e. Oxford: Wiley Blackwell.

Jiménez Cisneros, B.E., Oki, T., Arnell, N.W. et al. (2014). Freshwater resources. In: Climate Change 2014: Impacts, Adaptation, and Vulnerability. Part A: Global and Sectoral Aspects. Contribution of Working Group II to the Fifth Assessment Report of the Intergovernmental Panel on Climate Change (eds. C.B. Field, V.R. Barros, D.J. Dokken, et al.), 229–269. Cambridge, UK and New York, USA: Cambridge University Press.

Kelly, W.R. and Wilson, S.D. (2008). An evaluation of temporal changes in shallow groundwater quality in Northeastern Illinois using historical data. *Illinois State Water Survey, Scientific Report 2008–01*. Champaign, Illinois: Center for Groundwater Science.

Konikow, L. and Kendy, L. (2005). Groundwater depletion: a global problem. *Hydrogeology Journal* 13: 317–320.

Lanzoni, M., Darling, W.G., and Edmunds, W.M. (2018). Groundwater recharge in Sudan: an improved understanding of wadi-directed recharge. *Applied Geochemistry* 99: 55–64.

Lee, L. and Helsel, D. (2005). Baseline models of trace elements in major aquifers of the United States. *Applied Geochemistry* 20 (8): 1560–1570.

Li, L., Wing, B.A., Bui, T.H. et al. (2016). Sulfur mass-independent fractionation in subsurface fracture waters indicates a long-standing sulfur cycle in Precambrian rocks. *Nature Communications* 7: 13252.

Llamas, M. and Martínez-Santos, P. (2005). Intensive groundwater use: a silent revolution that cannot be ignored. *Water Science and Technology Series* 51 (8): 167–174.

MacDonald, A.M., Bonsor, H.C., O'Dochartaigh, B.E., and Taylor, R.G. (2012). Quantitative maps of groundwater resources in Africa. *Environmental Research Letters* 7: 024009. https://doi.org/10.1088/1748-9326/7/2/024009.

Margat, J. and van der Gun, J. (2013). Groundwater Around the World: A Geographic Synthesis. Boca Raton, FL: CRC Press (Taylor and Francis).

Morris, B., Lawrence, A., Chilton, J. et al. (2003). Groundwater and Its Susceptibility to Degradation: A Global Assessment of the Problem and Options for Management, Early Warning and Assessment Report Series, 126. Nairobi, Kenya: UNEP.

Ortiz-Zamora, D. and Ortega-Guerrero, A. (2010). Evolution of long-term land subsidence near Mexico City: review, field investigations, and predictive simulations. *Water Resources Research* 46 (1): W01513. https://doi.org/10.1029/2008WR007398.

Pitkänen, P., Partamies, S., and Luukkonen, A. (2004) *Hydrogeochemical interpretation of baseline groundwater conditions at the Olkiluoto site.* Posivia Report 2003–07. Olkiluoto, Finland: Posiva. http://www.posiva.fi/files/1247/Posiva_2003-07.pdf.

Portmann, F.Y., Döll, P., Eisner, S., and Flörke, M. (2013). Impact of climate change on renewable groundwater resources: assessing the benefits of avoided greenhouse gas emissions using selected CMIP5 climate projections. *Environmental Research Letters* 8: 0240023.

Post, V.E.A., Groen, J., Kooi, H. et al. (2013). Offshore fresh groundwater reserves as a global phenomenon. *Nature* 71: 71–78.

Ranjan, P., Kazama, S., and Sawamoto, M. (2006). Effects of climate change on coastal fresh groundwater resources. *Global Environmental Change* 16 (4): 388–399. https://doi.org/10.1016/j.gloenvcha.2006.03.006.

Scanlon, B.R., Healy, R.W., and Cook, P.G. (2002). Choosing appropriate techniques for quantifying groundwater recharge. *Hydrogeology Journal* 10 (1): 18–39.

Schmoll, O., Howard, G., Chilton, J., and Chorus, I. (2006). Protecting Groundwater for Health: Managing the Quality of Drinking-Water Sources, WHO Drinking Water Quality Series. Geneva/London: WHO/IWA Publishing.

Seibert, S., Burke, J., Faures, J.M. et al. (2010). Groundwater use for irrigation – a global inventory. *Hydrology and Earth System Science* 14: 1863–1880.

Selinus, O., Alloway, B., Centeno, J.A. et al. (eds.) (2013). Essentials of Medical Geology, Revised Edition. Dordrecht: Springer.

Smedley, P.L. (2008). Sources and distribution of arsenic in groundwater and aquifers. In: Arsenic in Groundwater: A World Problem (ed. T. Appelo), 4–32. Utrecht: International Association of Hydrogeologists. the Netherlands, IAH.

Stone, A.E.C. and Edmunds, W.M. (2014). Naturally-high nitrate in unsaturated zone sand dunes above the Stampriet Basin, Namibia. *Journal of Arid Environments* 105: 41–51.

Stone, A.E.C. and Edmunds, W.M. (2016). Unsaturated zone hydrostratigraphies: a novel archive of past climates in dryland continental regions. *Earth Science Reviews* 157: 121–144.

Tredoux, G., Englebrecht, J.P., and Talma, A.S. (2001). Nitrate in groundwater in southern Africa. In: New Approaches Characterising Groundwater Flow (eds. K. Seiler and S. Wohnlich), 663–666. Lisse, Netherlands: Swets and Zeitlinger.

van der Gun, J. (2012). Groundwater and Global Change: Trends, Opportunities and Challenges, United National World Water Assessment Programme Side Publication Series. Paris: UNESCO.

Wada, Y., Wisser, D., and Bierkens, M.F.P. (2014). Global modelling of withdrawal, allocation and consumptive use of surface water and groundwater resources. *Earth System Dynamics* 5: 15–40.

Wang, L., Stuart, M.E., Bloomfield, J.P. et al. (2012). Prediction of the arrival of peak nitrate concentrations at the water table at the regional scale in Great Britain. *Hydrological Processes* 26: 226–239.

WHO (2011). Guidelines for Drinking-Water Quality, 4e. Geneva: WHO Press http://www.who.int/iris/handle/10665/4458.

WHYMAP (2008). Maps and data from the World-wide Hydrogeological Mapping and Assessment Programme (WHYMAP). https://www.whymap.org/whymap/EN/Home/whymap_node.html (accessed 17 September 2018).

Zhou, Y. and Li, W. (2011). A review of regional groundwater flow modelling. *Geoscience Frontiers* 2 (2): 205–214.

4

Water Quality Modelling, Monitoring, and Management

Paul Whitehead[1], Michaela Dolk[2], Rebecca Peters[1], and Hannah Leckie[3]

[1] *School of Geography and the Environment, University of Oxford, UK*
[2] *Swiss Re, New York, USA*
[3] *Division of Climate, Biodiversity and Water, OECD, Paris, France*

4.1 Water Quality Modelling Background

4.1.1 Water Quality: The Problem

In addition to being major sources of water, river systems are often used as the principal disposal pathways for industrial, agricultural and domestic effluents. As demand for water increases and water quality deteriorates, there is a requirement for effective decision-making techniques that can be applied to solve water quality management problems.

In developing economies, the discharge of untreated wastewater is a serious and growing problem (see Chapter 16). The pervasive use of chemicals in agriculture and industry has led to a rising trend in pollution and increased threats to potable water supplies. Water pollution impacts human health and wellbeing, economic growth and freshwater ecosystems. The poor and vulnerable are often most at risk. In both developed and developing economies, water pollution from unregulated diffuse (non-point) sources of pollution from rural and urban areas continues to rise. Eutrophication, leading to hypoxia and algal blooms, is being caused primarily by agricultural runoff of excess nutrients, and this is a key water quality challenge globally (OECD 2017; UNESCO 2009). Factory waste from tanneries, the garment industry and metalworking is also a major issue, releasing dangerous levels of toxic metals and organics. Lakes and groundwaters are also impacted by pollution as agricultural or diffuse sources of pollution are flushed or seep into these water bodies. In addition, land use change in catchments and large-scale mining can create acid drainage problems that significantly damage the water quality and ecology of downstream waters. Unless attention is turned to managing both point and diffuse sources of pollution, further deterioration of water quality and freshwater ecosystems can be expected as human populations grow, industrial and agricultural production intensifies, and climate change causes significant alteration to the hydrological cycle.

Improving water quality is a critical element of the 2030 Sustainable Development Goals (SDGs), fulfilling an essential role in reducing poverty and disease and promoting

Water Science, Policy, and Management: A Global Challenge, First Edition. Edited by Simon J. Dadson, Dustin E. Garrick, Edmund C. Penning-Rowsell, Jim W. Hall, Rob Hope, and Jocelyne Hughes.
© 2020 John Wiley & Sons Ltd. Published 2020 by John Wiley & Sons Ltd.

sustainable growth. Target 6.3 aims to improve water quality by reducing pollution, eliminating dumping and minimizing the release of hazardous chemicals, and halving the proportion of untreated wastewater (United Nation 2015). There is a need for more effective decision-making and regulatory mechanisms that can be applied to reduce these problems nationally and internationally. Advances in water quality monitoring and modelling can offer a way forward by identifying pollution hotspots and pollution source priorities. By merging physical water quality models with economic models and sensitivity analysis, the efficiency and effectiveness of 'what-if' scenarios of various policy and infrastructure options can be tested without recourse to expensive testing in reality. Such a decision-support tool can save time and resources, assess the potential risks to stakeholders, and prioritize policy and management actions (OECD 2017).

4.1.2   Management Model Approaches and History

Over the past century, there has been increasing efforts worldwide devoted to the development of water quality models. As long ago as 1912 the Royal Commission on Sewage Disposal (1912) proposed a simple 1–8 ratio for effluent discharge to river flow in an attempt to clean up rivers in the United Kingdom. In the USA major developments by Streeter and Phelps (1925) led to the dissolved oxygen (DO) and biochemical oxygen demand (BOD) equations that enabled steady-state and dynamic modelling of oxygen balances in rivers (Cox and Whitehead 2009). Major studies in the area of water quality modelling were undertaken in the US (e.g. Great Lakes – Chapra 2018; Delaware River – Thomann and Mueller 1987), the UK (e.g. Bedford-Ouse River – Whitehead et al. 1981) and in Australia (e.g. Peel Harvey River and Estuary – Hornberger and Spear 1980). These historical approaches have in recent years been supplemented by increasingly complex models involving refined chemical processes and improved three-dimensional representations of rivers, lakes and groundwaters. In order to respond to water quality problems, environmental managers and policy-makers have considered a wide range of models, which can be classified into planning, design and operational models. Whilst higher-level steady-state models are used for long-term policy and planning analysis, stochastic models are used for the design of sustainable effluent treatment plants, and dynamic (hourly or daily) operational models to support the day-to-day management issues of pollution.

The purpose of a policy and planning model is to evaluate a large number of solutions and isolate a small number of near-optimal least-cost solutions. Techniques such as linear programming and dynamic programming have been used in combination with water quality models to evaluate alternative investment strategies (Loucks and van Beek 2017). Planning models generally use steady-state models, since they are based on average mass balance conditions. Steady-state models cannot account for natural fluctuations in water quality caused by changing hydrology, turbulence, reaction kinetics or other factors altering water quality. In the design phase, a small number of near-optimal least-cost solutions may be evaluated, and these are often considered in stochastic simulations to take into account the uncertainty associated with aquatic systems. Stochastic techniques such as simulation using Monte Carlo (MC) methods may be used to allow for the inherent uncertainties in both process knowledge and the driving variables (e.g. temperature, solar radiation, river flows). Warn (1981) and Whitehead and Young (1979) used MC analysis to address the problems of calculating effluent consent levels to meet in-stream river quality objectives and to evaluate the impact of effluents on downstream water quality. More recent stochastic modelling has been undertaken to address issues of uncertainty and how these can be built into water quality models (Wade et al. 2002a, b, c). The purpose of operational models is to assist in the efficient day-to-day (or hour-to-hour) management of a river system. An operational model uses data relating to the past and present state of the system to produce flow and water quality forecasts upon which management decisions can be taken. They also need to be process-based so that they can take into account the dynamics of river systems and correctly evaluate the impact of a discharge on the receiving river system. Operational models can be linked to telemetry systems monitoring flow and water quality in real time so that the model can be updated continuously (Boënne et al. 2014; Whitehead et al. 1984).

4.1.3 Generic Types of Water Quality Models

The concept of mass balance is fundamental to river, lake and groundwater quality modelling. In the formulation of the simplest mass balance model, it is assumed that chemicals and water quality determinands are conservative; that is, once they are discharged to the aquatic system they do not undergo any changes in mass, although they might be redistributed by dispersion or other physical processes. Key water quality variables include nutrients such as nitrogen and phosphorus, which undergo chemical change via processes such as nitrification, denitrification, adsorption on sediments, uptake by algae and macrophytes. Wetlands often act as a major sink of nutrients (see Chapter 5). Metals also react with the environment, creating metal complexes, adsorbing to suspended sediments and accumulating on river and lake sediments. Organics also undergo changes due to equilibrium processes where they transfer to the atmosphere, get adsorbed onto sediments or are taken up by fish in rivers and lakes.

The equations used to calculate the concentrations of these conservative substances are derived by considering a material balance over a length or area of the main water body and with mixing processes varying considerably. The degree of mixing provides a convenient means of distinguishing between various models currently available:

1) No mixing: a first-order differential equation model represented by

$$\frac{\partial c}{\partial t} = -u\frac{\partial c}{\partial x} - kc \tag{4.1}$$

2) Partial mixing: a second-order differential equation model represented by

$$\frac{\partial c}{\partial t} = -u\frac{\partial c}{\partial x} + E\frac{\partial^2 c}{\partial x^2} - kc \tag{4.2}$$

3) Complete mixing: a first-order mass balance equation over a well-mixed river reach

$$\frac{dc}{dt} = \frac{1}{T}(c_i - c) - kc \tag{4.3}$$

In these equations, t is time, x is distance along a river, c is concentration, u is the stream velocity, E is the dispersion coefficient, k is a chemical decay rate, d denotes the derivative and ∂ is the partial derivative, T is reach residence time, and subscript i refers to an upstream or input concentration in the river. Of course, in very large rivers, lakes, and groundwater systems, pollution is spread over three dimensions and this makes the modelling task much more difficult. The extension of the above equations into three dimensions of space is feasible, but solving the equations becomes complex and time-consuming from a computational perspective. The three equations are mathematical formulations of the different mixing processes occurring in river, lakes and groundwaters. No mixing means that there is just straight transport down a river system, whereas in fact in most rivers there is always dispersion operating on pollutants. Generally it takes time and distance before a chemical is completely mixed in the environment, and partial mixing is generally the normal situation immediately downstream of a discharge. However, this can become complete mixing at some point down the system after sufficient travel time has elapsed. The mixing differential equations get even more complicated when sources and sinks of pollutants are incorporated and when reaction kinetics and biological interactions are included.

4.1.4 Lumped Modelling Approaches

An alternative approach to the modelling of water quality in river and lake systems using a mass balance approach is the lumped parameter differential equation model (Chapra 2018). Time-varying water quality variables can be simulated using differential equations describing the mass balance over a short section of river or over a lake system when it is assumed that all the parameters and variables are uniform throughout the water body. In these circumstances, the models should be truly dynamic, able to deal with time-varying inputs and outputs, and of low order, with only the dominant modes of behaviour modelled, thereby retaining parametric efficiency (Young and Chotai 2004). Also the models need to be calibrated against real data collected from the aquatic systems and should try to account for the inevitable errors associated with laboratory analysis and sampling. Models need to address the issues of uncertainty that arise from the lack of knowledge of the inherent physical, chemical and biological mechanisms and interactions operating in the modelled system.

The mathematical form of the model is derived from a component mass balance across a reach of a river.

$$\frac{d}{dt}\mathbf{x}(t) = \frac{Q(t)}{V}\mathbf{u}(t) - \frac{Q(t)}{V}\mathbf{x}(t) + \mathbf{S}(t) + \zeta(t) \tag{4.4}$$

where

$\mathbf{u}(t)$ is the vector of input, upstream component concentration (mg/l)
$\mathbf{x}(t)$ is the vector of output, downstream component concentration (mg/l)
$\mathbf{S}(t)$ is the vector of component source and sink terms (mg/l/day)
$\zeta(t)$ is the vector of chance, random disturbances affecting the system (mg/l/day)
$Q(t)$ is the stream discharge (m^3/day)
V is the reach volume (m^3)

Chapra (2018) describes the development and application of lumped process-based models to river, lake and estuary systems, with these models proving to be very effective at delivering solutions to many water quality problems.

4.1.5 Case Study 1: Modelling of Metals Downstream of Mines in Transylvania

As an example of river pollution modelling in Eastern Europe, Case Study 1 deals with a long legacy of gold and silver mining in the Transylvania region of Romania. Old mines going back to pre-Roman times have been polluting the rivers for over 4000 years. Moreover, there is a desire to reopen and develop the mines both to clean up the acid mine drainage and extract metals, as well as to boost the local and national economy. The mine drainage is highly acidic, caused by rising mine waters dissolving oxides of sulphur to generate sulphuric acid. These acids dissolve mine minerals, flushing high concentrations of dangerous metals from the mine, such as arsenic, lead, chromium, zinc and copper. A dispersion pollution metals river model, as shown in Equation (4.2) above, has been applied to the Mures River (catchment area $30\,332\,km^2$, mean flow $184\,m^3/sec$) in Transylvania to simulate metals in the river system and to assess the environmental improvements from a clean-up operation at the mine (Chapra and Whitehead 2009). Figure 4.1 shows the effects of the installation of an effluent treatment plant at the mine to treat acid mine discharges which are toxic with a pH of 3.5 and high levels of arsenic, copper, chromium and zinc. The model has been used to assess the historic pollution and then to simulate the projected improvements in the river system downstream following the installation of an effluent treatment plant at the mine. The mine redevelopment and clean-up is still the subject of environmental dispute in Romania.

4.2 Water Quality Modelling at the Catchment Scale

4.2.1 Integrated Catchment Approach – A Brief Review

In recent years there has been an increasing focus on the development of an integrated approach to water quality modelling which takes into account the catchment scale. This is because pollution has become an issue concerning not only the management of point sources but also the pollution arising from diffuse sources. These diffuse sources include those from agriculture, groundwaters, and hydrocarbons in urban storm-water overflows. An additional source is from upland runoff carrying pollutants and excess sediments from felling of upstream forests and from air pollution, such as sulphur and nitrogen deposition. Water quality and water quantity should be managed in unison, as the two are interrelated and interdependent. For example, water scarcity reduces the capacity for dilution of point source pollution; high rainfall events may initiate diffuse pollution from land runoff (agricultural and urban) and combined sewer overflows into rivers; and poor water quality reduces the quantity of useable water and therefore exacerbates the problem of water scarcity.

There are relatively few models of whole catchments that incorporate the soil and groundwater components as well as river channel dynamics. This is somewhat surprising given that many problems are caused by non-point source or diffuse pollution. Where such models do exist they are often driven by overly complex hydrological models. In the USA there have been catchment model approaches based on the Soil and Water Assessment Tool (SWAT) model. SWAT (Srinivasan and Arnold 1994) has been developed over many years and has been designed as a catchment model to simulate the quality and quantity of surface and groundwater. The applications of the model have

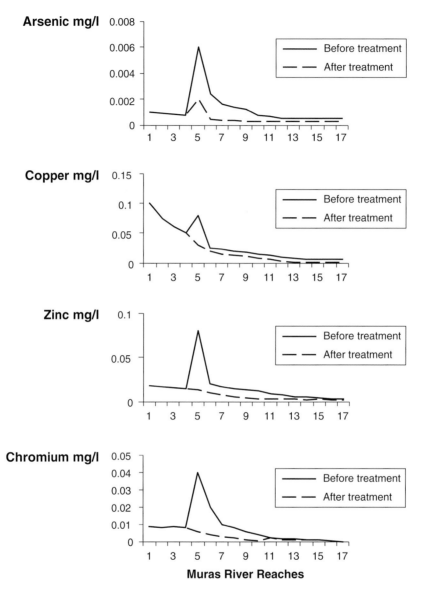

Figure 4.1 Simulation of metals along the Mures River system with and without mine drainage treatment, showing the elimination of high levels of pollution in close proximity to the polluted area and further reductions downstream. *Source:* Paul Whitehead.

focused on the environmental impact of land use change and land management practices, although this has extended to larger catchments and issues of climate change recently. SWAT is widely used in assessing soil erosion prevention and control, non-point source pollution control and regional management in catchments. Srinivasan and Arnold (1994) linked the water quality and flow equations to a geographical information system (GIS) and generated a valuable tool that has been widely used. However, SWAT still uses the older science of sediment transport based largely on US Department of Agriculture (USDA) empirical relationships rather than having any real process understanding. A much more sophisticated but more complex model is based on the MIKE 11 suite of programmes (https://www.mikepoweredbydhi.com), which aims to model in full detail the dynamics and transport mechanisms of water in a catchment (Estrup et al. 2006). MIKE SHE is a surface water and groundwater model that forms the basis of the MIKE 11 models and includes a full hydrological description of the key processes, such as overland flow, unsaturated flow, vegetation-based evapotranspiration, groundwater flow and fully dynamic channel flow. In recent years water quality components have been added into the models but, as might be expected with the full mathematical descriptions, the computational times are high, reflecting the nonlinear partial differential equations being solved using sophisticated numerical integration routines. Thus exploring model and system behaviour rapidly or addressing a set of scenarios is very difficult and very costly.

4.2.2 The Integrated Catchments (INCA) Model System

In an attempt to develop a process-based water quality model with hydrology and water quality modelled at the same level of complexity, Whitehead et al. (1998) developed the INCA model. The INCA model is dynamic (daily), process-based, and uses mass balance and reaction kinetic equations to simulate the principal mechanisms that are operating. Surface, soil and groundwater catchment zones are simulated together with leaching of water into a river or lake system. The land phase and river channels are modelled so that a semi-distributed description of water quality across the catchment can be obtained. The model can simulate up to six different land uses such as forest, arable, urban, surface vegetation (grazed or fertilized), and moorland. Sources of pollutants can be point sources, such as municipal sewage discharges or industrial effluents, or diffuse sources of pollutants from agriculture, atmospheric deposition and urban storm-water drainage.

INCA utilizes the lumped modelling approach described by Chapra (2018) and therefore is based on a set of ordinary differential equations where the flow and water quality equations are linked. The equations are solved using a fourth-order Runga Kutta method of solution with a Merson variable step length integration routine. This enables stable numerical integration of the equations and minimizes numerical problems. The advantage of this scheme is that scientific effort can be directed to ensuring correct process formulation and interaction rather than numerical stability problems. INCA enables the user to calculate water quality loads from different land uses, and information on annual and daily fluxes can be obtained. In addition, it is possible to evaluate scenarios of environmental change to assess impacts on flow, loads and water quality. Scenarios that can be tested may include changing effluent discharges, land use change (e.g. moorland to forest), changing atmospheric deposition (e.g. impacts of new atmospheric emission protocols), or changing agricultural practices (e.g. decreasing/increasing fertilizer applications). Over the past 16 years the INCA suite of models has evolved to include hydrology, nitrogen, ammonia, phosphorus, sediments, dissolved organic carbon (DOC), metals (including arsenic, lead, cadmium, copper, nickel, chromium, manganese, mercury and molybdenum), microplastics and pathogens (Futter et al. 2007; Nizzetto et al. 2016; Wade et al. 2002a, b, c; Whitehead et al. 2009, 2015a, 2016). Ecological interactions are also built into INCA by modelling the growth and death of macrophytes, epiphytes and phytoplankton in rivers and lakes. Again, a lumped differential modelling approach is used to simulate the interacting processes for ecology and the chemical and physical dynamics that affect it (Wade et al. 2002c; Whitehead and Hornberger 1984; Whitehead et al. 2015a). INCA has been applied to a wide range of issues and scales, from plot scale in forests to catchments as large as the Ganges, Brahmaputra and the Volta (Jin et al. 2018; Whitehead et al. 2015b).

INCA builds on all the history of water quality modelling and countless laboratory and field experiments to draw the best knowledge on water chemistry processes. It attempts to incorporate all this science into a coherent set of equations that can be solved efficiently on modern computing systems and also aims to have a user-friendly interface. It can be used rather like a computer game to explore all aspects of catchment behaviour, thereby providing the user, whether a scientist or manager, with a deep understanding of catchment dynamics.

The following two case studies demonstrate the application of the INCA model to simulate emerging contaminants in a large UK river system (the Thames) and extreme pollution in Bangladesh, and to explore potential restoration measures.

4.2.3 Case Study 2: Modelling Contaminants Using INCA – Metaldehyde in the Thames

Metaldehyde is becoming a serious problem for Thames Water in the UK, as the pesticide is increasingly used by farmers to control snail and slug pests on a range of crops. The pellets normally dissolve quickly in wet conditions and the chemical can then be rapidly flushed out during rainfall events. The high resulting concentrations in streams and the main Thames river system create problems for water abstraction and supply, as the concentrations often exceed the EU limits of 0.1 μg/l for organics in drinking water. The INCA Contaminants model (Lu et al. 2017) has recently been developed to assess organic pollutants in rivers and the model has been set up for the Thames catchment from the source at Cricklade to Teddington in London. Metaldehyde has been simulated in the river system and Figure 4.2a shows the simulated and observed metaldehyde concentrations at Datchet in the lower Thames. The model simulates the diffuse sources of metaldehyde assuming a farm application of the pesticide and can therefore be used to evaluate a set of application strategies. Figure 4.2b shows the peak concentrations of metaldehyde along the eight reaches of the Thames down the river under different application rates. The results suggest that an application rate no higher than 33 g/ha in the catchments is required to keep the metaldehyde down below the EU limit. The model is currently being used by Thames Water to assess best-practice mitigation strategies to address metaldehyde use and transport along the river system.

4.2.4 Case Study 3: Water Quality in the Turag-Balu River System, Dhaka, Bangladesh

There is a serious problem of water pollution in central Dhaka, Bangladesh, in the Turag-Balu river system. There is a very high organic pollution loading causing DO concentrations to fall to zero in the dry season, giving rise to so-called 'blackwater' conditions. In addition to low DO, the ammonia concentrations are very high at 10 mg/l, and large numbers of pathogens are present in the water. Factories along the river discharge pollutants with dangerous contaminants, runoff from waste tips is a serious

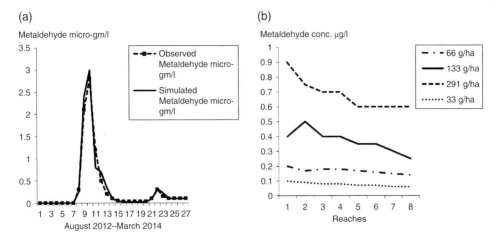

Figure 4.2 (a) Metaldehyde observed and simulated in the river Thames at Datchet (2012–2014). *Source:* Paul Whitehead. (b) Impacts of different application rates on metaldehyde concentration along the Thames reaches at eight locations from the source at Cricklade (in the Cotswolds) to Teddington in London. *Source:* Lu et al. (2017). Reproduced with the permission of the Royal Society of Chemistry (2017).

problem, and there is agricultural nutrient pollution from the upper reaches of the river. People using the water for drinking, washing clothes, bathing and vegetable production are under constant threat. There is wide-scale evidence of high levels of disease, skin infections and more serious illnesses impacting local people.

A baseline survey of water chemistry and pathogens has been undertaken by the Bangladesh University for Engineering and Technology, and the INCA models have been set up for the river system (Whitehead et al. 2018). INCA hydrology, nitrate, ammonia and pathogens models have been used to assess hydrobiochemical processes in the river and evaluate alternative strategies for the management of the pollution issues. The hydrology of the system is highly complex, and special attention needs to be paid to the additional flood flows and runoff coming from the upstream sections of the Turag generated by overland flow from the Brahmaputra. Having established the model, a series of management alternatives have been considered to address the water pollution issues. Three management strategies have been considered, namely, the introduction of effluent clean-up technologies for key discharges along the river, the augmentation of water flows in the upper Turag so as to increase the dilution effects in low flow conditions, and a combined strategy. Figure 4.3 shows the effects of these three strategies and indicates that clean-up of the effluents will significantly improve the water quality, reducing the mean ammonia concentrations from 10 to 3 mg/l. The augmentation flows are also very effective, reducing ammonia to 6 mg/l. Based on the model results, the best solution is to combine these policies, and this would reduce the ammonia levels to 1 mg/l, a relatively low concentration for river systems. However, other factors, such as the economic feasibility of the management strategies, need to be considered.

4.2.5 Model Uncertainty

There are always concerns about model limitations in any flow and water quality modelling study (Beck 1987). Models by definition are a simplification of reality, and hence

Ammonia mg/l

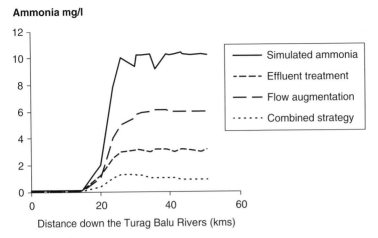

Distance down the Turag Balu Rivers (kms)

Figure 4.3 Simulated ammonia and control strategies along the Turag-Balu River System in Dhaka, Bangladesh. *Source:* Whitehead et al. (2018). Reproduced with permission of Elsevier.

there will always be uncertainty associated with model process equations and model parameters. Also, models are also only as good as the data with which they are driven and the data used for calibration and validation. Uncertainties in data, model processes and parameters will inevitably propagate to model outputs. Measuring hydrological and water quality variables is difficult, with at best 5% errors on flow measurement, with sampling and laboratory analytical techniques introducing further errors ranging from 5 to 20% (Whitehead et al. 1981), although chemical analysis techniques have improved significantly in recent years. Models are a simplification of reality and allow an approximation to any system to be evaluated (Chapra 2018). The fact that the system is dynamic and that chemical processes are kinetic in nature and interact with the hydrology makes modelling a very complex process. Information about the water quality impacts of climate change is limited, including their socioeconomic dimensions (Bates et al. 2008). Greater data collection and reliability can lead to continual improvements in the accuracy of models and their capacity to make informed policy decisions. For example, routine monitoring on a weekly or monthly basis gives little information on water quality dynamics, whereas daily or bi-daily sampling will give much better definition of the dynamic changes occurring and in-stream processes. A concerted scientific effort can reduce uncertainties in predicting water quality, and the trade-offs and benefits of current and alternative trends and scenarios. In the meantime, however, whilst large uncertainties continue to exist, characterization of these uncertainties is important, particularly when models are used to inform decision-making. One approach to resolving model uncertainty is to make use of Monte Carlo (MC) analysis. MC analysis is a technique in which multiple model simulations are performed using a random value for model parameters to generate model outputs. This procedure is repeated many times to generate multiple simulated outputs, with each ensemble member produced using a different parameter set. The ensemble averages and distributions of the outputs can be used to evaluate the uncertainty associated with the model. There have been extensive studies of uncertainty using such procedures (Beck 1987; Beven and Binley 1992, 2014; Whitehead and Young 1979).

One procedure is that of generalized sensitivity analysis (GSA), which has been used extensively to evaluate both parametric uncertainty in models and uncertainty in the data used to drive the model (Futter et al. 2007; Spear and Hornberger 1980). The GSA technique is based on the utilization of a simulation model together with a classification algorithm. The classification algorithm allows the model outputs to be identified as either representative or not representative of the observed behaviour. The idea is to inject uncertainty into the simulation model by selecting the parameters from specified probability distributions rather than from experimentally derived values. The simulation is repeated using different parameter sets, and the parameter set is classified as either producing or not producing a behaviour. Subsequent to these MC trials, statistical analysis of the parameter sets is used to identify the key parameters causing the model to reproduce the observed behaviour. The theory behind this statistical analysis is based on the separation between the cumulative probability distributions of two parameter sets, and a Kolmogorov–Smirnov two-sample test is utilized to test the separation. A knowledge of which parameters are controlling the system behaviour is particularly useful, as scientific effort can be deployed to determine the key processes represented by the parameters. For example, Wade et al. (2002a, b, c) found that the water quality models used in their studies were very sensitive to the parameters defining nitrogen dynamics such as denitrification processes, key hydrological factors such as base flow index, and growth rates in ecological models. The GSA technique has been applied in studies of a range of water quality problems, including eutrophication in the Peel–Harvey river and estuary system in Australia (Hornberger and Spear 1980) and algal growth in the River Thames (Bussi et al. 2016).

The GSA technique is one approach to uncertainty analysis, but other approaches include generalized likelihood uncertainty estimation (GLUE) by Beven and Binley (1992). This is a statistical method for quantifying the uncertainty of model predictions. The basic idea of GLUE is that that there are always different models that reproduce equally well an observed natural process, such as river flows or water quality. There is an implicit understanding that the models being used are approximations to what might be obtained from a thorough Bayesian analysis of the problem, if a fully adequate model of real-world hydrological processes were available. Again, a Monte Carlo approach is used to evaluate model performance and uncertainty statistics.

4.3 Monitoring Strategies Past and Present

Monitoring rivers, lakes and groundwaters has become a key task at local, regional, national and global levels over the past 50 years. This has been necessary to protect aquatic environments and ensure that pollution discharges have a minimal, or at least a managed, impact on the water systems. The datasets have been invaluable in assessing the state of the aquatic environment, detecting long-term trends, advancing the science of aquatic ecosystems, and as a tool to monitor the success of global initiatives such as the UN SDGs and European Directives such as the Water Framework Directive. In addition, monitoring datasets have been invaluable for establishing the system dynamics and providing a means of model development. Calibration and validation of models are crucial steps in proving that a model actually represents the river or catchment system, and so monitoring data is the key to successful model development.

4.3.1 Global Monitoring

The United Nations Global Environment Monitoring System (GEMS) freshwater programme is a major programme concerned with increasing the understanding of freshwater quality issues. This UN programme has many key objectives, including creating a worldwide monitoring system, strengthening monitoring protocols in developing countries, and generating understanding of water issues at global, national and regional levels. Over 80 countries have formally joined the GEMS/Water programme since its inception in 1978, with approximately 58 countries submitting data on a regular basis. The GEMS/Water website (https://www.unenvironment.org/explore-topics/water/what-we-do/monitoring-water-quality) offers exchange and sharing of information dealing with water and water management. The purpose of the website is to significantly improve communications and the flow of information between the UN, the World Health Organization (WHO) and regional, national and international agencies interested in water issues. Surface and groundwater quality monitoring data collected from the global GEMS/Water monitoring network is shared through the GEMStat information system (https://gemstat.org). Data are statistically treated and summarized for each individual station over a three-year period. Currently, there are 58 active countries with a total of 423 stations, and summaries are produced for physical, chemical and microbiological data plus globally significant variables such as metals and pesticides. GEMS supports the monitoring of six SDG indicators, and facilitates a standardized approach to monitoring the state of rivers globally.

4.3.2 National-scale Monitoring

In addition to global monitoring, there are national-scale monitoring systems in many countries. At the national level there is often primary legislation governing the management and monitoring of the environment. In the EU, for example, the EU Water Framework Directive, Nitrates Directive, Urban Wastewater Treatment Directive and the Habitats Directives seek to create a legal framework for all EU countries to ensure that they meet certain water quality and ecological standards. In fact, the aim of these directives is not just to monitor the environment but to restore damaged freshwater ecosystems across Europe. Data from the Environment Agency (EA) in the UK has been collected to meet these directives and used by researchers to understand the dynamics and processes controlling water quality and to assist with the development of water quality models. The data has provided a basis for catchment management and the setting of discharge standards for industrial and domestic effluents. In fact, prior to the EA, long-term datasets for water quality were collected by water conservation organizations and by water companies. The Thames is a good example, where water samples from the river have been analysed since the 1900s (Burt et al. 2014). These long-term datasets capture very interesting trends in water quality such as low levels of nutrients (N and P) in the freshwater sections of the Thames in the 1930s, pollution levels rising during the mid 1940s, as the Thames was ploughed up for food production, and then increasing nutrient levels in the 1960s and 1970s, reflecting the more intensive farming over these decades.

4.3.3 Long-term Monitoring of Key Scientific Sites

Perhaps some of the best time series of flow and water quality have arisen from long-term experiments such as the Hubbard Brook Study in the USA (Likens 2013) and the

Plynlimon study in Wales (Neal et al. 2013). These provide powerful datasets for improving scientific understanding of the key processes controlling water quality, such as processes controlling nutrient transport, hydrochemical controls on acidification, and impacts of land use change on water quality. Long-term data on DOC at Plynlimon shows increasing trends over the past 20 years, and this is generally thought to be a response of the carbon release mechanisms to reduced acid deposition. However, the spiky nature of the response also suggests a climate-driven component, with warmer conditions and increased precipitation beginning to affect water quality and drive peaks of DOC. Is this a first signature of climate change affecting water quality? Higher frequency datasets are also being used for new analysis such as fractal approaches developed to explain and qualify statistically detailed catchment dynamics (Kirchner and Neal 2013; Wade et al. 2012). There is still much debate about what these fractal patterns of behaviour mean for water quality analysis and how higher frequency data can be incorporated into modelling. Rapid water quality monitoring will always provide more information and show higher frequency patterns, so the fractal patterns are not such a surprise. Rather, it is more a case of how the information can be used to enhance models and understanding, especially when considering the underlying science or management alternatives. It is an exciting new area of research with great potential.

Long-term experiments also led to the development of spatial networks such as the Acid Waters Monitoring Networks covering sites across the USA, Canada and the UK (Jenkins et al. 1995). These extensive datasets provide long-term trends in acidity, with data being collected from the mid-1980s to the present. The datasets have been used extensively to support policy decisions concerning acid rain impacts and whether to stop releasing sulphur into the atmosphere from power stations in North America and Europe. Important models were developed for acidification processes based on the observed data (Cosby et al. 1986), and scenario analysis was used to arrive at major policy changes in the USA, Canada, the UK and across Europe. The models convinced the UK government that sulphur deposition was causing the acidification of rivers and lakes across the northern Europe. On the basis of this, the UK committed to reducing sulphur emissions by 60% in order to enable water quality to recover across the UK and Europe. This has led to a nearly 90% reduction in sulphur oxides to date, with significant recovery of rivers, lakes and catchments, largely solving the acid rain problem. However, new pressures from long-term nitrogen pollution pose an ongoing threat, enhancing the acidification process, and so there is still a need to maintain the long-term monitoring sites.

Monitoring has played a key role in understanding water quality and hydrology and has been crucial for scientific-based model development as well. Unfortunately, it is increasingly hard to fund monitoring programmes. The EA has been under financial pressure to reduce the routine monitoring of rivers and lakes in the UK, and research councils are extremely reluctant to fund monitoring, unless it is part of a highly focused scientific study. This means that less new information is available for understanding catchment response and for new model development. Crucial networks such as the Environmental Change Network (www.ecn.ac.uk) and the acid waters monitoring network (https://uk-air.defra.gov.uk/networks/network-info?view=aw) provide very important long-term records of changing environmental pollution and their impacts on soils, waters and terrestrial ecology.

4.3.4   Citizen Science Monitoring

Greater data collection and reliability can lead to continual improvements in the accuracy of models, and their capacity to inform policy decisions. An increasingly networked world offers opportunities for capturing new data, reducing uncertainties and engaging with the public. Remote and real-time sensing can generate new knowledge of the state of water quality, pollution sources, and options to address them. Earth observations and drones can be used to assess water quality in remote regions. Citizen science (i.e. mobile phone apps, online pollution reporting hotlines) and Earth observations can overcome challenges of inadequate data and data-sharing for transboundary management. For example, the Creek Watch iPhone App enables the US public to capture data on the quality and quantity of any water body at any point and time (https://researcher.watson. ibm.com/researcher/view_group.php?id=3240). The app enables new sources of data to be collected at little added cost, from which new insights can be derived and management decisions prioritized. Some challenges associated with using citizen science strategically include: integration and coherence of data gathered from various citizen and other sources; making them available to citizens, scientists, regulators and polluters in accessible and understandable ways; and ensuring that citizen science efforts and online platforms for accessing data are sustained beyond typical three-year project funding cycles. All of these new sources of data have the ability to reduce monitoring, compliance and enforcement costs. The new digital environment also offers opportunities for more collaborative and participatory relationships that allow relevant stakeholders to actively shape political priorities, collaborate in the design of public services, and participate in their delivery to provide more coherent and integrated solutions to complex challenges. For example, New Zealand has established a national public database (www. lawa.org.nz) to improve utilization of reliable real-time and historical water quality data by a range of users to inform business, recreational and environmental decisions. Sharing of data frees up significant overheads in delivering routine data requests, avoids double monitoring, and redirects effort into additional monitoring or policy work (OECD 2017).

4.3.5   Case Study 4: Monitoring and Modelling the Murray-Darling System in Australia

Australia's Murray-Darling Basin (MDB) covers an area of more than 1 million km^2, including parts of four states and the Australian Capital Territory. The river system fed by this catchment faces a number of water quality challenges, which require monitoring, modelling and management across jurisdictions. Amongst the major water quality issues are low DO, creating hypoxic black water conditions, salinity, and blue-green or cyanobacteria algal blooms. Water quality is monitored under the River Murray water quality monitoring programme, which is managed by the Murray-Darling Basin Authority on behalf of MDB governments. Data is used to detect changes in water quality, to provide a long-term record for research, and to inform water management decisions. Under the Murray-Darling Basin Plan, a series of water quality targets have been adopted, including:

- A DO target of at least 50% saturation;
- Salinity target values ranging from 580 to 1000 µS/cm EC (electrical conductivity) at five reporting sites, which should not be exceeded 95% of the time; and
- Recreational water cyanobacteria cell count target values.

Regular water quality sampling is required to monitor whether these targets are met.

Hypoxic 'blackwater' events occur when metabolism of large amounts of DOC (e.g. in returning floodwater) results in water column oxygen depletion, with potentially harmful consequences for aquatic organisms. For example, a large-scale 'blackwater' event in the southern MDB, which occurred due to a warm-season flooding in 2010–2011 (Whitworth et al. 2012), caused fish kills and negatively impacted crayfish populations (McCarthy et al. 2014). In New South Wales, a network of DO sensors is operated as part of an early warning system, with weekly information dissemination during high-risk periods (NSW Department of Industry 2017). In order to assist with low DO management, a desktop predictive model, the Blackwater Risk Assessment Tool (BRAT), has been developed by the Murray-Darling Freshwater Research Centre (Whitworth and Baldwin 2016). BRAT models low DO generation processes, such as inundation, DOC leaching from litter, DOC consumption, and oxygen consumption and re-aeration. The tool is being used to inform active management of forested floodplain inundation in the MDB.

Salinity issues in the Murray-Darling river system are a result of salt mobilization associated with elevated groundwater tables caused by land clearing and water-intensive agriculture. The issue rose to prominence in the 1960s, following a series of droughts. Since then, several strategies have been implemented to manage salinity, including the construction of salt interception schemes (SISs) and improved farming practices. Following a salinity audit in 1999, a Basin Salinity Management Strategy was adopted in 2001. Key aspects of this policy included the adoption of tributary and MDB outlet salinity targets, and accountability arrangements, including monitoring and reporting of salinity levels at key sites, including the MDB outlet at Morgan, South Australia. Modelling has been used to support the tracking of progress towards addressing water salinity issues. Comparison of modelled salinity levels (without salinity management strategies) and measured salinity levels at Morgan indicates that salinity management has been effective in helping to maintain salinity levels below the target of 800 µS/cm EC (Murray-Darling Basin Authority 2017a). Models were also used as part of the 2014 General Review of Salinity Management, to investigate salinity risks associated with different SIS operation scenarios (Murray-Darling Basin Authority 2017b). Following on from the Basin Salinity Management Strategy, the Basin Salinity Management 2030 policy was adopted in 2015.

Blue-green algae (cyanobacteria) blooms occur when conditions are favourable for rapid reproduction of cyanobacteria, such as warm water temperatures, sunlight and high nutrient concentrations. Blooms are a cause of concern, since some species of cyanobacteria produce toxins that are harmful to human and animal health and are difficult and costly to remove. A system of regular monitoring of phytoplankton and water quality in the Murray-Darling river system has been in place since 1980. The New South Wales Office of Water is currently working with the Commonwealth Scientific and Industrial Research Organisation (CSIRO) in Australia to develop an algal bloom early warning system that will utilize remote sensing technology (NSW Office of Water 2014). To help inform river operation practices to better manage water quality, a cyanobacteria bloom risk index model has been developed. This model has been developed as part of a prototype River Murray Decision Support System (DSS) for the South Australian region of the MDB (Overton et al. 2017). The DSS has been built using eWater Source, an integrated basin-scale modelling platform. The DSS also includes models for salinity and low DO risk, enabling an integrated analysis of river management scenario implications for water quality in the MDB.

4.4 Conclusions

The field of water quality processes, modelling, and management has expanded rapidly in the past 50 years as public expectations for cleaner, healthier rivers and better-quality water supplies have risen. Water quality modelling, from simple mass balance approaches to complex nonlinear dynamic models, provides a range of techniques to assist environmental managers to manage scarce resources effectively. The UN SDGs provide an international framework for assessing pollution control and water management. The targets of the SDGs, however, are unlikely to be met unless a concerted effort is made by national governments to apply the best techniques to clean up and restore aquatic environments.

Despite innovations in modelling methods across planning, design and operational systems, a substantial gap remains between theoretical and practical aspects of river management in developing countries. In many developed countries there is extensive legislation unpinning monitoring and management to force water companies, industry and environment agencies to work together to protect aquatic systems. However, in developing countries, the combination of water quality and ecological issues and socio-economic issues raises many challenges. The techniques of water quality modelling are likely to be further developed into the future to meet demands for modelling of new climatic conditions, previously unmodelled river systems, and new emergent contaminants, to address new ecological problems, and to incorporate multidisciplinary knowledge to assist with decision-making in the context of complex connected social, economic and environmental systems. There is a need to improve understanding and modelling of the impacts of climate change on water quality at scales relevant to decision-making, and of vulnerability to and ways of adapting to those impacts (IPCC 2014). Management approaches need to account for uncertainties around climate change projections and the impacts on water quality.

This chapter sought to highlight what can be achieved in water quality management and science with the support of appropriate models and monitoring. However, policy processes tend to be iterative and often reflect a shorter-term response to the most pressing problems of a particular period of government. Political groups are often obliged to put forward at least some populist policies in order to get elected, and must then attempt to deliver some of the promises against which they will be evaluated. In this respect, there are major issues to reconcile to enable effective river management in the context of conflicting political priorities. Many of these broader issues affecting water quality management cannot be resolved through water quality monitoring and modelling alone.

References

Bates, B.C., Kundzewicz, Z.W., Wu, S., and Palutikof, J.P. (eds.) (2008). *Climate Change and Water*, Technical Paper of the Intergovernmental Panel on Climate Change. Geneva: IPCC Secretariat.

Beck, M.B. (1987). Water quality modeling: a review of the analysis of uncertainty. *Water Resources Research* 23: 1393–1442.

Beven, K. and Binley, A. (1992). The future of distributed models: model calibration and uncertainty prediction. *Hydrological Processes* 6 (3): 279–298. https://doi.org/10.1002/hyp.3360060305.

Beven, K. and Binley, A. (2014). GLUE: 20 years on. *Hydrological Processes* 28 (24): 5897–5918. https://doi.org/10.1002/hyp.10082.

Boënne, W., Desmet, N., Van Looy, S., and Seuntjens, P. (2014). Use of online water quality monitoring for assessing the effects of WWTP overflows in rivers. *Environmental Science: Processes and Impacts* 16 (6): 1510–1518. https://doi.org/10.1039/c3em00449j.

Burt, T.P., Howden, N.J.K., and Worrall, F. (2014). On the importance of very long-term water quality records. *Wiley Interdisciplinary Reviews: Water* 1: 41–48.

Bussi, G., Whitehead, P.G., Bowes, M.J. et al. (2016). Impacts of climate change, land-use change and phosphorus reduction on phytoplankton in the River Thames (UK). *Science of the Total Environment* 507: 1507–1519. https://doi.org/10.1016/j.scitotenv.2016.02.109.

Chapra, S.C. (2018). *Surface Water Quality Modeling*. Illinois: Waveland Press Inc.

Chapra, S. and Whitehead, P.G. (2009). Modelling impacts of pollution in river systems: a new dispersion model and a case study of mine discharges in the Abrud, Aries and Mures River System in Transylvania, Romania. *Hydrological Research* 40 (2–3): 306–322.

Cosby, B.J., Whitehead, P.G., and Neale, R. (1986). A preliminary model of long term changes in stream acidity in South Western Scotland. *Journal of Hydrology* 84: 381–401.

Cox, B.A. and Whitehead, P.G. (2009). Impacts of climate change scenarios on dissolved oxygen in the River Thames, UK. *Hydrology Research* 40 (2–3): 138–152.

Estrup, H.E., Kronvang, B., Larsen, S.E. et al. (2006). Climate-change impacts on hydrology and nutrients in a Danish lowland river basin. *Science of the Total Environment* 365 (1): 223–237. https://doi.org/10.1016/j.scitotenv.2006.02.036.

Futter, M.N., Butterfield, D., Cosby, B.J. et al. (2007). Modeling the mechanisms that control in-stream dissolved organic carbon dynamics in upland and forested catchments. *Water Resources Research* 43 (2) https://doi.org/10.1029/2006WR004960.

Hornberger, G.M. and Spear, R.C. (1980). Eutrophication in Peel Inlet – I. The problem defining behaviour and a mathematical model for the phosphorus scenario. *Water Research* 14 (1): 29–42. https://doi.org/10.1016/0043-1354(80)90039-1.

IPCC (2014). *Climate Change 2014: Synthesis Report. Contribution of Working Groups I, II and III to the Fifth Assessment Report of the Intergovernmental Panel on Climate Change* [Core Writing Team: R.K. Pachauri and L.A. Meyer, eds.]. Geneva: IPCC.

Jenkins, A., Campbell, G., Renshaw, M. et al. (1995). Surface water acidification in the UK: current status, recent trends and future predictions. *Water, Air and Soil Pollution* 85: 565–570.

Jin, L., Whitehead, P.G., Addo, K.A. et al. (2018). Modeling future flows of the Volta River system: impacts of climate change and socio-economic changes. *Science of the Total Environment* 637-638: 1069–1080. https://doi.org/10.1016/j.scitotenv.2018.04.350.

Kirchner, J.W. and Neal, C. (2013). Universal fractal scaling in stream chemistry and its implications for solute transport and water quality trend detection. *Proceedings of the National Academy of Sciences of the United States of America* 110 (30): 12213–12218.

Likens, G.E. (2013). The Hubbard Brook ecosystem study: celebrating 50 years. *Bulletin of the Ecological Society of America* https://doi.org/10.1890/0012-9623-94.4.336.

Loucks, D.P. and van Beek, E. (2017). *Water Resource Systems Planning and Management*. Cham, Switzerland: Springer with Deltares and UNESCO-IHE.

Lu, Q., Whitehead, P.G., Bussi, G. et al. (2017). Modelling metaldehyde in freshwater systems: case study of the River Thames. *Environmental Science: Processes and Impacts* 19 (4): 586–595.

McCarthy, B., Zukowski, S., Whiterod, N. et al. (2014). Hypoxic blackwater event severely impacts Murray crayfish *Euastacus Armatus* populations in the Murray River, Australia. *Austral Ecology* 39: 491–500.

Murray-Darling Basin Authority (2017a). *Murray-Darling Basin Authority Annual Report 2016–17*. Canberra: Murray-Darling Basin Authority.

Murray-Darling Basin Authority (2017b). *Modelling to support the general review of salinity management in the basin*. Canberra: Murray-Darling Basin Authority.

Neal, C., Reynolds, B., Kirchner, J.W. et al. (2013). High-frequency precipitation and stream water quality time series from Plynlimon, Wales: an openly accessible data resource spanning the periodic table. *Hydrological Processes* 27 (17): 2531–2539.

Nizzetto, L., Bussi, G., Futter, M. et al. (2016). A theoretical assessment of microplastic transport in river catchments and their retention by soils and river sediments. *Environmental Science: Processes and Impacts* 18: 1050–1059.

NSW Department of Industry (2017). New South Wales Basin Plan Annual Report 2016/17. http://www.mdba.gov.au/sites/default/files/pubs/D18-9268-NSW-BP-implementation-report-2016-17.pdf [accessed 20 April 2018].

NSW Office of Water (2014). Algal Early Warning System under Development. http://archive.water.nsw.gov.au/__data/assets/pdf_file/0012/549498/media_release_141029_early_warning_algal_system_under_development_final.pdf [accessed 20 April 2018].

OECD (2017). *Diffuse Pollution, Degraded Waters: Emerging Policy Solutions*, OECD Studies on Water. Paris: OECD Publishing https://doi.org/10.1787/9789264269064-en.

Overton, I.C., Freebairn, A., Joehnk, K. et al. (2017). *River Murray Decision Support System: Prototype*. Technical Report Series No. 16/4. Adelaide: Goyder Institute for Water Research.

Royal Commission on Sewage Disposal (1912). *Eighth Report*. London: HMSO.

Spear, R.C. and Hornberger, G.M. (1980). Eutrophication in Peel Inlet – II. Identification of critical uncertainties via generalized sensitivity analysis. *Water Research* 14: 43–49.

Srinivasan, R. and Arnold, J.G. (1994). Integration of a basin-scale water quality model with GIS. *Water Resources Bulletin* 30 (3): 453–462.

Streeter, H.W. and Phelps, E.B. (1925). *A Study of the Pollution and Natural Purification of the Ohio River*. US Public Health Bulletin No. 146.

Thomann, R.V. and Mueller, J.A. (1987). *Principles of Surface Water Quality Modeling and Control*. New York: Harper Collins.

UNESCO (2009). *The Agenda 21 World Water Development Report 3: Water in a Changing World*. World Water Assessment Programme. Paris: UNESCO.

United Nations (2015). Resolution adopted by the General Assembly 22 September 2015: Transforming our world: the 2030 Agenda for Sustainable Development. A/RES/70/1.

Wade, A.J., Durand, P., Beaujouan, V. et al. (2002a). Towards a generic nitrogen model of European ecosystems: INCA, new model structure and equations. *Hydrology and Earth System Sciences* 6: 559–582.

Wade, A.J., Whitehead, P.G., and Butterfield, D. (2002b). The integrated catchments model of phosphorus dynamics INCA-P, a new approach for multiple source assessment in heterogeneous river systems: model structure and equations. *Hydrology and Earth System Sciences* 6: 583–606.

Wade, A., Whitehead, P., Hornberger, G. et al. (2002c). On modelling the impacts of phosphorus stripping at sewage works on in-stream phosphorus and macrophyte/epiphyte dynamics: a case study for the River Kennet. *Science of the Total Environment* 282: 395–415.

Wade, A.J., Palmer-Felgate, E.J., Halliday, S.J. et al. (2012). Hydrochemical processes in lowland rivers: insights from in situ, high-resolution monitoring. *Hydrology and Earth System Sciences* 16 (11): 4323–4342.

Warn, A.E. (1981). Calculating consent conditions to achieve river quality objectives. *Journal of Effective Water Treatment* 8: 152–155.

Whitehead, P.G. and Hornberger, G.M. (1984). Modelling algal behaviour in the River Thames. *Water Research* 18: 945–953.

Whitehead, P.G. and Young, P.C. (1979). Water quality in river systems: Monte-Carlo analysis. *Water Resources Research* 15: 451–459.

Whitehead, P.G., Beck, M.B., and O'Connell, E. (1981). A systems model of stream flow and water quality in the Bedford-Ouse river – 2. Water quality modelling. *Water Research* 15: 1157–1171.

Whitehead, P.G., Caddy, D.E., and Templeman, R.F. (1984). An on-line monitoring data management and forecasting system for the Bedford Ouse river basin. *Water Science and Technology* 16: 295–314.

Whitehead, P.G., Wilson, E.J., and Butterfield, D. (1998). A semi-distributed integrated nitrogen model for multiple source assessment in Catchments INCA: part I – model structure and process equations. *Science of the Total Environment* 210/211: 547–558.

Whitehead, P.G., Butterfield, D., and Wade, A.J. (2009). Simulating metals and mine discharges in river basins using a new integrated catchment model for metals: pollution impacts and restoration strategies in the Aries-Mures river system in Transylvania, Romania. *Hydrological Research* 40 (2–3): 323–346.

Whitehead, P.G., Bussi, G., Bowes, M.J. et al. (2015a). Dynamic modelling of multiple phytoplankton groups in rivers with an application to the Thames River system in the UK. *Environmental Modelling & Software* 74: 75–91. https://doi.org/10.1016/j.envsoft.2015.09.010.

Whitehead, P.G., Barbour, E., Futter, M.N. et al. (2015b). Impacts of climate change and socio-economic scenarios on flow and water quality of the Ganges, Brahmaputra and Meghna (GBM) river systems: low flow and flood statistics. *Environmental Science: Processes and Impacts* 17: 1057–1069. https://doi.org/10.1039/C4EM00619D.

Whitehead, P.G., Leckie, H., Rankinen, K. et al. (2016). An INCA model for pathogens in rivers and catchments: model structure, sensitivity analysis and application to the River Thames catchment, UK. *Science of the Total Environment* 527: 1061–1061. https://doi.org/10.1016/j.scitotenv.2016.01.128.

Whitehead, P.G., Bussi, G., Hossain, M.A. et al. (2018). Restoring water quality in the polluted Turag-Tongi-Balu river system, Dhaka: modelling nutrient and total coliform intervention strategies. *Science of the Total Environment* 631–632: 223–232.

Whitworth, K.L. and Baldwin, D.S. (2016). Improving our capacity to manage hypoxic blackwater events in lowland rivers: The Blackwater Risk Assessment Tool. *Ecological Modelling* 320: 292–298.

Whitworth, K.L., Baldwin, D.S., and Kerr, J.L. (2012). Drought, floods and water quality: drivers of a severe hypoxic blackwater event in a major river system (the southern Murray–Darling Basin, Australia). *Journal of Hydrology* 450-451: 190–198.

Young, P.C. and Chotai, A. (2004). Data-based mechanistic modelling and the simplification of environmental systems. In: *Environment Modelling: Finding Simplicity in Complexity*, 371–388. Chichester: Wiley.

5

Challenges for Freshwater Ecosystems

Jocelyne Hughes[1], Heather Bond[2], Clarke Knight[3], and Kieran Stanley[4]

[1] *School of Geography and the Environment, University of Oxford, UK*
[2] *Flood and Coastal Risk Management Directorate, Environment Agency, Wallingford, UK*
[3] *University of California at Berkeley, USA*
[4] *Institute for Atmospheric and Environmental Sciences, Goethe University Frankfurt, Frankfurt am Main, Germany*

5.1   How do Freshwater Ecosystems Work?

5.1.1   Structure and Function of Freshwater Ecosystems

It is a basic tenet in freshwater ecology that the overarching influences affecting energy flows and biotic processes within different freshwater ecosystems (rivers, lakes and wetlands are considered in this chapter) are driven by hydrology and catchment geology. Hydrology (e.g. water flow pathways, residence time, seasonality, extreme events) and geology (e.g. rock type, surface deposits, porosity, geochemistry) both influence the amount of water present in a water body at any one time, the water chemistry, and the substrate type (Moss 2018). This hydrogeological 'signature' gives rise to a range of freshwater ecosystem types, each with their own structure (species and habitat diversity, and productivity) and function (nutrient cycling processes, energy transfers, trophic interactions). For example, lotic or running waters such as rivers and streams host organisms that can cope with flows, shear stresses and disturbances; lentic or still/ enclosed waters such as lakes and ponds support intricately connected planktonic communities; and wetland habitats such as ombrotrophic (rain-fed) acid peatlands, minerotrophic (groundwater- and spring-fed) neutral to alkaline peatlands, and mineral-based, depositional marshes provide conditions for a range of emergent macrophyte or bryophyte species (Mitsch and Gosselink 2015). Within the broad context of a waterbody's hydrogeochemical signature, the structure and function of individual freshwaters will be influenced by the web of inter- and intra-species interactions; species traits and their ability to compete, survive and thereby affect their trophic level; isolation and the time taken for speciation; substrate type and the diversity of microorganisms tolerant of anoxic conditions; and disturbances or extreme events, such as floods, droughts or temperature extremes. Most significantly, the human influences on a catchment or waterbody may profoundly alter the baseline hydrogeochemical signature described above, with consequent sweeping impacts on the structure and function of freshwater

Water Science, Policy, and Management: A Global Challenge, First Edition. Edited by Simon J. Dadson, Dustin E. Garrick, Edmund C. Penning-Rowsell, Jim W. Hall, Rob Hope, and Jocelyne Hughes.
© 2020 John Wiley & Sons Ltd. Published 2020 by John Wiley & Sons Ltd.

ecosystems, such as land use and land cover affecting water and nutrient fluxes, nutrient enrichment causing eutrophication, release of a non-native aquatic species, destruction of wetlands for other land uses, the construction of impoundments, and the depletion of environmental flows.

Freshwater habitats cover less than 1% of the Earth's surface, but support an estimated 126 000 known species of plants and animals, representing about 10% of species described globally (Balian et al. 2008). Because freshwater habitats are insular, they support a high degree of endemism and species diversity. This high degree of specialization together with multiple interacting stressors (such as invasive non-native species (INNS), water extraction, pollution), leads to freshwater species facing a higher extinction risk than their terrestrial counterparts (Belgrano et al. 2015; Collen et al. 2014). The range of wetland and freshwater ecosystem types, their high species diversity (Collen et al. 2014) and the services and livelihoods that they provide to human populations (Figure 5.1; Ramsar Convention website; Russi et al. 2013) broadly drive and justify global freshwater conservation and restoration efforts and can be summarized into several key points.

Firstly, freshwaters, and particularly marshes, have the highest productivity (rate at which energy is stored as biomass over time) of any global ecosystem (Rocha and Goulden 2009) and provide life-sustaining provisioning services to human populations globally, such as food, fuel and water (Figure 5.1; Mitsch and Gosselink 2015; Moss 2018). Yet wetland and freshwater ecosystems are the fastest declining of any global habitat, as measured by species loss and habitat reduction (IUCN Freshwater Biodiversity Unit website; Gardner et al. 2015; WWF 2018). For example, between 1970 and 2014, 83% of biodiversity in freshwaters was lost compared with 38% in terrestrial systems and 36% in marine systems. Conserving and managing this remarkable species diversity (and species traits) is the cornerstone to maintaining the services and livelihoods that freshwaters provide. Secondly, conceptualizing or modelling the impacts of anthropogenic disturbances and climate change on the functional diversity of freshwater ecosystems is critical to securing the functional services that humans rely upon (ecological and monetary), for instance, water purification and water quality, sustainable food systems, flood alleviation and aquifer recharge. The effective management of freshwater functions and services requires multidisciplinary approaches (Tharme et al. 2019), enabling natural scientists to work with social scientists, and policy-makers (top-down) with community groups and educators (bottom-up). Thirdly, the role of microbial communities in freshwaters, including bacteria, viruses, microbial fungi, cyanobacterial symbionts and microbial eukaryotes, is key to energy transfer and biogeochemical cycles in freshwater ecosystems. These cannot be seen with the naked eye, and unless they are a health risk or become a nuisance through proliferation, they are largely overlooked by policy-makers and conservationists. Yet without diverse and productive communities of microorganisms supplying energy at the base of trophic webs or driving nutrient cycling and biogeochemical reactions (e.g. nitrogen-fixing diazotrophs, methanotrophs, denitrifiers), wetlands and freshwaters cannot provide vital regulating functions (Bodelier and Dedysh 2013; Debroas et al. 2017; Sigee 2005).

5.1.2 Key Challenges in Freshwater Ecology

Freshwater ecosystem diversity provides vital provisioning and regulating services for human populations. The challenges faced by managers and conservationists in

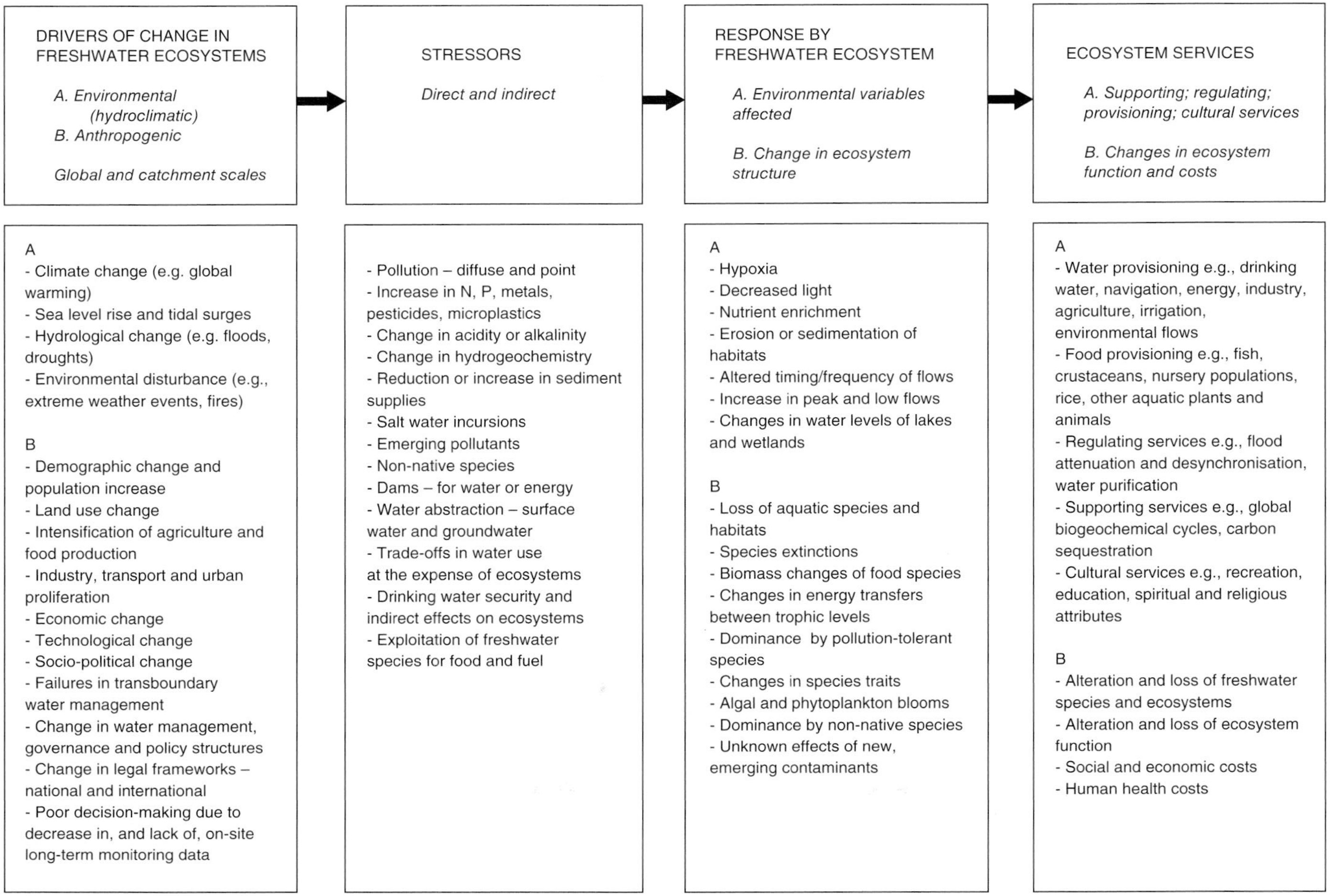

DRIVERS OF CHANGE IN FRESHWATER ECOSYSTEMS

A. Environmental (hydroclimatic)
B. Anthropogenic

Global and catchment scales

A
- Climate change (e.g. global warming)
- Sea level rise and tidal surges
- Hydrological change (e.g. floods, droughts)
- Environmental disturbance (e.g., extreme weather events, fires)

B
- Demographic change and population increase
- Land use change
- Intensification of agriculture and food production
- Industry, transport and urban proliferation
- Economic change
- Technological change
- Socio-political change
- Failures in transboundary water management
- Change in water management, governance and policy structures
- Change in legal frameworks – national and international
- Poor decision-making due to decrease in, and lack of, on-site long-term monitoring data

STRESSORS

Direct and indirect

- Pollution – diffuse and point
- Increase in N, P, metals, pesticides, microplastics
- Change in acidity or alkalinity
- Change in hydrogeochemistry
- Reduction or increase in sediment supplies
- Salt water incursions
- Emerging pollutants
- Non-native species
- Dams – for water or energy
- Water abstraction – surface water and groundwater
- Trade-offs in water use at the expense of ecosystems
- Drinking water security and indirect effects on ecosystems
- Exploitation of freshwater species for food and fuel

RESPONSE BY FRESHWATER ECOSYSTEM

A. Environmental variables affected

B. Change in ecosystem structure

A
- Hypoxia
- Decreased light
- Nutrient enrichment
- Erosion or sedimentation of habitats
- Altered timing/frequency of flows
- Increase in peak and low flows
- Changes in water levels of lakes and wetlands

B
- Loss of aquatic species and habitats
- Species extinctions
- Biomass changes of food species
- Changes in energy transfers between trophic levels
- Dominance by pollution-tolerant species
- Changes in species traits
- Algal and phytoplankton blooms
- Dominance by non-native species
- Unknown effects of new, emerging contaminants

ECOSYSTEM SERVICES

A. Supporting; regulating; provisioning; cultural services

B. Changes in ecosystem function and costs

A
- Water provisioning e.g., drinking water, navigation, energy, industry, agriculture, irrigation, environmental flows
- Food provisioning e.g., fish, crustaceans, nursery populations, rice, other aquatic plants and animals
- Regulating services e.g., flood attenuation and desynchronisation, water purification
- Supporting services e.g., global biogeochemical cycles, carbon sequestration
- Cultural services e.g., recreation, education, spiritual and religious attributes

B
- Alteration and loss of freshwater species and ecosystems
- Alteration and loss of ecosystem function
- Social and economic costs
- Human health costs

Figure 5.1 Challenges and threats (Drivers and Stressors) to maintaining the structure and function of freshwater ecosystems (Response and Services). *Sources:* partially modified from Grizzetti et al. (2016). Licenced under CC-BY-4.0; Jocelyne M.R. Hughes.

maintaining these services are numerous (Figure 5.1). At the global scale, they include the impacts of an increasing human (9.8 billion by 2050; 2017 estimate from UN DESA website) and urban population (68% by 2050; 2018 estimate from UN DESA website), and the consequent competing interests for water by humans, food and energy systems, and ecosystems; and an increase in extreme hydrologies and temperatures leading to water insecurity and reduced water quality because of changing climates. At the catchment or local scale there are challenges, for example, from industrial and agricultural wastewater causing pollution and nutrient-rich waters, especially increases in nitrates, phosphates and acidification; the proliferation of dams or water abstraction points altering river flows and groundwater levels and their impacts on environmental flows, water quality and livelihoods; the pervasive spread of non-native species into freshwaters via ballast water, the aquarium, horticultural and pet trades, and water sports, and their impact on provisioning services and ecosystem processes; and from contaminants of emerging concern such as microplastics, pharmaceuticals and personal care products and their impact on aquatic organisms, food webs and human health. In the following sections we will evaluate a number of examples of freshwater ecosystem challenges and discuss technical, economic or institutional solutions. We will also examine how it is possible to create or engineer freshwater habitats to deliver ecosystem services at the site scale.

5.2 The Challenge of Water Quality Management: Linking Freshwater Ecosystems to Water Quality

5.2.1 'The Kidneys of the Landscape'

Wetlands and freshwaters can naturally improve water quality by storing and recycling organic matter, nutrients and human waste, and have been described as 'the kidneys of the landscape' (Mitsch and Gosselink 2015; Moss 2018; Russi et al. 2013). The mechanisms for improving water quality are both physical and biochemical: water flows slowly through the wetland, allowing solids to settle. Denitrification, volatilization and plant uptake are the mechanisms that remove nitrogen (N) from the system, whereas ammonification and nitrification processes transform N in the sediment. Kadlec and Wallace (2009) describe the key pathways for phosphorus (P) removal in wetlands through plant uptake and soil sorption. Water volume, incoming load and concentration of pollutants, depth of the wetland, vegetation type and distribution, microorganism diversity, substrate properties, light, temperature and hydraulic efficiency are all factors that contribute to determining the water treatment capacity and efficiency of a wetland.

5.2.2 Constructed Wetlands

Based on the above properties of natural wetlands for water quality remediation, artificial or engineered wetlands, called constructed wetlands (CWs), have been designed to mimic these natural processes for a specific remediation purpose and load volume. These include vertical-flow, horizontal-flow and floating CWs, and each uses a different design of emergent or aquatic plants and substrate types to achieve a particular water treatment function. For more than half a century, CWs have successfully been used around the world to treat municipal and industrial wastewater. For example, 3500 vertical-flow CWs have been built across France in villages and isolated communities for raw wastewater treatment since the 1970s. In comparison with conventional wastewater treatment (see Chapter 16), CWs are cheaper to build, require less energy, less maintenance, and accommodate pulse flows, such as high seasonal volumes of runoff (Kadlec and Wallace 2009). More recently, CWs have been expanded into the treatment of agricultural wastewaters and runoff (Payne et al. 2009). This option for reducing nutrient loading from livestock farms is relatively low in construction and operation costs and is effective at reducing nutrient levels entering water bodies (Box 5.1; Gottschall et al. 2007). Additional benefits from CWs include the creation of wildlife habitats and aesthetically appealing green space, particularly when combined with urban green infrastructure projects.

Box 5.1 Constructed Wetlands for Reducing Nutrients Within the Eastern Ontario Beef Industry

The eutrophication of surface waters with associated decrease in water quality is a ubiquitous global freshwater challenge. Agricultural regions (particularly arable and livestock) contribute significantly to nutrient loadings of freshwaters by producing runoff rich in nitrates and phosphates. On-site solutions to this problem come at varying scales (e.g. re-siting of silage heaps, reduced applications of fertilizers, planting riparian buffers along rivers), and may be incentivized by policy (e.g. EU Water Framework Directive, Australia's National Water Quality Management Strategy) and finance (e.g. subsidies, offsetting). The impact of a site-scale technical solution to reduce nutrient loadings to a water course was evaluated for a farm in eastern Canada (Bond 2016). This multidisciplinary project aimed to assess the technical and social viability of using a CW system to treat manure runoff from livestock farms in Ontario. Focus was on the Eastern Ontario beef industry, where outdoor manure storage is common and CWs are a new practice, but the results can be applied to livestock farms globally.

The CW was located on Pemdale farm, a small beef farm (30 cattle) near Winchester, Eastern Ontario, where cattle manure is stored in an outdoor concrete confinement (Figure 5.2). Previous studies of CWs on cattle farms had determined reductions of 48–98% in N levels and 35–96% in P levels (Hunt and Poach 2001); whilst a review two decades ago of a database of 300 wastewater wetlands concluded that in general nitrogen was reduced by more than 50% while phosphorus reduction was less efficient due to saturation from prolonged loading (Knight et al. 1992). The technical performance of a free water surface CW system at Pemdale farm was tested by analysing water samples taken at various stages across the three ponds (Figure 5.2). Laboratory analysis of the water samples confirmed these trends, with reduction efficiencies of 50% total phosphorus, 51% total Kjeldahl nitrogen, and 82% ammonium. Interviews with regional beef farmers and other stakeholders in Eastern Ontario suggested that CWs could be successfully deployed for treating wastewater on small- to medium-sized farms with outdoor livestock confinement. Furthermore, it was found that perceptions of the water quality problem and awareness of CW remediation processes strongly influenced farmers' interest in the technology; but economic incentives were needed for further uptake, and a lack of technical expertise limited CW adoption by farmers.

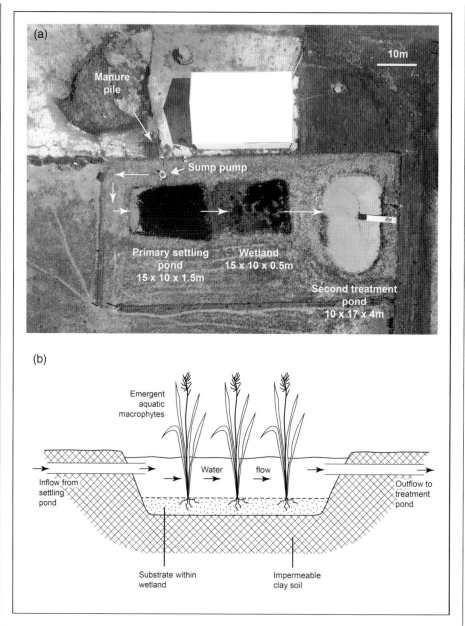

Figure 5.2 (a) Aerial view of the free water surface constructed wetland (FWSCW) system on Pemdale farm, Ontario, Canada, showing the manure pile, the settling pond, wetland, and treatment pond comprising the FWSCW; and direction of main runoff and flow pathways (white arrows). *Source:* courtesy of Larry Bond. (b) Schematic cross-section of the wetland within the FWSCW system showing the flow pathways (arrows), substrate and plants (not drawn to scale). *Source:* chapter authors. (*See color representation of this figure in color plate section*).

A final example focuses on the *Living-Filter*, a floating reedbed planted hydroponi-cally with three species of emergent macrophyte, together with baffles, that was installed at the water abstraction point in Farmoor Reservoir, Oxfordshire, UK. This novel CW used several mechanisms to reduce the phytoplankton loading to the water treatment works: by acting as a biofilter, where algal cells were physically trapped by the hydro-ponic roots of the plants growing down through the water column; through biomanipu-lation, in which the plant roots created a suitable habitat or refuge for the proliferation of zooplankton and other microorganisms that consumed the phytoplankton; by pro-viding a physical surface for the attachment of secondary grazers; and by releasing alle-lochemicals from the roots in response to environmental stressors (Castro-Castellon et al. 2016). In this CW a combination of processes, including biomanipulation, caused chlorophyll-*a* (a measure of phytoplankton biomass) removal efficiency of up to 45% during the first seven weeks of operation.

5.2.3   Managing Freshwater Ecosystems for Water Quality Enhancement

The successful management of freshwater ecosystems (rivers, lakes and wetlands) for water quality remediation has been demonstrated for countries, regions or catchments where there are effective policy, governance, participatory and financing structures in place (Rollason et al. 2018). By far the most cost-effective approach is to ensure that policies and incentives are in place to manage and conserve the natural functions of *existing* wetlands and freshwaters (rural or urban). For example, the proliferation of Catchment Partnerships (CPs) in the UK enable stakeholders to achieve a common aim of water quality enhancement and flood management through ecosystem stewardship and monitoring (see Evenlode CP website; Catchment-Based Approach [CaBA] web-site). Both CaBA (stakeholder-driven) and integrated water management approaches (policy-driven) have been successfully used to manage freshwater ecosystems and enhance water quality in a variety of physical settings, including the Tweed River Basin in Scotland where 'trusted intermediaries' have built a rapport between local communi-ties and government (Rouillard and Spray 2017). In urban areas, the construction of *new* wetlands can significantly enhance the quality of urban runoff and reduce flood peaks by storing storm water. For example, in Melbourne, Australia, it has required new governance and policy structures to apply integrated (urban) water management via water-sensitive design schemes to the management of all components of urban water quantity, quality and usage (City of Melbourne 2014). Part of the approach for manag-ing the chemical and nutrient loading of storm water entering Melbourne's waterways and Port Philip Bay has required Melbourne Water to construct over 600 wetlands across the city. These wetlands treat storm water and are rich environments for wildlife; and their functioning, efficiency and management is based on modelling changing urban land uses in the catchment in relation to the underlying geology. The integration of policy- and stakeholder-driven solutions for water quality management, with scien-tific monitoring and data gathering, is essential in order to evaluate outcomes and secure sustainable financing mechanisms (Wheater and Gober 2015).

In Philadelphia, USA, novel approaches for financing habitat creation for water qual-ity remediation and storm water storage come from the public and private sector. The Philadelphia Water Department and Philadelphia Industrial Development Corporation have created two Incentives Programmes for subsidizing the creation of wetlands and green infrastructure across different property types with the aim of reducing pollution to the city's sewers and waterways and enhancing water quality. Similarly, in Canberra, Australia, millions of dollars have been invested into creating wetlands, rain gardens and ponds to improve water quality in the catchment (low dissolved oxygen, high sediment loads, high nitrate, phosphorus and phytoplankton loads coupled with extreme hydrological fluctuations) over a 30-year period (ACT Healthy Waterways Project). The project is part of the Murray-Darling Basin Plan, and is a joint federal and state initiative with strong community involvement. By joining up decision-making and creating efficient management structures involving all stakeholders, effective networks of natural and engineered wetlands can be deployed at the landscape scale as innovative and cost-effective water quality enhancement solutions.

5.3 The Challenge of Invasive Non-native Species: Impacts on Diversity and Ecosystem Function

5.3.1 The Spread of Non-native Freshwater Species

Biological invasions and the spread of non-native species (or alien species) are significant drivers of global ecological change and deterioration (see UN Convention on Biological Diversity (CBD) website). At the highest level, globalization has led to the proliferation of INNS, resulting in biodiversity loss in all ecosystems as well as negative impacts on livelihoods and economic activities (Simberloff 2013). Freshwater ecosystems are particularly vulnerable to the spread of non-native species compared with terrestrial ecosystems due to a high degree of natural isolation and endemism (Richter et al. 1997). It must be noted, however, that not all non-native species are invasive in the sense that they are detrimental to ecosystems, or proliferate and spread; the impact of a novel species could be undetectable or benign, especially at the earliest stage of introduction. INNS are often introduced into freshwater systems unintentionally through the pet trade, and recreational activities such as sport fishing and boating. International shipping also contributes to the spread of non-native species when ships discharge ballast water that contains organisms transported from other ports. Connected to shipping and trade, movement along canals has created opportunities for novel species introductions. For instance, because the River Thames estuary is a major port and is linked to inland waterways, the Thames has become one of the world's most highly invaded freshwater ecosystems, with about 100 non-native aquatic species (Jackson and Grey 2013; Box 5.2).

5.3.2 Impacts of INNS

There are numerous examples from all continents on the effects of INNS on native populations and communities of freshwater species (Box 5.2; Coetzee et al. 2019; Simberloff 2013), with non-native freshwater species introductions represented from all animal and plant phyla, including microorganisms. For example, the introduction in the 1950s of Nile perch (*Lates niloticus*) as a commercial food source into Lake Victoria, the world's largest tropical lake, is a dramatic and well-documented example of ecosystem change with the consequent extinction of approximately 200 endemic cichlid fish

Box 5.2 What Drives Non-native Amphipod Distributions in the River Thames?

The distribution of three species of amphipod as well as habitat and human influences were mapped across 84 sites in the upper Thames catchment (Knight et al. 2017) to assess spatial patterns and environmental predictors: recently arrived non-native demon shrimp (*Dikerogammarus haemobaphes*; Figure 5.3), established non-native Northern River crangonyctid (*Crangonyx pseudogracilis*), and a native freshwater amphipod species (*Gammarus pulex*). The findings showed widespread distribution and density of *G. pulex* relative to *D. haemobaphes*, suggesting that the full impact of the current spread has yet to be felt since its 2012 introduction. Habitat partitioning occurred, with both *D. haemobaphes* and *C. pseudogracilis* occupying vegetative habitats, and native *G. pulex* occupying pebble/gravel habitats. From preliminary samples taken from the exterior hulls of narrowboats, the study showed that effective biosecurity might be best focused on boat traffic in the Thames and Cherwell rivers.

5 mm

Figure 5.3 *Dikerogammarus haemobaphes* is native to the Ponto-Caspian region and has been recorded in the Thames catchment since 2012. This specimen measures 8 mm. *Source:* courtesy of J. Keble-Williams and J. Hughes. (*See color representation of this figure in color plate section*).

species. This example also illustrates how trade-offs in freshwater ecosystem management can have huge impacts on species conservation (endemic fish in this case) for commercial gain (non-native Nile perch fishery). Globally, the aquarium and ornamental industry is growing at 14% annually, with faster trends in many developing countries, including China (Chen et al. 2017). A large-scale study in Guangzhou, Shenzhen and Hong Kong conducted in 2000–2003 showed that almost a million individual turtles comprising 157 species were bought and sold at market. The highly invasive red-eared slider turtle (*Trachemys scripta elegans*), a native of the United States, is farmed and traded in five provinces across China, with an estimated population of 25 million. They are considered a threat to native turtles due to their early sexual maturity, large size and physical aggression, and are banned in Australia, the EU and Japan. There are many examples of INNS aquatic plants in all biomes. In tropical regions, invasive water hyacinth (*Eichhornia crassipes*) has proliferated in many water bodies, typically blocking waterways to boat traffic, reducing water quality by blocking sunlight and causing low oxygen levels, dominating native aquatic macrophyte communities through competition and tolerating nutrient-rich waters. This species, along with countless others, has spread via the ornamental plant trade and unintentional transport of plant fragments via sporting equipment. A number of aquatic plant INNS are now banned from importation to some countries and attempts have been made to raise awareness of INNS of concern (e.g. European Union's 2014 Regulation on Invasive Alien Species).

The ecological impacts of INNS on inland water systems are wide-ranging, and biodiversity loss poses serious concerns as one of the many expected consequences. In general, INNS decrease the abundance of native freshwater species through competition for resources, predation, parasitism or hybridization, and therefore compromise ecosystem function. Ciruna et al. (2004) summarizes the ecological impacts including changes to the hydrological regime, water chemistry, and physical habitat; habitat connectivity; ecosystem structure, processes and function; as well as impacts to the biological community ranging from the population to the genetic level. Hybridization may, for example, diminish the genetic variability of a native species. Non-native species such as the invasive quagga mussel (*Dreissena bugensis*) are profound 'ecosystem engineers', altering the environment they invade and physically blocking pipes and outflows through colonization and proliferation. In particular, invasive mussels affect nutrient mineralization rates, oxygen levels and sedimentation rates, and they become dominant organisms in some freshwater habitats.

The severity of ecological impacts depends on several factors. Of chief importance are the characteristics or traits of the non-native species, the extent of the invasion, the amount of propagule pressure, and the vulnerability of the particular ecosystem (Ciruna et al. 2004). For instance, impacts increase if the non-native species is particularly well-suited to its new environment, or if there are synergistic interactions with other processes that threaten native species. In the UK, Ponto-Caspian species (from the Caspian, Azov and Black Sea regions of southeastern Europe), are more likely to colonize compared with other non-natives due to the climatic and habitat similarities between the native and invaded regions, and therefore represent a severe threat to inland water systems. In addition to direct ecological impacts, INNS harm economies, although the two are closely related. Damage caused by INNS is estimated at almost 5% of the world economy (see review of economic impacts on the Nature Conservancy website). In the UK, the impact of invasive species amounts to £1.7 billion annually, based on an evaluation of direct impacts to a range of industries (Roy et al. 2014). In freshwaters, the direct economic impacts relate to infrastructure fouling, commercial species loss, water quality remediation, and so on. These costs are largely reflected in detection, eradication and quarantine, and are reported to increase with spread and establishment of INNS.

5.3.3 What Can be Done About the Problem?

The CBD is a legally binding international policy instrument that requires ratifying nations 'to work to prevent the introduction, spread, and export of all types of invasive species'. The guiding framework for global INNS management is the precautionary principle which urges protective management decisions in the face of scientific uncertainty, especially when damage or degradation may be irreversible. In this light, the

CBD suggests a three-tiered approach for international INNS management. The first and most cost-effective measure is prevention; secondly, rapid containment and eradication; and lastly, if INNS become established, control and containment (see examples in Wong et al. 2017).

Improving coordination in international policy is an essential goal for effective INNS management. For example, within the EU, each country approaches invasive species management differently; the Netherlands has a national policy to address invasive species, but other nations do not have coherent policies, compromising containment efforts for transboundary systems such as waterways. Even with international agreements and domestic regulations, invasions are often unintentional (Vander Zanden and Olden 2008). The customs and quarantine practices that safeguard a country against human disease and agricultural pests are often inadequate protection for freshwater ecosystem integrity. Numerous entry pathways illustrate this vulnerability. At the macro-scale, shipping vessels transport organisms from different ecosystems, providing a constant source of propagule pressure into ports (see 'Ballast Water Convention' on the International Maritime Organisation website). The protection of freshwaters from INNS will depend on identifying those risks and vectors that facilitate the spread of non-native species; raising awareness of their potential impacts amongst stakeholders, such as the Clean-Check-Dry initiative by the UK Environment Agency; using available technologies such as environmental DNA (eDNA) to track and trace INNS; and ensuring that data and information about INNS is shared nationally, internationally, and across management organizations.

5.4 The Challenge of Environmental Change: Managing Biogeochemical Cycles and Water Security in Freshwaters

5.4.1 Impacts of Warming and Changing Atmospheric GHGs on Freshwaters

Increasing global temperatures and atmospheric greenhouse gas (GHG) concentrations (or mole fractions) due to human activities has led to increased research and awareness of temperature and elevated GHG effects on freshwater ecosystems. Many freshwater species have evolved physiologies linked to temperature and water quality to survive within specific habitat niches (Ficke et al. 2007). Alterations to water temperature and dissolved CO_2 mole fractions have wide-ranging effects, including altered organism growth, sexual development, reproduction and mortality.

The magnitude of regional temperature increases linked to climate change (and associated changes in water quantity) are correlated with latitude, with the largest increases in temperature predicted at higher latitudes (Ficke et al. 2007). Many freshwater species, such as fish and amphibians, are exotherms, which results in their body temperatures being identical to that of their environment. Biochemical processes within freshwater aquatic organisms are often a function of body temperature, and this parameter affects an organism's growth, sexual development and reproduction (Ficke et al. 2007). To compensate for changes in temperature, many species thermoregulate behaviourally, and alterations in temperatures within freshwater ecosystems may result in a latitudinal or altitudinal shift in species' populations as a means of adaptation.

Temperature changes may also cause alterations in local community population size and dynamics, causing community spectrum shifts towards smaller organisms and negatively impacting freshwater food webs. Hogg et al. (1995) showed that an increase in temperature within a first-order stream in Ontario, Canada, by 2.0–3.5°C over a two-year period had an effect on population genetics of two invertebrate species: *Hyalella azteca* (stonefly) and *Nemoura trispinosa* (amphipod). The study found that increased temperature within the stream resulted in lower population densities but increased growth rates for both invertebrate species.

Temperature also has significant effects on dissolved oxygen and GHG mole fractions within aquatic ecosystems. As temperature increases within the aquatic column, solubility of dissolved oxygen and other GHGs decreases (Figure 5.4). Reduced oxygen mole fractions can cause hypoxic stress or mortality in many aquatic species that depend on dissolved oxygen to breathe. Concurrently, increased temperatures within aquatic environments can lead to an increase in CO_2, CH_4 and N_2O production through biogeochemical processes such as autotrophic respiration, fermentation, methanogenesis and denitrification. Depending on aquatic temperature (driving gas solubility in water) and the partial pressure difference at the air–water interface, vertical degassing from the supersaturated water body to the atmosphere (termed evasion) occurs. CO_2, CH_4 and N_2O evasion has been measured in a number of different freshwater environments, including streams and wetland water bodies, with many of these studies showing increased rates of GHG evasion with temperature.

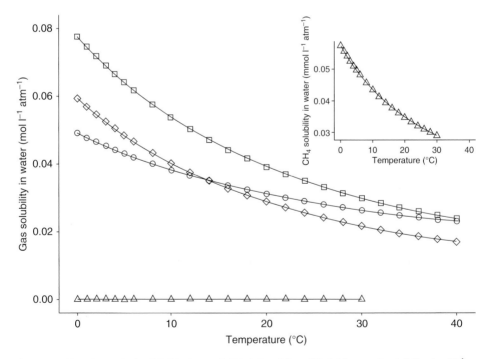

Figure 5.4 Variation in O_2 (circle), CO_2 (square), CH_4 (triangle), and N_2O (diamond) solubility (mol l^{-1} atm^{-1}) in freshwater with temperature at 1 atm. Inset panel shows CH_4 solubility as mmol l^{-1} atm^{-1} (1000 times less than main panel) due to CH_4 being sparsely soluble in water. *Source:* compiled from data in Weiss *(1970, 1974)*, Yamamoto et al. *(1976)*, and Weiss and Price *(1980)*.

Increasing atmospheric CO_2 mole fractions and higher dissolved organic carbon inputs to the aquatic environment, driven by human activities, have resulted in the weak acidification of freshwater ecosystems by increasing the partial pressure of CO_2 (pCO_2). Weak acidification is caused by the addition of weak acids, such as carbonic acid or acetic acid, and occurs when the acid only partially dissociates within the aquatic environment leading to changes of environmental pH of ~1 or 2 (Hasler et al. 2018). Freshwater ecosystems have a wide range of natural pH levels, ranging from <2 to 12; however, weak acidification is a stressor on freshwater taxa and has been linked to alterations in aquatic organism development and biomass. Primary producers, such as macrophytes and phytoplankton, have been shown to have an increase in growth rate and biomass when exposed to weak acidification. However, consumers, such as fish, amphibians and macroinvertebrates, show a reduction in growth rates in response to weak acidification, likely to result from the utilization of energetically expensive ion and acid–base regulatory mechanisms to counter the acidification of fluids and tissues (Hasler et al. 2018). These regulatory mechanisms have energetic costs associated with maintaining homeostasis, and freshwater animals only have finite energy budgets, often resulting in reduction of growth rate, calcification, and body mass, as a consequence of chronic physiological stress (Hasler et al. 2018).

Sub-lethal injuries and mortality have also been observed in freshwater organisms due to weak acidification. Fish exposed to weak acidification linked to high CO_2 exposure have been shown to have vertebrae abnormalities, gill lesions, nephrocalcinosis (deposition of calcium salts in the stomach or kidneys) and liver inflammation, which can lead to premature death. Freshwater snails, crayfish, clams and amphipods have been found to be acid-sensitive (e.g. diminished populations of freshwater snails in acidified Norwegian lakes), demonstrating that small decreases in pH have been shown to cause direct mortality in freshwater taxa (Hasler et al. 2018). Finally, alterations in behavioural responses have been recorded in fish and invertebrates exposed to weak acidification, such as negative impacts to olfaction, alarm cue responses and loss of predator avoidance behaviours (Hasler et al. 2018; Weiss et al. 2018). Elevated pCO_2 can cause hypercarpnia (CO_2 retention in the body) and has a narcotic effect on fish and invertebrates. Weiss et al. (2018) studied the effects of elevated pCO_2 on *Daphnia* (a planktonic crustacean) and noted a reduction in the ability of *Daphnia* to sense predators, as well as form adequate inducible defences to predators, driven by increasing pCO_2 but not by alterations in pH.

5.4.2 Environmental Flows

Changes in global water storages and fluxes due to climate change have created water security challenges for many freshwater habitats. Ecosystem structure and function is compromised when there is insufficient water to maintain basic ecosystem processes. This situation is further compounded when dams or impoundments are constructed on rivers without due consideration for downstream environmental flows. The Brisbane Declaration on Environmental Flows and Wetland Water Regimes (2007 and 2017) and Action Agenda (2018) set out the action needed to protect river flows globally. Specifically it states that 'environmental flows describe the quantity, timing, and quality of freshwater flows and levels necessary to sustain aquatic ecosystems which, in turn, support human cultures, economies, sustainable livelihoods, and well-being'. Water security is fundamental to maintaining freshwater ecosystem structure and function (Box 5.3), and

---

Box 5.3 Managing Water Security in a Hillside, Calcareous Fen: Marley Fen, Oxfordshire, UK

Marley Fen is a hillside calcareous fen (peatland) located in Wytham Woods, Oxfordshire (Stanley 2011). The >1 ha site is surrounded by mixed woodland and grassland, and pollen analysis on a 4.5 m core suggests a peat basal age of 11 000 years BP. The site was formed and is maintained by groundwater from upwellings in the underlying Corallian limestone bedrock. The predominant water source into the fen originates from an upwelling in the catchment above the site, feeding the site with nutrient- and base-rich water that maintains diverse communities of calcareous fen flora and fauna typical of this temperate zone (Stanley 2011), including a number of nationally rare and Red Data Book *Diptera* species. The fen had been drying over the past 50 years due to changing climate patterns and a lack of catchment water management; the plant communities have been compromised due to a lack of *in situ* disturbance (traditionally, grazing), with a consequent and significant loss of plant species. A portion of the major water source flowed through a conifer plantation which decreased groundwater flows to the fen, increased the groundwater NO_3^- mole fractions, and therefore affected the structure and productivity of the plant communities. The principle of maintaining water security (in this case, groundwater flow) for ecosystem function is the same in Marley Fen as for other freshwaters, the difference being that the wetland does not support livelihoods but is managed for its biodiversity value.

In 2010 a management plan to restore the fen was adopted with the aim of cutting back the plantation and woodland tree species in order to increase groundwater flow to the peat, with a subsequent reduction in peat accretion and loss of peat by oxidation, and a decrease in shade on the fen plant communities. Increasing water levels and the potential for re-wetting creates resilience in wetlands that are in water-scarce catchments. In 2012, the number of trees in the conifer plantation higher up in the fen catchment was reduced by 60% to decrease NO_3^- inputs into the fen. Management of trees near to the fen and in the catchment has led to increased water security in the fen: increased water levels and re-wetting to promote peat accretion; reduced nitrogen stress on fen flora and fauna communities; and an increase in plant species richness and abundance (five new species over five years). It is predicted, via yearly ongoing monitoring of the vegetation, that seasonal *in situ* cutting and removal of dominant *Phragmites australis* that began in 2017 will further enhance the biodiversity and restore the fen structure and function.

---

to be achieved requires creative governance, management and financial structures. A number of countries have environmental flows incorporated into policy (e.g. EU Water Framework Directive), legal frameworks (e.g. USA Endangered Species Act requires habitat protection for endangered species; South African National Water Act), and water trading (e.g. Australia and western USA) (Katz 2012).

There are, however, numerous examples globally where environmental flows have not been managed in favour of freshwater ecosystem services (e.g. Mekong, Nile). Ibáñez et al. (2010), in a study of sediment dynamics in the Ramsar-listed Ebro River delta (marshes and wetlands) in Spain, used surface elevation tables, marker horizons and lead-210 dating to demonstrate surface marsh subsidence and accretion dynamics as a result of reduced freshwater sediment loadings from the Ebro River. The reduced water flows and sediment loadings from the high Ebro catchment are due to a long-standing programme of dam construction (108 dams). Together with the competing needs of downstream stakeholders including irrigators, variable precipitation patterns and water scarcity, and sea-level rise, the Ebro delta is shrinking with consequent loss of wetland function. There are, however, a number of positive examples of rivers where environmental flows have been maintained or reinstated. These are reviewed, together with the policy mechanisms involved, by Harwood et al. (2017) for the Ganga River, Savannah River, Crocodile River and others.

5.5 Approaches to Tackling the Challenges of Freshwater Ecosystem Conservation and Management

5.5.1 Technical Innovations

5.5.1.1 Environmental DNA

Freshwater ecological management requires knowledge of species distributions within the ecosystem. Environmental DNA techniques (eDNA), or metabarcoding, represent a significant innovation in freshwater monitoring and management (Taberlet et al. 2018), which has traditionally relied on visual detection and counting techniques – a practice that is often time-consuming and requires taxonomic expertise. eDNA refers to nuclear or mitochondrial DNA that is released from an aquatic organism (e.g. faeces, skin, hair, gametes, pollen, short fragments of genetic material) and can be extracted from environmental samples such as sediment or water. The detection of eDNA allows for the monitoring of plant and animal species in aquatic settings and indicates a recent presence of that species in the sample (DNA in older samples are destroyed by ultraviolet light and microbial activity).

The benefits and potential of eDNA monitoring for freshwater management are widespread. The continual improvement of technology for identifying and analysing eDNA in aquatic habitats means that target species can be detected within a few hours of taking a water sample (Taberlet et al. 2018). This is especially useful for monitoring species that are difficult to detect using conventional methods, or sensitive species that are endangered or threatened. Additionally, eDNA is proving effective for the discovery and monitoring of INNS, including the early stages of an invasion in a new habitat or at low population densities. eDNA techniques can be applied to a diversity of taxa and habitats, including fish, amphibians, plants and benthic arthropods; and in any freshwater ecosystem such as a controlled mesocosm, CW, pond, river and large lake. For example, in the Laurentian Great Lakes of the USA and Canada, invasive Asian carp have been monitored upstream of the lakes using eDNA. A limitation of genetic monitoring techniques is that a significant fraction of DNA sequencing remains unassigned or belongs to unknown taxa, although recent research has shown that using supervised machine learning (SML) can build robust predictive models of eDNA sequences (Cordier et al. 2017).

5.5.1.2 Remote Sensing Methods and Databases

With the numerous challenges facing freshwater ecosystems, there is a need for innovative techniques and solutions for the conservation and management of aquatic biodiversity and wetland functions. Part of the solution requires rigorous monitoring, identification and documentation of the biotic and abiotic processes that affect ecosystem dynamics, as well as continued monitoring of global patterns of biodiversity change, including freshwater biodiversity. Hughes et al. (2019) review the methods and applications of remote sensing techniques in freshwater ecology, including the deployment of unmanned aerial vehicles (or drones) for monitoring the extent of wetland habitat and freshwater bodies. Remote sensing methods are cost-effective and allow large and inaccessible areas to be surveyed. The publication of, and open access to, field and remotely sensed data is crucial and empowering for decision-makers. Large open-access datasets comprising ecological monitoring information provide the opportunity to use data for advanced comparative and synthetic analyses, policy-making, as well as ecological assessments and management (Jiang et al. 2016).

Global databases are allowing greater ease in sharing information and tracking changes in biological diversity. However, there are challenges associated with the development of data management systems for ecological monitoring, as well as in the integration of data systems, due to the complexity and diversity of ecological data collection (Reichman et al. 2011). To be useful at a global level, this information needs to be accessible, digital and integrated into one standardized format. A number of database management programmes meet these requirements and have proven to be useful in sharing and analysing global biological diversity information.

For example, OSAEM, the Open, Sharable and Extensible Data Management System for Aquatic Ecological Monitoring, was designed to share datasets from the Korean National Aquatic Ecological Monitoring Program (NAEMP), and provides an open, sharable and intuitive database sharing option; and the Atlas of Living Australia is a national-level open database. These examples represent one general approach to ecological monitoring database systems which use data-level and defined data models for monitoring. These data models allow for the development of software programs to process and analyse datasets automatically (Jiang et al. 2016).

The second approach to biodiversity database systems uses a metadata-level data-sharing approach that creates and shares metadata meeting a certain standard. Metadata provide the context for data collection and describe all aspects of a specific dataset to allow a better understanding of its content and background, which is critical in making appropriate re-use of biological data. DataOne and Ecological Metadata Language (EML) are examples of these systems, as well as the global database management system Darwin Core. This is a widely distributed biodiversity data-sharing platform, with approximately 300 million Darwin Core-formatted records published by more than 340 organizations in 43 countries using the Global Biodiversity Information Facility. It is primarily based on taxa and their occurrence in the environment, with information related to modern biological specimens, their spatiotemporal occurrence, and supporting evidence. In order to standardize this data, Darwin Core uses a glossary of terms and standards, and the mapping of accumulated knowledge is referred to as the Darwin Cloud. Another key advantage of systems such as Darwin Core is the global scope of its data – a necessity in monitoring and assessing macroecological studies.

Organizations such as Biodiversity Observation Networks (BONs) contribute to efforts in mobilizing biodiversity information, and the Global Wetland Observation System (GWOS) is part of BONs. The concept of essential biodiversity variables (EBVs) was introduced by the Group on Earth Observation. EBVs are the key measurements required to report on changes in biodiversity, and include genetic composition, species populations, species traits, community composition, ecosystem structure and ecosystem function. A thorough review of the main freshwater-related information systems and secondary data sources has been undertaken by De Wever et al. (2019) and Schmidt-Kloiber and De Wever (2018). The next step for making use of data in large database management systems is in developing decision support tools linked to these systems, to facilitate management recommendations for large complex freshwater ecosystems.

5.5.2   Social Science Innovations

It has been demonstrated via a number of flagship examples (see Water Stewardship on the WWF website) that the successful management and conservation of freshwater ecosystems relies on a multidisciplinary approach involving local communities and stakeholders as well as scientists, social scientists and policy-makers (Tharme et al. 2019). Successful outcomes rely on a mixture of top-down approaches including national and international legislation and policy for water resource and ecosystem management, and international conventions for species protection and the movement of species; effective governance structures that match the needs of water and ecosystem management usually at the catchment scale; and bottom-up approaches involving local communities and stakeholders.

Freshwater ecology management benefits from innovative engagement with the public and involving local communities and stakeholders in management and decision-making activities. Included in this engagement are citizen scientists who generate new data and knowledge; they are not only an added resource for project managers, but their involvement in wetland creation and restoration projects through public engagement and education contributes to their success. Community engagement is an iterative process that works towards broad societal and ecological goals. FreshWater Watch (FWW) is an example of a global citizen science programme from Earthwatch that has more than 22 000 global water quality samples (as of mid-2019). Individuals and communities can participate in data gathering and water monitoring, contribute to the extensive database, and gain an understanding of freshwater ecosystem processes and functions. FWW has a number of different projects within its scope, all of which are consistent in their methodologies, data upload and educational approaches, ensuring quality control, comparable data and reduced costs (Thornhill et al. 2016). Another example of a successful wetland conservation and restoration project involving numerous partnerships and stakeholders is the Big Fen Project in the UK. This is the largest wetland restoration project in Europe (a landscape of 3700 ha) and has involved working with farmers, landowners, agencies, local councils, volunteers and local communities to create a wetland landscape for wildlife, water quality enhancement and flood alleviation.

A final example illustrates the impact of media and social media in raising awareness of pollution issues. Labelled as the 'Attenborough Effect', the Blue Planet TV series shown in 2017 revolutionized public opinion about one-use plastics. After the final episode, the response on Twitter was huge – a public outcry at the over-use and disposal of plastics. The effect of the Blue Planet series on public opinion and behaviour has been pivotal in changing habits and forcing governments and industry to gradually dispense with one-use plastics. There are lessons to be learnt from this phenomenon for the conservation and management of freshwater ecosystems: media and social media could be effectively used to raise awareness of species diversity, and the functions, services and livelihoods provided by freshwater ecosystems upon which humans depend.

References

Balian, E.V., Segers, H., Lévêque, C., and Martens, K. (2008). The freshwater animal diversity assessment: an overview of the results. *Hydrobiologia* 595: 627–637.

Belgrano, A., Woodward, G., and Jacob, U. (eds.) (2015). *Aquatic Functional Biodiversity: An Ecological and Evolutionary Perspective*. Cambridge, MA: Academic Press.

Bodelier, P.L.E. and Dedysh, S.N. (2013). Microbiology of wetlands. *Frontiers in Microbiology* 4: 79. https://doi.org/10.3389/fmicb.2013.00079.

Bond, H. (2016). *An Assessment of Constructed Wetlands for Reducing Nutrients in Manure Runoff within the Eastern Ontario Beef Industry*. MSc thesis. University of Oxford.

Castro-Castellon, A.T., Chipps, M.J., Hankins, N.P., and Hughes, J.M.R. (2016). Lessons from the "living-filter": an in-reservoir floating treatment wetland for phytoplankton reduction prior to a water treatment works intake. *Ecological Engineering* 95: 839–851.

Chen, Y., Sun, C., and Zhan, A. (2017). Biological invasions in aquatic ecosystems in China. *Aquatic Ecosystem Health and Management* 20: 402–412.

Ciruna, K.A., Meyerson, L.A., and Gutierrez, A. (2004). *The ecological and socio-economic impacts of invasive alien species in inland water ecosystems*. Report to the Conservation on Biological Diversity on behalf of the Global Invasive Species Programme, Washington, DC.

City of Melbourne (2014). *Total Watermark: City as a Catchment Strategy*. City of Melbourne, Australia: https://www.melbourne.vic.gov.au/about-council/vision-goals/eco-city/Pages/total-watermark-city-catchment-strategy.aspx.

Coetzee, J., Hill, M.P., Hussner, A. et al. (2019). Invasive aquatic species. In: *Freshwater Ecology and Conservation: Approaches and Techniques*, Techniques in Ecology and Conservation Series (ed. J.M.R. Hughes), 338–357. Oxford: Oxford University Press.

Collen, B., Whitton, F., Dyer, E.E. et al. (2014). Global patterns of freshwater species diversity, threat and endemism. *Global Ecology and Biogeography* 23: 40–51.

Cordier, T., Esling, P., Lejzerowicz, F. et al. (2017). Predicting the ecological quality status of marine environments from eDNA metabarcoding data using supervised machine learning. *Environmental Science & Technology* 51: 9118–9126. https://doi.org/10.1021/acs.est.7b01518.

De Wever, A., Schmidt-Kloiber, A., Bremerich, V., and Freyhof, J. (2019). Secondary data: taking advantage of existing data and improving data availability for supporting freshwater ecology research and biodiversity conservation. In: *Freshwater Ecology and Conservation: Approaches and Techniques* (ed. J.M.R. Hughes). Oxford: Oxford University Press.

Debroas, D., Domaizon, I., Humbert, J.-F. et al. (2017). Overview of freshwater microbial eukaryotes diversity: a first analysis of publicly available metabarcoding data. *FEMS Microbiology Ecology* 93: –fix023. https://doi.org/10.1093/femsec/fix023.

Ficke, A.D., Myrick, C.A., and Hansen, L.J. (2007). Potential impacts of global climate change on freshwater fisheries. *Reviews in Fish Biology and Fisheries* 17: 581–613.

Gardner, R.C., Barchiesi, S., Beltrame, C. et al. (2015). *State of the World's Wetlands and Their Services to People: A Compilation of Recent Analyses*. Ramsar Briefing Note No. 7 (31 March 2015). Gland: Ramsar Convention Secretariat. doi:10.2139/ssrn.2589447.

Gottschall, N., Boutin, C., Crolla, A. et al. (2007). The role of plants in the removal of nutrients at a constructed wetland treating agricultural (dairy) wastewater, Ontario, Canada. *Ecological Engineering* 29: 154–163.

Grizzetti, B., Lanzanova, D., Liquete, C. et al. (2016). Assessing water ecosystem services for water resource management. *Environmental Science and Policy* 61: 194–203. https://doi.org/10.1016/j.envsci.2016.04.008.

Harwood, A., Johnson, S., Richter, B. et al. (2017). *Listen to the River: Lessons from a Global Review of Environmental Flow Success Stories.* Woking: WWF- UK.

Hasler, C.T., Jeffrey, J.D., Schneider, E.V.C. et al. (2018). Biological consequences of weak acidification caused by elevated carbon dioxide in freshwater ecosystems. *Hydrobiologia* 806: 1–12.

Hogg, I.D., Williams, D.D., Eadie, J.M., and Butt, S.A. (1995). The consequences of global warming for stream invertebrates: a field simulation. *Journal of Thermal Biology* 20: 199–206.

Hughes, J.M.R., Clarkson, B.R., Castro-Castellon, A.T., and Hess, L.L. (2019). Wetland plants and aquatic macrophytes. In: *Freshwater Ecology and Conservation: Approaches and Techniques*, Techniques in Ecology and Conservation Series (ed. J.M.R. Hughes), 173–206. Oxford: Oxford University Press.

Hunt, P.G. and Poach, M.E. (2001). State of the art animal wastewater treatment in constructed wetlands. *Water Science and Technology* 44: 19–25.

Ibáñez, C., Sharpe, P.J., Day, J.W. et al. (2010). Vertical Accretion and Relative Sea Level Rise in the Ebro Delta Wetlands (Catalonia, Spain). *Wetlands* 30: 979. https://doi.org/10.1007/s13157-010-0092-0.

Jackson, M.C. and Grey, J. (2013). Accelerating rates of freshwater invasions in the catchment of the River Thames. *Biological Invasions* 15: 945–951.

Jiang, M., Jeong, K., Park, J.H. et al. (2016). Open, sharable, and extensible data management for the Korea national aquatic ecological monitoring and assessment program: A RESTful API-Based approach. *Water* 8: 201.

Kadlec, R.H. and Wallace, S.D. (2009). *Treatment Wetlands.* Boca Raton, FL: CRC Press.

Katz, D. (2012). Cash flows: water markets and environmental flows in theory and practice. In: *Water Trading and Global Water Scarcity: International Perspectives* (ed. J. Maestu), 214–232. Washington, DC: Resources for the Future (RFF Press)/Routledge–Taylor and Francis.

Knight, R.L., Rubles, R.W., Kadlec, R.H., and Reed, S.C. (1992). Wetlands for wastewater treatment performance data base. In: *Constructed Wetlands for Wastewater Treatment* (ed. G. Moshiri), 35–58. Boca Raton, FL: Lewis Publishers.

Knight, C.A., Hughes, J.M.R., and Johns, T. (2017). What drives non-native amphipod distributions in the River Thames? The role of habitat and human activity on species abundance. *Crustaceana* 90: 399–416.

Mitsch, W.J. and Gosselink, J.G. (2015). *Wetlands*, 5e. Chichester: Wiley.

Moss, B. (2018). *Ecology of Freshwaters: Earth's Bloodstream*, 5e. Hoboken N.J.: Wiley.

Payne, V.W.E. et al. (2009). Constructed wetlands. In: *National Engineering Handbook*, Part 637. 210-VI-NEH. Washington, DC: USDA, Natural Resources Conservation Service.

Reichman, O.J., Jones, M.B., and Schildhauer, M.P. (2011). Challenges and opportunities of open data in ecology. *Science* 331: 703–705.

Richter, B.D., Braun, D.P., Mendelson, M.A., and Master, L.L. (1997). Threats to imperiled freshwater fauna. *Conservation Biology* 11: 1081–1093.

Rocha, A.V. and Goulden, M.L. (2009). Why is marsh productivity so high? New insights from eddy covariance and biomass measurements in a Typha marsh. *Agricultural and Forest Meteorology* 149: 159–168.

Rollason, E., Bracken, L.J., Hardy, R.J., and Large, A.R.G. (2018). Evaluating the success of public participation in integrated catchment management. *Journal of Environmental Management* 228: 267–278.

Rouillard, J.J. and Spray, C.J. (2017). Working across scales in integrated catchment management: lessons learned for adaptive water governance from regional experiences. *Regional Environmental Change* 17: 1869–1880.

Roy, H.E., Peyton, J., Aldridge, D.C. et al. (2014). Horizon scanning for invasive alien species with the potential to threaten biodiversity in Great Britain. *Global Change Biology* 20: 3859–3871.

Russi, D., ten Brink, P., Farmer, A. et al. (2013). *The Economics of Ecosystems and Biodiversity (TEEB) for Water and Wetlands*. London and Brussels: IEEP; Gland: Ramsar Secretariat.

Schmidt-Kloiber, A. and De Wever, A. (2018). Biodiversity and freshwater information systems. In: *Riverine Ecosystem Management* (eds. S. Schmutz and J. Sendzimir), 391–412. Cham: Springer.

Sigee, D.C. (2005). *Freshwater Microbiology: Biodiversity and Dynamic Interactions of Microorganisms in the Aquatic Environment*. Chichester: Wiley.

Simberloff, D. (2013). *Invasive Species: What Everyone Needs to Know*. Oxford: Oxford University Press.

Stanley, K.M. (2011). *The Physicochemical regime of Marley Fen, Oxfordshire, UK: Impacts on plant growth and distribution*. MSc thesis. University of Oxford.

Taberlet, P., Bonin, A., Zinger, L., and Coissac, E. (2018). *Environmental DNA – For Biodiversity Research and Monitoring*. Oxford: Oxford University Press.

Tharme, R., Tickner, D., Hughes, J.M.R. et al. (2019). Approaches to freshwater ecology and conservation. In: *Freshwater Ecology and Conservation: Approaches and Techniques*, Techniques in Ecology and Conservation Series (ed. J.M.R. Hughes), 20–47. Oxford: Oxford University Press.

Thornhill, I., Loiselle, S., Lind, K., and Ophof, D. (2016). The citizen science opportunity for researchers and agencies. *BioScience* 66: 720–721.

Vander Zanden, M.J. and Olden, J.D. (2008). A management framework for preventing the secondary spread of aquatic invasive species. *Canadian Journal of Fisheries and Aquatic Sciences* 65: 1512–1522.

Weiss, R.F. (1970). The solubility of nitrogen, oxygen and argon in water and seawater. *Deep Sea Research and Oceanographic Abstracts* 17: 721–735.

Weiss, R.F. (1974). Carbon dioxide in water and seawater: the solubility of a non-ideal gas. *Marine Chemistry* 2: 203–215.

Weiss, R.F. and Price, B.A. (1980). Nitrous oxide solubility in water and seawater. *Marine Chemistry* 8: 347–359.

Weiss, L.C., Pötter, L., Steiger, A. et al. (2018). Rising pCO_2 in freshwater ecosystems has the potential to negatively affect predator-induced defenses in *Daphnia*. *Current Biology* 28: 327–332.

Wheater, H.S. and Gober, P. (2015). Water security and the science agenda. *Water Resources Research* 51: 5406–5424.

Wong, W.H., Piria, M., Collas, F.P.L. et al. (2017). Management of invasive species in inland waters: technology development and international cooperation. *Management of Biological Invasions* 8: 267–272.

WWF (2018). *Living Planet Report – 2018: Aiming Higher* (eds. M. Grooten and R.E.A. Almond). Gland, Switzerland: WWF.

Yamamoto, S., Alcauskas, J.B., and Crozier, T.E. (1976). Solubility of methane in distilled water and seawater. *Journal of Chemical and Engineering Data* 21: 78–80.

6

Water and Health

A Dynamic, Enduring Challenge

Katrina J. Charles[1], Saskia Nowicki[1], Patrick Thomson[1,2], and David Bradley[1,3,4]

[1] *School of Geography and the Environment, University of Oxford, UK*
[2] *Smith School of Enterprise and the Environment, University of Oxford, UK*
[3] *London School of Hygiene and Tropical Medicine, UK*
[4] *Department of Zoology, University of Oxford, UK*

6.1 Introduction

Health outcomes have been a primary motivator for many developments in the water sector, particularly in water supply, sanitation and water resources development. The evolution of water research over past centuries in Europe and North America, where reliable and abundant quantities of water are widely accessible, made water quality and acute health outcomes a primary concern. The priorities of the sector have shifted since then as research expanded into different geographies and development challenges. In recent decades, the sector was influenced by the Millennium Development Goals (MDGs) to focus on delivering improved water facilities. Currently, it is being strongly directed by the Sustainable Development Goals (SDGs; see Chapter 1), and the dialogue that preceded the goals, to consider levels of service provision.

In this chapter, we first explore in Section 6.2 the established approaches for classification and measurement of water-related health outcomes, with an emphasis on infectious diseases. In Section 6.3, we provide some examples of shifts in the water sector where new innovations (in measurement and treatment methods) and knowledge (particularly evidence of emerging diseases) have substantially impacted public health. In Section 6.4, we explore how knowledge of water and health linkages is being enriched by research into chronic drinking water related diseases. And finally, in Section 6.5 we issue a challenge to future water researchers, from all disciplines, framing three major areas where innovation is needed to overcome seemingly intractable water-related health problems.

6.2 Classifying and Measuring Health Outcomes

Infectious diseases were the initial driver for health-related water research taking place in areas with relatively abundant water, in Europe and North America. This focus on infectious diseases was extended to other parts of the world, and in 1972, as part of the

Water Science, Policy, and Management: A Global Challenge, First Edition. Edited by Simon J. Dadson, Dustin E. Garrick, Edmund C. Penning-Rowsell, Jim W. Hall, Rob Hope, and Jocelyne Hughes.
© 2020 John Wiley & Sons Ltd. Published 2020 by John Wiley & Sons Ltd.

seminal Drawers of Water study in East Africa, a functional water-related disease classification was created (White et al. 1972; Table 6.1). This classification, which is sometimes referred to as the Bradley classification, for the first time considered diseases not by their causal microbes or symptoms of infection, but by their transmission pathways. This was done to enable consideration of the impact of interventions. It was also intended to highlight the importance of water quantity in the communities of East Africa, where the study was based. The classification was designed to differentiate the water-related diseases that are associated with water resource developments (such as dams and irrigation canals), including the insect vector-borne disease of malaria and water-based diseases such as schistosomiasis.

Our understanding of the transmission cycles of many water-related diseases has greatly improved in the decades since this classification was first developed. Transmission is a more complex and sophisticated process than we imagined – for example, the ability of *Legionella* to proliferate intracellularly in free-living freshwater amoebae was only recently understood. It is scarcely possible to capture this complexity and retain a simple and robust categorization. Nevertheless, the original four-category system is now inadequate and many advances in our understanding of water-related disease, several of which are discussed by Bartram and Hunter (2015), must be considered in modern classifications. For example, pathogens in water can directly enter the human body via the respiratory tract as well as via the gastrointestinal route. Indeed, *Legionella*, which now accounts for half the reported waterborne outbreaks in the USA, often enters via the respiratory tract in aerosols. Secondly, the role of handwashing with soap in reducing transmission of conventional respiratory pathogens, such as the influenza virus and other highly infectious diseases, creates a further subsection of water-washed diseases. As raised by Gerba and Nichols (2015), the categorization of water-based diseases

Table 6.1 The Bradley classification of water-related diseases.

Category	Definition	Example
I Waterborne a) Classical b) Nonclassical	Where water acts as a passive vehicle for an infecting agent (e.g. faecal-oral diseases).	Typhoid Infectious hepatitis
II Water-washed a) Superficial b) Intestinal	Where infections decrease as a result of increasing the volume of available water, irrespective of water quality.	Trachoma, scabies Shigella dysentery
III Water-based a) Water-multiplied percutaneous b) Ingested	Where a necessary part of the life-cycle of the infecting agent takes place in an aquatic animal such as a snail.	Schistosomiasis Guinea worm
IV) Water-related insect vectors a) Water-biting b) Water-breeding	Where infections are spread by insects that breed in water or bite near it.	Gambian sleeping sickness Onchocerciasis

Source: Adapted from White et al. (1972).

merits further attention as well. With the guinea-worm approaching extinction owing to its successful control, a better subdivision of the water-based diseases may be between diseases with an aquatic intermediate host(s) and pathogens that also live non-parasitically in freshwater.

To further complicate classification attempts, gains in knowledge have somewhat blurred the boundary between classical waterborne disease agents and freshwater microbes that can become pathogenic, especially in immunologically compromised hosts. The increase in people experiencing deliberate immunosuppression (e.g. for transplanted organ survival) or suffering immunological damage (e.g. from diseases such as HIV) has widened the range of opportunistic pathogens (Bartram 2015). Additionally, some free-living microorganisms that proliferate greatly in water under certain conditions can produce toxins of public health importance (e.g. some cyanobacteria that cause 'algal blooms' produce neurotoxins). These toxins sit more comfortably within a classification system that includes chemically toxic metals (e.g. arsenic). It remains important, however, to distinguish between infective replicating agents and the vast array of toxic chemicals. The health impacts of toxic chemicals are better understood now (as discussed in Section 6.4) than they were when the Bradley classification was originally developed. At that time, toxic chemicals in water were not considered a critical issue compared with acute infectious disease.

Diarrhoea is an indicator of acute infectious waterborne diseases, and self-reported diarrhoea remains a key method used to measure health risk associated with water supply and sanitation, despite many limitations. Diarrhoea is caused by a range of pathogens, with the majority of childhood deaths from diarrhoea associated with rotavirus, shigella, adenovirus, cholera and *Cryptosporidium*, according to the Global Burden of Disease study (GBD 2018). Diarrhoea remains a major cause of disease internationally: it is the eighth leading contributor to mortality, causing an estimated 1.6 million deaths in 2016. Unsafe water and sanitation are amongst the leading risk factors for diarrhoea, with 72% and 56% of diarrhoea deaths in children younger than five years attributed to unsafe water and poor sanitation, respectively.

6.3 Politics and Innovation in Water and Health

The water and health field changed rapidly over the previous century, constantly evolving to deal with new evidence, changing risks and environments, but also in response to changing societal concerns and political agendas (see Chapters 8 and 19). The drivers of implementation in any sector are strongly political, and water and health is no different, but the sector's work has also been heavily influenced by disruptive innovations. In water-related health research (and health-related water research), innovations in the science of detection and treatment and changes in knowledge have rapidly shifted academic discussions into action. They have changed the way that researchers and practitioners in the sector work. Whilst there have been many innovations in the broad field of water and health (Bartram 2015), in this section we draw from David Bradley's experience of working in the sector across six decades to explore some of his key reflections on the politics and innovations that have driven agendas and shaped the sector. Understanding these drivers of change is helpful for reflecting on how impact can be achieved.

6.3.1 Measurement: Understanding the Role of Malnutrition and Infection in Diarrhoea

In the 1920s, research into infectious diarrhoeal disease and nutrition began to overlap. The role of nutrition in modifying host resistance to infection was recognized and initially overestimated, drawing considerable research attention in the following decades (Scrimshaw et al. 1959). Research focused on the impacts of nutrition on infections, with comparatively few studies examining the reverse – how infections impact nutrition. By the 1960s, the complexity of infection–malnutrition interactions was better understood, but the relative importance of infection versus nutrition was still debated. It was recognized that in cases '[w]here both malnutrition and infection are serious … success in control of either condition commonly depends on the other' (Scrimshaw et al. 1959, p. 396). Yet, research and interventions continued to focus more on nutrition, in part because determination of causal agents (pathogens) remained out of reach for many diarrhoeal cases.

In this case, disruptive innovation took the form of increased measurement capability, which enabled the detection and quantification of faecal–oral pathogens in faeces. Technological innovations increased detection of aetiological agents: first of bacteria with better culture methods, then with electron microscopes viruses were identified on and in gut tissue. Innovations in tissue culturing techniques for detection in later decades helped to expand virus vaccine programmes (College of Physicians of Philadelphia 2019). Some of these changes were very rapid: *Cryptosporidium* was first detected in Bangladesh immediately following news of a paper that reported a new method for staining the parasite cysts (David Bradley, personal communication). These innovations were supported by collaborations between microbiologists and nutritionists, such as the ground-breaking work in Guatemala tracking the impact of infections on growth faltering in malnourished children (Mata et al. 1972). Identifying pathogenic causes of diarrhoea helped to balance the debate about the relative importance of infection versus nutrition and pushed the conversation towards treatment and environmental interventions.

6.3.2 Treatment: Oral Rehydration Therapy (ORT)

Cholera is one of the most researched diarrhoeal diseases. It can spread rapidly through a population, causing severe fluid loss through diarrhoea and vomiting, and leading to death if not treated. Before the nineteenth century, if you contracted cholera there was a 70% chance you would die. Cholera originated in South Asia and travelled to Europe and North America as trade increased. More recently, the 'seventh pandemic' has spread the disease to many areas across the globe. Large cholera outbreaks in Europe and North America have been identified as drivers of the sanitation revolution in the industrialized world (Guerrant et al. 2003). With the recognition that dehydration was the cause of death from cholera infections in the 1830s, treatments started to include intravenous fluid replacement (Guerrant et al. 2003), reducing the mortality rate to 40%. This required many litres of sterile intravenous fluid, however, with a significant production and storage chain to maintain during epidemics.

In the 1960s, new research identified the importance of glucose in enabling the uptake of sodium in cholera patients, sparking the development of ORT, with the first trials taking place in what is now Bangladesh in 1970. ORT use rapidly went from trial to implementation as the 1971 Bangladesh war of independence resulted in cholera treatment being needed for high numbers of refugees – higher than could be supplied by intravenous methods. Mortality was reduced with ORT down to 3.6% (Guerrant et al. 2003), with intravenous treatment largely replaced except in cases with vomiting. Over time the success of ORT as a low-cost intervention, capable of being delivered in settings with limited medical facilities, has enabled the transformation of cholera from a disease endemic in Bangladesh, to one that has been observed to be underreported in specific areas where there is good access to ORT as most cases can be treated at home.

Cholera remains a very serious disease, with regular outbreaks across the world. It is considered an 'indicator of inequity' because it typically impacts the poorest who are malnourished and lack adequate water and sanitation. Research on ORT and cholera has continued, broadening the solution from sugars to carbohydrates more generally. But whilst the timing of outbreaks remains difficult to forecast, and environmental or human reservoirs of the disease are debated, ORT has enabled rapid and cost-effective responses that have saved countless lives. In this case innovation occurred rapidly, driven by a case overload associated with displacement to poor living conditions during the war (Guerrant et al. 2003).

6.3.3   Knowledge: Emerging Health Issues

Health policy narratives are influenced by the available knowledge. We highlight this here as it is a major challenge to delivering progress in water-related health, but also to inspire future researchers about the dynamic nature of the challenge. Climate change has been an area that has strongly influenced health policy and research in recent years. To determine future needs for health systems and technologies, we now need to understand much more about the climatic factors that influence health outcomes through changing environmental conditions, human behaviour, and vector habitats. These changes include the emergence (and re-emergence) of infectious diseases, such as the Zika virus outbreaks in South America in 2015, and environmental transmission of mosquito-borne diseases re-emerging in Europe. Determinants of disease are understood in the context of the climate conditions and behaviours that spread them.

Here the system has been disrupted by the changing environment; however, it is the disruption of the sector motivated by this knowledge that we are highlighting. The new evidence on climate and health has driven policy responses and research. This process of knowledge influencing priorities does not always rely on external disruption – changes in public health patterns have themselves led to new knowledge that is influencing policy and ongoing research. As we succeed in treating acute diseases and our life expectancy increases, our understanding of health vulnerabilities changes. Chronic health conditions gain relative importance, and policies and research become more focused on long-term exposures and immune system changes.

Although scientific knowledge is important for influencing sector priorities, it is crucial to acknowledge that political context and the potential for success are equally if not more important factors determining policy formation, traction and effectiveness. Accordingly, in the next section we give an example where politics rather than science led disruptive change in the water sector.

6.3.4   Politics and the Pace of Disruption

Not all disruption is rapid, nor scientifically driven. Some ideas historically have taken much longer to move from academia to practical implementation. One commonly taught to students, in many different disciplines, is John Snow's theory of waterborne disease, which stemmed from London's 1854 Broad Street pump cholera outbreak. Snow's work is now recognized as ground-breaking epidemiology. When he initially reported that cholera was waterborne, however, it took longer for his theory to be widely accepted than it did for 'the most advanced and elaborate sewage system in the entire world' to be constructed and largely operational in London by 1866 (Johnson 2006, p. 208). The delay to sewer construction was largely political, rather than due to a lack of knowledge that clean water was important.

The prevailing belief at the time was that disease spread through bad air (the miasma theory), but this belief and belief in the importance of clean water were not mutually exclusive. Long before science provided an explanation, societies were aware that water and illness were linked. That awareness was reason enough for Scotland and northern British industrial towns to construct extensive waterworks by 1850 to combat shortages and provide cleaner water (Soloman 2010). In London, a filtration system was introduced for Chelsea in 1828 – preceding Snow's theory of waterborne cholera by decades.

After the first cholera outbreak in England, a commentator in a medical journal stated that 'there was never a real panic' because it was quickly realised that the great majority of cholera victims were amongst 'the poorest of the poor' (Inglis 1971, p. 272). London was experiencing a sanitary crisis and it was a reflection of social inequality. It was also 'an early manifestation of an inherent dilemma in the industrial market economy': the lack of an internal mechanism to ensure environmental sustainability (Soloman 2010, p. 260). Furthermore, there was ongoing debate over the rightful role of London's government as the administrative class and industrial magnates had risen to prominence, heralding a shift from self-serving aristocratic ideals towards Victorian ideals of 'social conscience and civic pride' (Plumb 1963, p. 97). So, London's eventual sanitary revolution was not driven as much by scientific understanding of waterborne disease as it was by political motivations (e.g. the Great Stink) and ideals of civic pride.

Even as the science of waterborne disease has advanced in the many decades since, these historic sociopolitical debates are still reflected in current efforts to tackle cholera and other waterborne diseases amongst the urban poor. The persistence of these debates underscores the importance of political will for enabling positive change in the water sector. Innovation and advances in knowledge are important, but not sufficient to move the sector forward without political backing.

6.4 Beyond Outbreaks: The Underreported Health Burden of Inadequate Water Supplies

Historically, research tackled the acute health burden of infectious disease, as discussed in Section 6.2. Health problems resulting from long-term exposure to infectious diseases received substantially less attention and remain underreported. Similarly, salinity and chemical contaminants such as arsenic, fluoride, lead and manganese, which typically have significant health effects only after long-term exposure, have been overlooked in traditional classifications of water-related health issues. To date, there has been no global assessment of the health burden from chemical contaminants in water, although there is enough evidence to know that the impacts are varied and widespread.

In recent years, research has increasingly directed attention towards the health burden of long-term exposure to both infectious diseases and chemical contaminants rather than solely to their acute effects. This new focus on chronic conditions is changing the way researchers and practitioners think about water and health. As the collection of research expands, researchers and practitioners in the water sector gain a better understanding of the importance and complexity of health outcomes from long-term consumption of contaminated drinking water. In this section we explore five growing areas of research: the chronic impacts of infectious diseases through enteric environmental dysfunction (EED); visible physical effects of chemical exposure such as skin and skeletal damage (e.g. due to hyperkeratosis, peripheral neuropathy, and fluorosis); less visible physical effects of chemical exposure such as hypertension and cancer; cognitive impairment; and psychosocial distress. These are all relatively young areas of research in which additional studies have high impact potential.

6.5 Enteric Environmental Dysfunction

Increasingly in recent years, the chronic impact of infectious diseases has been recognized. EED is the term used to describe inflammation of the gut from chronic exposure to pathogens associated with poor drinking water quality, poor hygiene, and high faecal contamination from humans and animals in the environment. This inflammation leads to malabsorption of food and malnutrition, which increases both susceptibility to further infections and the risk of stunting (Figure 6.1). The impact of EED on the gut is also associated with lower effectiveness of oral vaccines, such as the oral rotavirus vaccine (Korpe and Petri 2012). Whilst the relationship between individual infections and stunting was first demonstrated in Guatemala by Mata et al. (1972), researchers have only recently begun to realise the cumulative impact of clinical and subclinical asymptomatic infections via EED and stunting.

Stunting affects an estimated 165 million children under five years of age, many in areas with poor access to water and sanitation (Prendergast and Humphrey 2014). Although height is a linear measure, stunting is also associated with growth retardation in other areas of the body. Consequently, it can result in increased mortality and morbidity, including reduced cognitive development. Furthermore, it is associated with chronic non-communicable diseases, including metabolic disorders, such as diabetes, and cardiovascular diseases, as well as adult obesity (Guerrant et al. 2013).

Stunting has lifelong implications for those directly affected; it also has intergenerational impacts. Impacts on cognitive development reduce the potential for educational attainment and employment. Increased health costs and risks place economic burdens on families. These reinforce an intergenerational cycle of poverty, which is exacerbated because women who were stunted in childhood are biologically more likely to have stunted offspring (Prendergast and Humphrey 2014).

6.5.1 Visible Disease from Chemical Exposure

Arsenic and fluoride are two of the best-known environmental contaminants of concern for drinking water, in part because they cause visible physical damage. Hazardous concentrations of these elements are generally associated with groundwaters in the

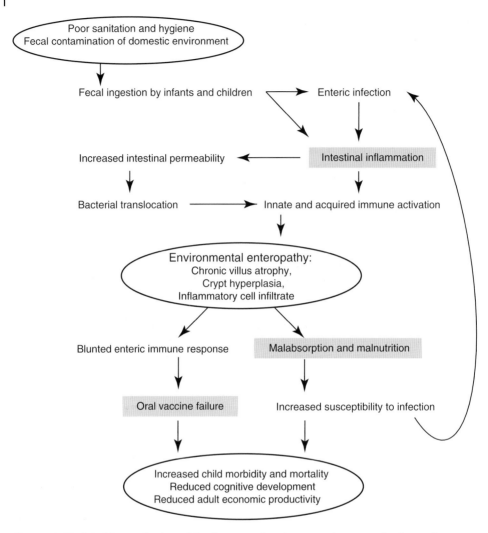

Figure 6.1 Model of the mechanism of development of environmental enteropathy. *Source:* Korpe and Petri (2012). Reproduced with permission of Elsevier Ltd. (*See color representation of this figure in color plate section*).

Indian subcontinent, China, Central Africa and South America, although the geology in most parts of the world can create localized areas of high concentration. Both arsenic and fluoride are readily absorbed through the human gastrointestinal tract (WHO 2017).

The most-documented health impacts of absorbed arsenic are dermal lesions in the form of changed pigmentation, hyperkeratosis (overproduction of keratin in the skin causing thickening in places and resulting in a sandpaper-like texture of the skin), and peripheral neuropathy (which results in weakness, numbness and pain due to peripheral nerve damage). Arsenic poisoning can also cause peripheral vascular disease (PVD) – a blood circulation disorder that produces pain and fatigue as well as visible symptoms such as reddish blue or pale colouration of legs and arms, and wounds or ulcers that will not heal. PVD also increases blood clot formation, which can lead to organ damage and loss of fingers, toes or limbs (WHO 2017). There are no known benefits of consuming arsenic, and considerable uncertainty remains 'over the actual risks at low concentrations' (WHO 2017, p. 315). As a result, concentrations in drinking water should ideally be zero (the WHO gives a provisional guideline of 0.01 mg/l as a concentration that can reasonably be achieved by treatment and detected with current analytical methods).

Unlike arsenic, there is clear evidence that fluoride may be a beneficial nutrient. At low concentrations (0.5–2 mg/l), it has been shown to improve dental health (WHO 2017), leading to fluoridation of some municipal water supplies and fluoride being added to dental products. High fluoride intake is detrimental, however, especially for children during their development. It can cause fluorosis – a condition marked by dis-coloured and pitted teeth, weakened bones (which are more susceptible to fracture) and immobilized joints. Dental fluorosis is associated with long-term consumption of water with concentrations as low as 0.9 mg/l, whereas skeletal fluorosis is associated with con-centrations around 3–6 mg/l, with crippling cases usually resulting from greater than 10 mg/l (WHO 2017). The severity of fluorosis is determined by the cumulative absorp-tion of fluoride over time, but fluorosis symptoms can improve if fluoride consumption stops or is reduced. Food and airborne particulates can also be significant sources of fluoride, so the safe amount of fluoride in drinking water is dependent on the volume consumed and the contribution of fluoride from food and air. Although the WHO pro-vides a drinking water guideline for fluoride of 1.5 mg/l based on epidemiological evi-dence, the organization also emphasizes the need for context-specific water quality standards that account for total fluoride intake from water as well as food, airborne particulates, and dental care.

6.5.2   Hypertension and Cancer

Beyond very visible physical effects from arsenic and fluoride intake, drinking unsafe water that contains high levels of salinity and heavy metals can increase the likelihood of developing hypertension and cancer. It is widely recognized that high dietary sodium chloride intake raises blood pressure and can lead to hypertension (Institute of Medicine 2005), which increases the likelihood of preeclampsia and cardiovascular and renal dis-eases. Despite this recognition, the WHO has not developed a health-based guideline for salinity in drinking water for two main reasons (WHO 2017). Firstly, there is an assumption that salt intake from food substantially outweighs intake from water. Since salt is detectable by taste, acceptability limits are generally reached before salinity becomes a health concern. Consequently, people will elect not to drink water that is hazardously salty, assuming that they have another option. The second reason that no health-based drinking water guideline is available for salinity is that '[n]o firm conclu-sions can be drawn concerning the possible association between sodium in drinking water and the occurrence of hypertension' (WHO 2017, p. 416). There have been rela-tively few studies examining the relationship between blood pressure and sodium in drinking water, and most were conducted in developed countries.

One exception is a small collection of studies from coastal Bangladesh that have examined the impact of high-salinity drinking water on blood pressure. Scheelbeek et al. (2017), for example, found that drinking water was a significant source of daily sodium intake and that the concentration of salinity in drinking water was highly associated with blood pressure and the likelihood of hypertension in consumers. The assumption that taste-based acceptability thresholds direct people to drink water with limited salt content holds true in many places, particularly in the developed world; however, there are water-insecure coastal zones and dryland, groundwater-dependent regions where people – especially impoverished people – have few options besides drinking saline water.

Salinity is not the only component of drinking water quality that can impact cardiovascular health: stunting from EED is associated with cardiovascular problems and a growing number of studies have reported associations between hypertension and intake of environmental toxins such as arsenic, cadmium, mercury (Martins et al. 2018) and lead (USEPA 2013). For arsenic in particular, studies have reported associations between concentrations in drinking water and hypertension (Abhyankar et al. 2012) and cardiovascular disease (Moon et al. 2012), although a dose–response relationship has not been established.

In addition to being a risk factor for cardiovascular and renal diseases, 'there is overwhelming evidence that consumption of elevated levels of arsenic through drinking water is causally related to the development of cancer' in the skin, lungs, bladder and kidneys (WHO 2017, p. 316). Studies have also pointed to a possible association between fluoride in drinking water and cancer, but the evidence is considered inconclusive and more data are needed – particularly for bone cancer, given the impact of fluorosis on bone development (WHO 2017). Whilst low-cost water treatment is available for arsenic and fluoride, as well as pathogens, there is little evidence of sustainable use at the household or community level in low-income settings (Waddington and Snilstveit 2009).

6.5.3 Cognitive Impairment

Chronic exposure to contaminated drinking water can also impact brain function. Stunting from EED and intake of heavy metals such as arsenic, lead, and possibly manganese can affect cognitive development. These impacts have been found to be most severe for children, although the mechanisms of cognitive impairment are not well understood. For arsenic, studies have demonstrated an inverse relationship between arsenic intake and intellectual function amongst children and adults (Wasserman et al. 2004; Bryant et al. 2011).

For lead, the USEPA (2013) recognizes that even low concentrations in the blood of children can damage their central and peripheral nervous systems and cause lower intelligence quotient scores, behavioural problems and learning disabilities. Unlike arsenic and fluoride, which are naturally occurring environmental contaminants, lead in drinking water usually leaches out of lead piping in municipal distribution systems. This was the cause of the water crisis in Flint, Michigan, that began in 2014 when the city changed the raw water source and treatment practices for their municipal supply, resulting in lead leaching from pipes into their drinking water.

In addition to arsenic and lead, a few epidemiological studies have pointed to an association between manganese in drinking water and learning difficulties in children – more work is needed to strengthen these findings and demonstrate a causal relationship (WHO 2017). Food is usually a more important source of manganese than water, but some surface and groundwater sources can be significant if they have sustained anaerobic conditions (under which manganese is more soluble). As with salinity, elevated manganese in drinking water (more than 0.1 mg/l) can cause an undesirable taste that may discourage people from drinking the water if they have a better-tasting alternative (WHO 2017).

6.5.4 Psychosocial Distress

In addition to the aforementioned physical impacts, people with chronic health conditions often have their suffering compounded by loss of ability – ability to be productive, to earn an income, to fit a social role. This loss of ability can have immediate and long-term impacts on individuals, families and communities. Both the experience of the disease itself and its follow-on effects have important psychological consequences. A few studies have examined the psychosocial dimensions of inadequate water supplies. For example, Wutich and Ragsdale (2008) found that emotional distress was significantly associated with water insecurity in a Bolivian squatter settlement; Brinkel et al. (2009) reviewed literature on the social and mental health effects of arsenic exposure and found them to be numerous; and Thomas and Godfrey (2018) identified emotional distress associated with intrahousehold conflict around limited water services in rural Ethiopia. Psychosocial distress is a growing area of public health research. When applied to the water sector, acknowledging this health dimension further emphasizes the importance of securing reliable, safe and affordable water supplies.

6.5.5 Revisiting the Water-Related Burden of Disease

Chronic conditions resulting from long-term exposure to bacteriological and chemical contamination in drinking water create a far-reaching, often intergenerational health burden. This burden is unevenly distributed, with geography, wealth, age, gender, race, and other axes of inequality being important factors. Those with education, and who can afford more costly alternatives, and those with access to better nutrition can reduce their exposure or moderate the impacts – for example, a diet high in calcium can somewhat mitigate the impact of fluoride and lead ingestion (USEPA 2013; WHO 2017), and a high potassium diet can moderate the blood pressure increase from high sodium intake (Institute of Medicine 2005). The chronic conditions described in the preceding sections are generally more severe if people are exposed as children, although elderly people are more at risk from blood pressure increases. Brinkel et al. (2009) reported that Bangladeshi women suffer more from arsenic-related ostracism than do men. In the United States, African-Americans are reported to have above average prevalence of salt sensitivity and therefore higher risk of hypertension (Institute of Medicine 2005). These many inequalities add another layer of complexity for understanding and combating the health burden of inadequate water supplies.

This burden has considerable implications for meeting development aims, including the SDGs. The extent and impacts of EED and chemical contamination demonstrate that a more nuanced understanding of the complexities of water and health interactions is needed. Focusing on diarrhoea in children under five as the primary outcome of water, sanitation and hygiene (WASH) interventions does not adequately represent the complex benefits of good WASH. Although a shift has been made to focus on anthropometric measures, such as stunting, this still does not capture the full burden of poor WASH nor the potential benefits of improving WASH. It is worth noting that cost–benefit studies still rely heavily on diarrhoea risk reductions as the main source of data on health benefits and health cost-savings. A more holistic understanding of the water-related burden of disease may greatly increase our ability to implement effective water interventions for improved long-term, multigenerational benefits.

6.6   Water and Health Challenges in the SDG Period

Improvements in water-related health have been significant over the past decades. We know much more about infectious diseases, how to interrupt their transmission pathways, and how to treat them. These gains in knowledge have enabled substantial reductions in diarrhoeal diseases, malaria, schistosomiasis and other water-related diseases. Additionally, as highlighted in Section 6.4, we are beginning to understand the importance of chronic water-related health problems and the need for approaches that better reflect the complexity of water-health systems. In light of this complexity and the ambitious SDGs, addressing water-related health raises important interdisciplinary challenges.

In this section we frame three current research challenges where innovative thinking is needed: meeting higher service levels for WASH; developing new technologies for testing water quality; and building institutional capacities to decrease inequalities in global water supply. These challenges are central to achieving the water SDG, but they are not all-encompassing. The future will also see water and health researchers focusing on other important challenges, for instance: identifying and managing emerging diseases and contaminants including endocrine-disrupting chemicals; dealing with the increasing use of human and animal antibiotics and the resulting rise of antimicrobial resistance; addressing declining water infrastructure; and tackling previously unrecognized inequalities in developing and developed countries.

The greatest conceptual difference between the water MDG and the water SDG is the shift from thinking about infrastructure to thinking about services – a shift from considering things to considering people. The drinking water SDG is Target 6.1:

> By 2030, achieve universal and equitable access to safe and affordable drinking water for all.

'Safely managed' is the gold standard by which progress towards Target 6.1 will be judged (Table 6.2). The ladder of service levels is presented rather than a single standard, so that monitoring can reflect important intermediate steps as well as progress in achieving the gold standard (WHO/UNICEF 2015). Whilst the focus here is on water quality, it is important to note that the new category of 'safely managed' also addresses quantity, a critical factor for ensuring health gains, through reducing the burden of collection that often physically restricts water use, and through ensuring water is available when needed.

Explicit inclusion of water quality in the definition of 'safely managed' is an important aspect of the shift in emphasis from infrastructure towards more holistic service provision. The old infrastructure approach was often characterized by discrete investments with little follow-up after construction; in contrast, ensuring good water quality in the long term requires ongoing management to maintain and verify water safety measures. In the MDG period, progress towards safe water provision was measured only by the

Table 6.2 The WHO/UNICEF Joint Monitoring Programme (JMP) monitoring levels for drinking water.

Service level	JMP definition
Safely managed	Drinking water from an improved[a] water source that is located on premises, available when needed and free from faecal and priority chemical contamination.
Basic	Drinking water from an improved[a] source, provided collection time is not more than 30 minutes for a round trip, including queuing.
Limited	Drinking water from an improved[a] source for which collection time exceeds 30 minutes for a round trip, including queuing.
Unimproved	Drinking water from an unprotected dug well or unprotected spring.
Surface water	Drinking water directly from a river, dam, lake, pond, stream, canal, or irrigation canal.

Source: Adapted from WHO/UNICEF (2017).
[a] Improved sources are supposed to be protected from faecal contamination by nature of their design/construction. The category includes: piped water, boreholes or tubewells, protected dug wells, protected springs, rainwater, and packaged or delivered water.

prevalence of 'improved' water sources. The move to include water quality explicitly was informed by several studies, including a meta-analysis by Bain et al. (2014) that showed the inadequacy of 'improved' infrastructure as a proxy for safe water. These studies indicated that the achievement of the water MDG (Target 7c), which was celebrated in 2010, was less of a triumph than initially thought because 'improved' infrastructure is not consistently free from faecal contamination. For the SDGs, safe water will be judged by water quality tests and, in recognition of the importance of chronic health issues, priority chemical contamination will be considered in addition to faecal contamination (priority in this case depends on context, but arsenic, fluoride and lead are high-profile contaminants). Thus, the SDG focus on services is especially pertinent to the relationship between water and health.

6.6.1 Improving Service Levels

The move to the service-oriented SDGs and the addition of the JMP's service ladder has coincided with more nuanced investigation into the health outcomes associated with different service levels and the health benefits achieved by different interventions. It has also enabled water and health research to be situated in a clearer policy context. A rigorous WASH randomized control trial recently conducted in rural Kenya and Bangladesh tried to unpack some of the uncertainties around WASH effectiveness, focusing in particular on impacts on infants. These trials found no significant additional benefit from WASH interventions being added to nutrition programmes (Luby et al. 2018; Null et al. 2018). This is consistent with earlier evidence from Fewtrell et al. (2005), who conducted a number of reviews and meta-analyses of studies of water, sanitation and hygiene interventions that suggested multipronged interventions (i.e. those addressing a combination of water supply, quality, sanitation and hygiene aspects) did not have a greater impact on health than did individual interventions.

A meta-analysis by Wolf et al. (2018a) showed that having high-quality piped water leads to a much greater reduction in diarrhoea than interim service levels. Increasingly, studies are highlighting problematic limitations and variability in WASH service levels. The benefits of piped water supply can be realised only if supplies are reliable (e.g. Majuru et al. 2016). For non-piped supplies, service levels vary based on intermittent water availability due to poor maintenance and related governance issues (see Chapter 9). Additionally, users often choose to switch between different water sources based on seasonality and other factors, often switching to less microbially safe surface water sources during rainy seasons (Thomson et al. 2019). It is clear that the relationship between health outcomes and a water source type or category of infrastructure is heavily mediated by the level of service supplied.

Wolf et al. (2018b) cautioned against interpreting a minimal reduction in diarrhoea as indicative of an ineffective intervention, as a particular intervention may only target one of many pathways of faecal–oral transmission, making the individual intervention a necessary but not sufficient condition for reducing diarrhoea morbidity. Unless these multiple pathways are targeted simultaneously, the faecal exposure is often unlikely to be reduced sufficiently to result in a reduction in enteric disease (Robb et al. 2017). The relationship between higher levels of service and better health outcomes is positive but complex and nonlinear. Further research to understand this complexity is vital for designing WASH programmes and interventions that ensure the most effective use of inevitably limited resources.

6.6.2   Improving Water Quality Testing Methods

The purpose of drinking water quality testing is to reduce uncertainty about the wholesomeness of water for consumption, thereby promoting health-protective management of supplies and prioritization of interventions. The water SDG, in particular the 'safely managed' gold standard, has highlighted the importance of water quality testing; however, generating water quality information is costly, time-intensive and rife with uncertainty. It can be particularly difficult in low-resource settings and where organizational structures are not in place to conduct monitoring programmes or make use of water quality data.

There are many different pathogens – bacteria, protozoa and viruses – that spread by faecal contamination of water supplies. Although methods have significantly advanced, many pathogens remain difficult to detect without intensive, expensive molecular methods – even then, it is difficult to determine which individual micro-organisms are infectious. For the last century, instead of testing for pathogens directly, monitoring programmes have used culturable coliform bacteria as indicators of faecal contamination risk. The preferred indicator, *Escherichia coli*, continues to be widely used, but it has some important limitations:

- It takes 18–24 hours for test results to be available.
- Traditional culturing methods require experienced technicians as well as sterile working conditions and access to a power supply for refrigeration and incubation of growth media and samples.
- The cost of consumables may prohibit frequent sampling where budgets are constrained, but frequent sampling is important to understand and manage temporal and spatial changes in water-related health risks.
- Viruses and protozoa are physiologically distinct from *E. coli* and thus have different environmental transport and survival patterns, so absence of *E. coli* does not guarantee safety (Leclerc et al. 2001).

- *E. coli* can survive and reproduce in soils, sediments, water, and biofilms on infrastructure (for example Brennan et al. 2010) – so their presence is not necessarily indicative of recent faecal contamination.

In addition to these limitations, the purpose of *E. coli* testing is often misunderstood: instead of interpreting results within a risk assessment framework, there is a tendency to equate *E. coli* in general (a common enteric species, the majority of which are non-pathogenic) with *E. coli* in its pathogenic form, and therefore to interpret tests as direct assessments of pathogen presence.

In recent years, new approaches have been developed to get around some of the limitations of traditional *E. coli* sampling. For example, simple methods for most probable number assessment of culturable *E. coli* have been developed and marketed for application in low-resource and emergency settings by companies such as Aquagenx (https://www.aquagenx.com). Moving further from the traditional approach, emergence of a new paradigm in DNA sequencing has the potential to lower barriers to direct analysis of pathogens (Jain et al. 2016) – although clinical applications are more prevalent than environmental monitoring applications to date. A third approach has been to use a well-established form of measurement (fluorimetry) in a new way. This approach attempts to use tryptophan-like fluorescence, an indicator of microbial activity in water, as a measure of faecal contamination risk (e.g. Nowicki et al. 2019). Although each of these approaches has certain strengths, they all have limitations and do not provide universally applicable solutions for microbial water quality monitoring.

We have focused here on microbial water quality monitoring, but innovation to improve chemical water quality monitoring capacities could also have widespread implications. Especially with increasing attention on water quality testing in light of the drinking water SDG, research into new approaches to testing for chemical and pathogenic contamination has relevance and far-reaching impact potential.

6.6.3 Leaving No One Behind

The SDGs are accompanied by the oft-repeated imperative to leave no one behind and to prioritize those who are most difficult to reach. For SDG 6.1, those who are most difficult to reach are poor inhabitants of rural areas. As of 2015, it was estimated that only 24% of people in sub-Saharan Africa (SSA) had safely managed water and most of them were living in urban areas. One of the criteria for safely managed water is for it to be supplied on premises. This is a difficult standard to meet in rural areas where supplies are typically decentralized, and low-density settlements make extensive piped networks prohibitively expensive under current cost–benefit considerations. When only the water quality criterion is accounted for, the estimate for total population with good quality water rises to 42%, but there is insufficient data to make an estimate for the rural areas.

Identification methods exist for the pathogens and chemicals discussed in this chapter; however, even where tests are affordable, in low-resource settings useful water quality data are often not readily available to managers of non-utility supplies, whether at the household, community or government level. In rural areas of SSA, where community-led models of water management predominate, drinking water quality is rarely tested. In SSA, rural areas will require the most 'attention and additional resources to achieve regulatory compliance for water quality monitoring' (Peletz et al. 2016).

Monitoring is only part of the challenge. Once water quality data are generated, water safety planning processes must be in place to act on them. But the prevalence of decentralized community management is a barrier to increasing water safety because community-level water managers rarely have the training or access to resources that are required for water quality testing and designing and implementing water safety plans.

Safe, community-managed water supplies require a dual approach where water quality information and community management are both priorities. To align these priorities, communities must be supported in understanding and managing the safety of their drinking water. As elaborated in Chapter 9, recent literature promotes institutional pluralism as a strategy for improving rural water service provision. The challenges for achieving the SDG targets for sanitation, which is necessary for reducing stunting and the infectious disease burden, are similarly complex (see Chapter 16). The core concept is that appropriate sharing of responsibilities between the government, private sector, communities, families and individuals can take advantage of the differing strengths of each domain. Indeed, even as 'leaving no one behind' points to the needs and rights of everyone for water, so too there are responsibilities towards the management of water at every level of society, and 'everyone playing their part' points to the responsibilities that accompany rights. Such an approach could provide an opportunity to include an effective, health-protective water quality monitoring element as part of the wider management of rural drinking water services.

6.7 Conclusions

Despite many advances, and a much deeper understanding of the complexity of intersections between water and health, it remains a relevant area and a critical challenge for delivering the SDGs. In this chapter we have reflected on historical context and the current state of knowledge, framing three challenges for future scholars that underpin the delivery of the SDGs: achieving high service levels, safe water, and universal access. To achieve these, we need disruption. It has been clear from the start of the SDGs that these ambitious targets of universal coverage of basic water and sanitation will not be achieved within the time-frame (and available finances; see Chapters 15 and 17) of current approaches; it is even less likely that access for all will be achieved with a high level of service.

Disruption in the water sector is not always rapid. Change has come from different pathways as a result of innovation and learning, of crisis, and of political will. Reflecting on these historical developments helps us to consider how change happens, and that opportunities for innovation can be shaped by processes far removed from technological advancement and scientific understanding.

Acknowledgements

This work is an output from the REACH programme (www.reachwater.org.uk) funded by UK Aid from the UK Department for International Development (DFID) for the benefit of developing countries (Aries Code 201880) and the Gro for GooD project (UPGro Consortium Grant: NE/M008894/1 (NERC, ESRC and DFID). However, the views expressed, and information contained in it are not necessarily those of or endorsed by DFID, which can accept no responsibility for such views or information or for any reliance placed on them.

References

Abhyankar, L.N., Jones, M.R., Guallar, E., and Navas-Acien, A. (2012). Arsenic exposure and hypertension: a systematic review. *Environmental Health Perspectives* 120 (4): 494–500.

Bain, R., Cronk, R., Wright, J. et al. (2014). Fecal contamination of drinking-water in low- and middle-income countries: a systematic review and meta-analysis. *PLoS Medicine* 11 (5) https://doi.org/10.1371/journal.pmed.1001644.

Bartram, J. (ed.) (2015). *Routledge Handbook of Water and Health*. London and New York: Routledge.

Bartram, J. and Hunter, P. (2015). Bradley Classification of disease transmission routes for water-related hazards. In: *Routledge Handbook of Water and Health* (eds. J. Bartram, R. Baum, P.A. Coclanis, et al.). London and New York: Routledge.

Brennan, F.P., O'Flaherty, V., Kramers, G. et al. (2010). Long-term persistence and leaching of escherichia coli in temperate maritime soils. *Applied and Environmental Microbiology* 76 (5): 1449–1455. https://doi.org/10.1128/AEM.02335-09.

Brinkel, J., Khan, M.H., and Kraemer, A. (2009). A systematic review of arsenic exposure and its social and mental health effects with special reference to Bangladesh. *International Journal of Environmental Research and Public Health* 6 (5): 1609–1619. https://doi.org/10.3390/ijerph6051609.

Bryant, S.E.O., Edwards, M., Menon, C.V. et al. (2011). Long-term low-level arsenic exposure is associated with poorer neuropsychological functioning: a project FRONTIER study. *International Journal of Environmental Research and Public Health* 8 (3): 861–874. https://doi.org/10.3390/ijerph8030861.

College of Physicians of Philadelphia (2019). Early Tissue and Cell Culture in Vaccine Development, The History of Vaccines: An Educational Resource by the College of Physicians of Philadelphia. https://www.historyofvaccines.org/content/articles/early-tissue-and-cell-culture-vaccine-development (accessed 20 February 2019).

Fewtrell, L., Kaufmann, R.B., Kay, D. et al. (2005). Water, sanitation, and hygiene interventions to reduce diarrhoea in less developed countries: a systematic review and meta-analysis. *The Lancet. Infectious Diseases* 5 (1): 42–52. https://doi.org/10.1016/S1473-3099(04)01253-8.

GBD 2016 Diarrhoeal Disease Collaborators (2018). Estimates of the global, regional, and national morbidity, mortality, and aetiologies of diarrhoea in 195 countries: a systematic analysis for the Global Burden of Disease Study 2016. *The Lancet. Infectious Diseases* 18 (11): 1211–1228. https://doi.org/10.1016/S1473-3099(18)30362-1.

Gerba, C.P. and Nichols, G.L. (2015). Water-based disease and microbial growth. In: *Routledge Handbook of Water and Health* (eds. J. Bartram, R. Baum, P.A. Coclanis, et al.). London and New York: Routledge https://doi.org/10.4324/9781315693606.

Guerrant, R.L., Carneiro-Filho, B.A., and Dillingham, R.A. (2003). Cholera, diarrhea, and oral rehydration therapy: triumph and indictment. *Clinical Infectious Diseases* 37 (3): 398–405. https://doi.org/10.1086/376619.

Guerrant, R.L., DeBoer, M.D., Moore, S.R. et al. (2013). The impoverished gut – a triple burden of diarrhoea, stunting and chronic disease. *Nature Reviews Gastroenterology and Hepatology* 10 (4): 220–229. https://doi.org/10.1038/nrgastro.2012.239.

Inglis, B. (1971). *Poverty and the Industrial Revolution*. London: Hodder and Stoughton Limited.

Institute of Medicine (2005). *Dietary Reference Intakes for Water, Potassium, Sodium, Chloride, and Sulfate*. Washington, USA: The National Academies Press https://doi.org/10.17226/10925.

Jain, M., Olsen, H.E., Paten, B., and Akeson, M. (2016). The Oxford Nanopore MinION: delivery of nanopore sequencing to the genomics community. *Genome Biology* 17 (1): 239. https://doi.org/10.1186/s13059-016-1103-0.

Johnson, S. (2006). *The Ghost Map: A Street, A City, An Epidemic and the Hidden Power of Urban Networks*. London: Penguin Group.

Korpe, P.S. and Petri, W.A. (2012). Environmental enteropathy: critical implications of a poorly understood condition. *Trends in Molecular Medicine* 18 (6): 328–336. https://doi.org/10.1016/j.molmed.2012.04.007.

Leclerc, H., Mossel, D.A., Edberg, S.C., and Struijk, C.B. (2001). Advances in the bacteriology of the coliform group: their suitability as markers of microbial water safety. *Annual Review of Microbiology* 55: 201–234. https://doi.org/10.1146/annurev.micro.55.1.201.

Luby, S.P., Rahman, M., Arnold, B.F. et al. (2018). Effects of water quality, sanitation, handwashing, and nutritional interventions on diarrhoea and child growth in rural Bangladesh: a cluster randomised controlled trial. *The Lancet Global Health* 6 (3): e302–e315. https://doi.org/10.1016/S2214-109X(17)30490-4.

Majuru, B., Suhrcke, M., and Hunter, P.R. (2016). How do households respond to unreliable water supplies? A systematic review. *International Journal of Environmental Research and Public Health* https://doi.org/10.3390/ijerph13121222.

Martins, A.d.C. Jr., Carneiro, M.F.H., Grotto, D. et al. (2018). Arsenic, cadmium, and mercury-induced hypertension: mechanisms and epidemiological findings. *Journal of Toxicology and Environmental Health, Part B, Critical Reviews* 21 (2): 61–82. https://doi.org/10.1080/10937404.2018.1432025.

Mata, L.J., Urrutia, J.J., Albertazzi, C. et al. (1972). Influence of recurrent infections on nutrition and growth of children in Guatemala. *American Journal of Clinical Nutrition* 25 (11): 1267–1275. https://doi.org/10.1093/ajcn/25.11.1267.

Moon, K., Guallar, E., and Navas-Acien, A. (2012). Arsenic exposure and cardiovascular disease: an updated systematic review. *Current Atherosclerosis Reports* 14 (6): 542–555. https://doi.org/10.1007/s11883-012-0280-x.

Nowicki, S., Lapworth, D., Ward, J.S.T. et al. (2019). Tryptophan-like fluorescence as a measure of microbial contamination risk in groundwater. *Science of the Total Environment* 646: 782–791. https://doi.org/10.1016/j.scitotenv.2018.07.274.

Null, C., Stewart, C.P., Pickering, A.J. et al. (2018). Effects of water quality, sanitation, handwashing, and nutritional interventions on diarrhoea and child growth in rural Kenya: a cluster-randomised controlled trial. *The Lancet Global Health* 6 (3): e316–e329. https://doi.org/10.1016/S2214-109X(18)30005-6.

Peletz, R., Kumpel, E., Bonham, M. et al. (2016). To what extent is drinking water tested in sub-saharan Africa? A comparative analysis of regulated water quality monitoring. *International Journal of Environmental Research and Public Health* 13 (3): 275. https://doi.org/10.3390/ijerph13030275.

Plumb, J.H. (1963). *England in the Eighteenth Century*. London: Penguin Group.

Prendergast, A.J. and Humphrey, J.H. (2014). The stunting syndrome in developing countries. *Paediatrics and International Child Health* 34 (4): 250–265. https://doi.org/10.1179/2046905514Y.0000000158.

Robb, K., Null, C., Teunis, P. et al. (2017). Assessment of fecal exposure pathways in low-income urban neighborhoods in Accra, Ghana: rationale, design, methods, and key findings of the Sanipath study. *American Journal of Tropical Medicine and Hygiene* 97 (4): 1020–1032. https://doi.org/10.4269/ajtmh.16-0508.

Scheelbeek, P.F.D., Chowdhury, M.A.H., Haines, A. et al. (2017). Drinking water salinity and raised blood pressure: evidence from a cohort study in coastal Bangladesh. *Environmental Health Perspectives* 125 (5): 057007. https://doi.org/10.1289/EHP659.

Scrimshaw, N.S., Taylor, C.E., and Gordon, J.E. (1959). Interactions of nutrition and infection. *The American Journal of the Medical Sciences* 237: 367–403.

Soloman, S. (2010). *Water: The Epic Struggle for Wealth, Power, and Civilization*. New York: HarperCollins Publishers.

Thomas, V. and Godfrey, S. (2018). Understanding water-related emotional distress for improving water services: a case study from an Ethiopian small town. *Journal of Water Sanitation and Hygiene for Development* 8 (2): 196–207. https://doi.org/10.2166/washdev.2018.167.

Thomson, P., Bradley, D., Katilu, A. et al. (2019). Rainfall and groundwater use in rural Kenya. *Science of the Total Environment* 649: 722–730. https://doi.org/10.1016/j.scitotenv.2018.08.330.

USEPA (2013). *Integrated Science Assessment for Lead*. North Carolina: Office of Research and Development, National Center for Environmental Assessment.

Waddington, H. and Snilstveit, B. (2009). Effectiveness and sustainability of water, sanitation, and hygiene interventions in combating diarrhoea. *Journal of Development Effectiveness* 1 (3): 295–335. https://doi.org/10.1080/19439340903141175.

Wasserman, G.A., Liu, X., Parvez, F. et al. (2004). Water arsenic exposure and children's intellectual function in Araihazar, Bangladesh. *Environmental Health Perspectives* 112 (13): 1329–1333. https://doi.org/10.1289/ehp.6964.

White, G., Bradley, D., and White, A. (1972). *Drawers of Water. Domestic Water Use in East Africa*. Chicago: University of Chicago Press.

WHO (2017). *Guidelines for Drinking-Water Quality: Fourth Edition Incorporating the First Addendum*, 4e. Geneva: WHO Press http://apps.who.int/iris/bitstream/handle/10665/254637/9789241549950-eng.pdf;jsessionid=2E1B7E42B16F03AB88E253B119976D63?sequence=1.

WHO/UNICEF (2015). *Progress on Sanitation and Drinking Water – 2015 Update and MDG Assessment*. Geneva: World Health Organization (WHO) and United Nations Children's Fund (UNICEF) https://washdata.org/reports.

WHO/UNICEF (2017). *Progress on Drinking Water, Sanitation and Hygiene – 2017 Update and Sustainable Development Goal (SDG) Baselines*. Geneva: World Health Organization (WHO) and United Nations Children's Fund (UNICEF) https://washdata.org/reports.

Wolf, J., Hunter, P.R., Freeman, M.C. et al. (2018a). Impact of drinking water, sanitation and handwashing with soap on childhood diarrhoeal disease: updated meta-analysis and meta-regression. *Tropical Medicine & International Health* 23 (5): 508–525. https://doi.org/10.1111/tmi.13051.

Wolf, J., Johnston, R., Hunter, P.R. et al. (2018b). A Faecal Contamination Index for interpreting heterogeneous diarrhoea impacts of water, sanitation and hygiene interventions and overall, regional and country estimates of community sanitation coverage with a focus on low- and middle-income countries. *International Journal of Hygiene and Environmental Health* https://doi.org/10.1016/j.ijheh.2018.11.005.

Wutich, A. and Ragsdale, K. (2008). Water insecurity and emotional distress: coping with supply, access, and seasonal variability of water in a Bolivian squatter settlement. *Social Science & Medicine* 67 (12): 2116–2125. https://doi.org/10.1016/j.socscimed.2008.09.042.

7

Monitoring and Modelling Hydrological Processes

Simon J. Dadson[1], Feyera Hirpa[1], Patrick Thomson[1,2], and Megan Konar[3]

[1] *School of Geography and the Environment, University of Oxford, UK*
[2] *Smith School of Enterprise and the Environment, University of Oxford, UK*
[3] *University of Illinois at Urbana-Champaign, USA*

7.1 Modelling Hydrological Systems: Current Approaches

Hydrological models are powerful tools for understanding catchment processes and have been used in several practical applications such as flood and drought forecasting (e.g. Chapter 11), water resources planning (e.g. Chapter 14), climate change risk assessment (e.g. Chapters 2 and 4), and studies of the human impact on hydrological systems (e.g. Thielen et al. 2009; Alfieri et al. 2013; Hirabayashi et al. 2013; Haddeland et al. 2014; Schewe et al. 2014). Since its early stages of development in the 1960s and 1970s (e.g. Freeze and Harlan 1969; Burnash et al. 1973; Beven and Kirby 1979), the field of hydrological modelling has seen the proliferation of many different approaches, from using simple, catchment-scale, conceptual models to more complex, physically-based, distributed models.

The goal of any hydrological modeller is to represent the catchment processes, and thereby reproduce observed hydrological variables (such as streamflow and soil moisture) with as much accuracy as necessary given the requirements of the problem to be solved. To achieve this goal, various modelling groups follow different approaches (see the review by Hrachowitz and Clark 2017). There are three major classes of hydrological models based on the spatial resolution at which they operate: lumped, distributed and semi-distributed (Figure 7.1). Lumped models consider the catchment as a single homogeneous unit with uniform (area-averaged) state variables and parameter sets. By contrast, distributed models divide the catchment into smaller elements (often uniformly spaced grids) to represent the spatial heterogeneity within the model domain. They have the advantage of producing spatially distributed predictions across the model domain; however, this comes with an additional complexity of distributed model parameterization often with inadequate measurements to understand the processes operating in sufficient detail at each of the elemental units (e.g. Beven and Cloke 2012; Clark et al. 2017). Semi-distributed models are the most widely applied since they are simpler and faster than distributed models. They are essentially aggregates of several lumped models

Water Science, Policy, and Management: A Global Challenge, First Edition. Edited by Simon J. Dadson, Dustin E. Garrick, Edmund C. Penning-Rowsell, Jim W. Hall, Rob Hope, and Jocelyne Hughes.
© 2020 John Wiley & Sons Ltd. Published 2020 by John Wiley & Sons Ltd.

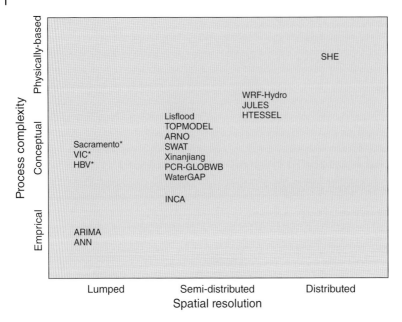

Figure 7.1 Schematic classification of commonly used models (see Table 7.1) based on process complexity and spatial resolution. The models marked with * have their recent versions operating at a semi-distributed resolution. Some models (e.g. JULES, WRF-hydro) have a mix of process complexity and spatial resolution. *Source:* authors.

applied over small homogeneous units based on concepts of hydrological similarity (Beven and Kirby 1979), hydrological response units (Arnold et al. 1998) or sub-basin boundaries (Todini 1996).

Models can also be categorized by the complexity of the physical process they represent: empirical, conceptual and physically-based models. Empirical models, commonly referred to as black-box models, are based on statistical relations between input and output data without an explicit consideration of the physical processes (see Xu et al. 2017). Examples of such models include recursive estimation (Ng and Young 1990), autoregressive integrated moving average models (ARIMA) (Montanari et al. 1997) and artificial neural networks (e.g. Dawson and Wilby 2001). Conceptual models provide simplified representations of physical processes (e.g. infiltration, evaporation, surface and subsurface flows) and use empirical equations to describe the interactions between the processes (e.g. Burnash et al. 1973; Beven and Kirby 1979; Lindstrom et al. 1997). Physically-based models explicitly represent hydrological processes (evaporation, transpiration, surface flow, infiltration, subsurface flow, and percolation) and provide a detailed description of the physical principles controlling the processes. These models require a detailed specification of model parameters to represent catchment properties (e.g. Freeze and Harlan 1969; Abbott et al. 1986a, b; Refsgaard et al. 2010; Clark et al. 2017).

The choice of appropriate model complexity and resolution depends on the details of a particular application, including the level of historical data available on fluxes and stores of water in the catchment. Scholars argue that 'different modelling strategies are complementary rather than mutually exclusive as they have different strengths and are thus suitable for different purposes' (Hrachowitz and Clark 2017). For example, detailed process-based models are commonly used alongside data-intensive monitoring campaigns to understand catchment response and to test hypotheses about hydrological processes. In contrast, large-scale conceptual models are more suitable for real-time applications (e.g. flood early warning) in data-sparse regions. There is a growing number of basic and applied research works aimed at developing and using hydrological modelling tools to assist decision-makers in areas of flood forecasting, seasonal drought outlooks and climate risk assessments. Hydrological models are increasingly being used to quantify hydrological risks and to inform societal risk reduction strategies at a range of scales (see Chapter 14).

7.1.1 From Local Catchment Models to Global Hydrological Studies

Hydrological models have been applied across ranges of scales, from local catchment scale to river basin, continental and global scales. Local catchment models in highly instrumented watersheds are commonly used to test hypotheses and new theoretical representations of hydrological processes. These approaches have been particularly valuable when conducted in parallel with long-term instrumentation programmes, for example in studies at major experimental watersheds including Plynlimon ($19\,km^2$) in the UK, and the Goodwin Creek ($21\,km^2$), Little Washita ($611\,km^2$) and Walnut Gulch ($150\,km^2$) watersheds in the US (see review by Tetzlaff et al. 2017).

With the improved availability of global meteorological datasets (e.g. Beck et al. 2017), digital elevation models (DEMs) (Lehner et al. 2008), lake and reservoir levels (Crétaux et al. 2011; Gao et al. 2012), soil moisture (Wagner et al. 1999), evapotranspiration (e.g. Miralles et al. 2011) and land cover data (https://www.esa-landcover-cci.org), and advances in computational power, there has been a notable growth in global-scale hydrological modelling in recent years, taking up the challenge to produce wide-area projections of hydrological responses to future climate and other environmental changes. There are currently several high-resolution (e.g. sub-daily and ~10 km) global-scale hydrological models (Bierkens 2015).

In common with all physical models, hydrological models must adhere to the basic principles of mass, energy and momentum conservation (see Beven (2012) for a thorough introduction). However, whilst most rainfall-runoff models focus on solving the mass balance equation for water, they often neglect the other two components (i.e. energy and momentum). The water balance equation partitions precipitation into root-zone soil water storage, evapotranspiration, surface and subsurface runoff, and drainage to groundwater. Rainfall-runoff models estimate runoff by combining a solution to the mass balance equation at each model time-step and spatial unit, with additional relations to describe fluxes between model stores. A significant challenge in specifying a model appropriate for large-scale applications arises from the fact that hydrological fluxes, and the properties of the stores through which they flow, themselves vary across at least seven orders of magnitude in time and space (Figure 7.2). Large-scale models that are widely applied in water resources modelling (e.g. WaterGAP: Döll et al. 2003; PCR-GLOBWB: van Beek et al. 2011; mHM: Samaniego et al. 2010) and river flow forecasting (e.g. Lisflood: De Roo et al. 2000; E-Hype: Lindström et al. 2010) use the water balance to estimate surface water availability and runoff.

By contrast, land surface models use mass and energy balance equations to model the interactions between the land surface and atmosphere, often because these models

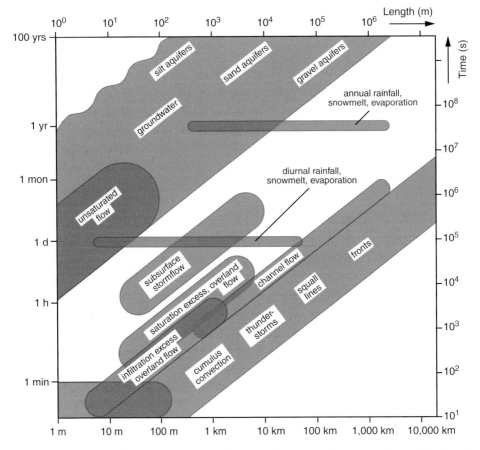

Figure 7.2 Hydrological processes at a range of characteristic space–time scales. *Source:* Blöschl and Sivapalan (1995). Reproduced with permission of John Wiley and Sons Ltd.

serve as the land-surface schemes in meteorological models and therefore required mass, energy and momentum fluxes to be exchanged with the overlying atmospheric model. Soil moisture content, soil thermal properties and vegetation control both the water balance and the partitioning of the available energy between latent and sensible heat components at the land surface (Seneviratne et al. 2010). Since in many regions there is a strong coupling between soil moisture and precipitation (Koster et al. 2004), a good representation of the land surface is needed to correctly predict rainfall using weather and climate models (Balsamo et al. 2009). Land surface models used in hydrological applications include VIC in African drought monitoring (Sheffield et al. 2014), HTESSEL in global flood forecasting (Alfieri et al. 2013), JULES for modelling the complete terrestrial water cycle (Best et al. 2011), and WRF-hydro for supporting hydrological predictions in the United States (Gochis et al. 2018).

Whilst such approaches conform more closely to physical principles, this does not always translate into improved model performance when evaluated by reference to observed measures (e.g. streamflow), particularly when processes operating below the model grid resolution are parametrized (cf. Figure 7.2). Notwithstanding recent developments, there are still recognized challenges in making 'hydrological models of everywhere locally relevant' as outlined in Bierkens et al. (2015). The challenges include the inability of the current generation of macro-scale models to represent subgrid processes, the lack of adequate measurements to parametrize high-resolution models, and the computational cost of completing operational simulations within a reasonable timeframe (Beven and Cloke 2012).

Despite their limitations, global-scale models are being used for both research and operational purposes in large parts of the developing world where local hydrological models do not exist (Alfieri et al. 2018). The Global Flood Awareness System (GloFAS [www.globalfloods.eu]: Alfieri et al. 2013) and the Global Flood Monitoring System (GFMS [http://flood.umd.edu]: Wu et al. 2014) are examples of operational flood forecasting systems producing real-time flood information at fully global scales. Proof-of-concept global drought monitoring and prediction systems (Yuan and Wood 2013; Hao et al. 2014) and applications of hydrological models in continental-scale operational flood forecasting in Europe, North America, Australia and Africa are reviewed by Emerton et al. (2016).

Moreover, the application of global-scale models has provided quantitative support, including studies of water use and abstraction (e.g. Wada et al. 2014), groundwater modelling (e.g. Haddeland et al. 2014), water availability estimation (e.g. Schewe et al. 2014), and climate risk assessment (e.g. Hirabayashi et al. 2013; Haddeland et al. 2014; Schewe et al. 2014). Whilst these models are often helpful for policy development and awareness-raising around pressing issues of national and international concern, they have to date been of limited use to water managers, who typically employ tailored catchment or water supply system models to the situation of interest. Examples of such models include operational flood forecasting systems in the Danube river basin (European Flood Awareness System: Thielen et al. 2009), water resources management in the Nile river basin (e.g. Wheeler et al. 2016), environmental impact studies in the Mekong basin (e.g. Kite 2001), and climate change impact studies in the Colorado River basin (e.g. Barnett and Pierce 2009).

7.1.2 Validation, Verification, and Confirmation in Hydrological Modelling

Since models are mathematical approximations of reality, they are inevitably subject to uncertainty. The uncertainties arise from incomplete representations of the physical processes in the model structure, uncertainty in model parameters, and errors in the input data and initial state (Beven and Freer 2001). Hydrological models are evaluated based on the degree to which they reproduce an observable variable (e.g. streamflow). Several evaluation metrics are used to validate the model outputs (Zappa et al. 2013; Brown et al. 2014). Many hydrological models contain parameters that are either unobservable or hard to observe with the required accuracy and precision. In such cases, a process of model calibration is used to assign optimal model parameter values in such a way as to satisfy the conditions of an objective function (Sorooshian and Gupta 1995). A single or multi-objective function can be used for measuring the degree to which model output matches observations. The selection of an appropriate objective function depends on the application and data availability, and may impact the model performance (reviewed in Pechlivanidis et al. 2011). Several parameter optimization techniques are in current use. They are primarily based on gradient descent algorithms, simulated annealing, and genetic algorithms (see reviews in Sorooshian and Gupta 1995; Franchini 1996; Beven 2012).

Scholars have different philosophies for quantifying and reducing model hydrological uncertainty. Beven and Binley (1992) developed the generalized likelihood uncertainty estimation (GLUE) framework in which multiple simulations are performed, and the likelihood of each model is estimated with respect to the variable of interest. Then, based on the likelihood measure, multiple models and parameter sets with acceptable behaviour in reproducing the variable are selected (Beven and Freer 2001). This philosophy is rooted in the concept of equifinality, in which it is argued that a single optimal parameter set that uniquely represents reality cannot be identified (Beven and Freer 2001), and therefore multiple parameter sets should be simultaneously considered where their performance is determined by the analyst to be satisfactory.

Conversely, other scholars argue that the GLUE approach fails to identify the best model for operational use (e.g. flood forecasting) and cannot separately quantify the model structural error and the model parameter uncertainty, which is sampled from a uniform parameter space (Stedinger et al. 2008; Clark et al. 2011). Moreover, it is suggested that the GLUE likelihood function is not necessarily based on the actual statistical distribution of model errors. Clark et al. (2011) advocate the use of 'multiple-hypothesis frameworks', in which model development and improvements are made based on testing hypotheses about hydrological processes and catchment properties. This approach would help to systematically evaluate alternative modelling approaches and reduce the predictive uncertainty arising from model structure and parameter errors.

7.1.3 Representing Human-managed Water Systems

Amongst the most pressing challenges shared by hydrological modellers, water managers and policy-makers is the need to make predictions of the future availability of water resources and of the threats to human livelihood and wellbeing from water-related extremes. Humanity relies on vast quantities of water for economic and cultural activities (Marston et al. 2018; Roobavannan et al. 2018). Over the last hundred years, global water withdrawals have increased by almost an order of magnitude, from 500 to 4000 km^3/year (Oki and Kanae 2006). Human water use leads to changes in the global hydrological cycle through the construction of dams, diversion of natural water flows, consumption of irrigation water supplies in agriculture, and abstraction of groundwater (Sivapalan et al. 2012, 2014; Wada et al. 2017). Many other human activities, such as land use and land cover changes, can impact the hydrological cycle; and water scarcity or extremes accentuate water storages and fluxes. This makes it imperative to understand where and when water demand outstrips available water supply.

Models can help water managers to understand and predict mismatches in water supplies and demand. Yet until recently, hydrological models have primarily focused on physical processes, ignoring human modifications to the hydrological cycle. Starting in the late 1990s, a new generation of global hydrology and water resources models (GHWMs) were developed to explicitly account for human activities in order to assess water security (Bierkens 2015). Initial attempts to reconcile hydrological models with human water use focused on comparisons with national water resources statistics

Table 7.1 Commonly used hydrological and land surface models.

Model	Model type	Typical domain/ resolution	Reference	Citations[a] (Google Scholar)	Citations[a] (scopus.com)
Sacramento	Rainfall-runoff	Watershed/ sub-basin, variable	Burnash et al. (1973)	882	463
TOPMODEL	Rainfall-runoff	Watershed/ sub-basin, variable	Beven and Kirby (1979)	6126	3396
SHE	Rainfall-runoff	Variable/ variable, variable	Abbott et al. (1986a)	1500	786
Xinanjiang	Rainfall-runoff	Watershed/ sub-basin, daily	Zhao (1992)	679	438
VIC	Land surface model	Global/0.5°, daily	Liang et al. (1994)	2375	859
ARNO	Rainfall-runoff	Watershed/ sub-basin, daily	Todini (1996)	595	323
HBV	Rainfall-runoff	Watershed, variable,	Lindstrom et al. (1997)	847	515
SWAT	Rainfall-runoff	Watershed, variable	Arnold et al. (1998)	5343	3156
INCA	Rainfall-runoff/water quality	Watershed, variable	Whitehead et al. (1998)	376	257
Lisflood	Rainfall-runoff	Global/0.1°, daily	De Roo et al. (2000)	279	130
WaterGAP	Rainfall-runoff	Global/0.5°, daily	Döll et al. (2003)	827	517
HTESSEL	Land surface model	Global/variable, 3-hourly	Balsamo et al. (2009)	387	273
JULES	Land surface model	Global/0.5°, 3-hourly	Best et al. (2011)	462	285
PCR-GLOBWB	Rainfall-runoff	Global/0.5°, daily	van Beek et al. (2011)	214	152

[a] based on the number of citations for the document listed in the reference section as of 28 April 2018. Note that many local and regional models exist and represent human activities, but are not included here due to the focus on global hydrology models.

(Vörösmarty et al. 2000). Over time, human water use has been explicitly added to global hydrology models through GHWMs, many of which are also listed in Table 7.1.

The major novelty of GHWMs is their representation of anthropogenic water use. As GHWMs have evolved, the water use component of the model has also increased in sophistication and complexity. Initial GHWMs parameterized water use (industrial and domestic) with national population and GDP data. Starting in the early 2000s, maps of global irrigated areas (Döll and Siebert 2000) enabled irrigation water requirements to be estimated. Since 2000, new developments in these models have included reservoir operations (Hanasaki et al. 2006), urban water infrastructure and demands (McDonald et al. 2014), and determining intra-annual variability in water supply and demand (Hanasaki et al. 2008a, b). These advances have been used to better understand flood hazards and risks (Hirabayashi et al. 2013; Ward et al. 2013), groundwater depletion (Gleeson et al. 2010, and the contribution of human modifications of water stocks to global sea-level change (Pokhrel et al. 2012; Wada et al. 2012).

Challenges remain in our understanding (and modelling) of interactions between human water use and the hydrological cycle. Most GHWMs represent human actions as responses to physical variables. However, human actors are much more complex and typically respond to economic and policy settings, in addition to physical constraints (Konar et al. 2013, 2016). As such, an important area for future research is to better understand human decision-making related to water use and then incorporate these responses into GHWMs. For this reason, multidisciplinary collaborations between hydrologists, water resources engineers, economists, sociologists and anthropologists are likely to be particularly promising (Sivapalan et al. 2012, 2014).

7.2   Monitoring Hydrological Systems

7.2.1   Monitoring the Global Water Cycle Across Scales

The earliest known examples of hydrological monitoring come from Egypt around 3000 BCE: water levels were measured using marks on cliffs along the side of the river, such as at the second cataract at Semna – now submerged by Lake Nasser (Biswas 1970). Horizontal markings on three stone statues from around 200 BCE were used to measure the canal levels in the Dujiang Weir Project, a large agriculture project undertaken in Sichuan Province in China during the Warring States Period (Guowei 2001). As technology advanced in Egypt, water was diverted from the main river channel prior to measurement. An elaborate example of this practice was the construction of gauging stations – known as Nilometers – such as found at Rhoda Island in Cairo, which provides a water level record from 622 CE into the early twentieth century. Flows from the Nile were diverted into a deep shaft with a surrounding staircase, in the centre of which stood a measuring column. Priests would monitor the water level against the column to determine flood and drought conditions critical to agricultural planning and taxation (Biswas 1970; Eliasson 2013).

Like these early examples of staff gauges, manual rain gauges have existed in more or less their current form for many centuries. Sanskrit texts from the fourth century BCE include the design of rain gauges and what weights of rain were to be expected in different parts of what is now India, and rather like the river level gauging on the Nile, these levels determined taxation (NIH 1990). In the way that the Nilometers diverted flows so that water level would not be measured in turbulent conditions, rainfall measurement techniques have been refined to reduce errors: conical rain gauges are recorded in China from 1247 CE; bronze cylinders set on stone pillars were used to measure rainfall in Korea in 1441 (Biswas 1970).

Modern techniques are now available to measure a suite of hydrological fluxes, including precipitation, evaporation and flow within soils and vegetation. Precipitation, in the form of rain or snow, can be measured using field instruments of many different designs. Innovation in the design of rain gauges over the past century has typically concentrated on the development of cumulative gauges with logging and telemetry, and on the reduction of undercatch bias (5–25%) caused by interference from wind and topography (Beven 2012). Ground-based radar systems promise better precipitation estimates, albeit with substantial uncertainties. Evaporative fluxes have traditionally been measured using evaporation pans, from which daily evaporation is calculated by measuring the change in water volume. The use of eddy-covariance instruments to calculate the surface exchange of fluxes of trace gases including water vapour has grown in its application in the past 20 years. Such measurements contain their own uncertainties (see Baldocchi et al. 2001) but provide new datasets for coupled land–atmosphere exchange studies. Nonetheless, most hydrological models still rely on potential evaporation estimated from meteorological drivers (radiation, temperature, wind and humidity), which are more straightforward to measure. Promising techniques are available for measuring the quantity of water stored in soils and groundwater, although the fact that these processes operate over at least seven orders of magnitude (cf. Figure 7.2) introduces considerable uncertainty.

Gauging of river flows remains the mainstay of hydrometry. The techniques employed are tailored to fit the stream in question, but most river gauges measure the time-varying flow depth and relate that to discharge through the use of rating relationships that must be calibrated from time to time. Innovation has been seen in the sensors used to measure depth (whether they be manual records from stage boards or automatic stage recorders attached to pressure transducers) and in the equipment available to survey cross-sections (including new acoustic Doppler current profilers; ADCPs), but the basic principles remain the same. Where channel geometry acts as a constraint on measurement accuracy, engineered structures with a regularized geometry such as weirs can be introduced to the channel (see Strangeways 2003 for a comprehensive account).

7.2.2 Decline of *In Situ* Monitoring

Whilst sensor technology has developed, albeit slowly, the deployment and maintenance of meteorological and hydrometric infrastructure has in many countries suffered a sharp decline. The number of river gauging stations whose data are held by the Global Run-off Data Centre (GRDC) is currently in decline, having dropped to below 3000 stations in 2013 from a peak of 7909 in 1979 (GRDC 2018). Several factors responsible for this decline include funding and logistical constraints as well as war and political changes in the countries collecting data. A similar contemporaneous decline in *in situ* meteorological monitoring has been observed, particularly in Africa and South America.

Whilst there is some hope that new Earth observation (EO) techniques might be able to supplement the gauged network, and it is certainly the case that EO has greatly accelerated progress in some areas of meteorology (Slingo and Palmer 2011), the application of EO techniques to hydrology has so far only been able to address the largest-scale questions (Famiglietti et al. 2015).

Current and future policy and management responses to the decline of monitoring networks warrant careful attention both to new and existing technologies, but also to funding and governance models to promote the availability of widespread, real-time environmental informatics. Fekete et al. (2015) call for an '*in situ* renaissance' whilst

Rogers and Tsirkunov (2013) devote an entire book to the importance of developing national meteorological and hydrological services, specifically in reference to resilience to climate change. Recognizing the economic and human cost associated with severe and unforeseen weather events, the World Bank launched the Africa Hydromet Program in 2015, the first phase of which will invest around USD 600 million over eight years into the modernization of hydrometeorological services in 15 African countries, strengthening early warning systems to enhance resilience to extreme weather events.

7.2.3 The Role of EO

The first EO satellite was launched in 1959 in the form of the United States Navy's Vanguard 2, which made 244 cloud cover observations over a few weeks before its batteries expired. It remains in orbit to this day. Later that year the Explorer 6 took the world's first satellite photograph, a grainy image of clouds over the Pacific Ocean. These were followed by a number of further launches though the 1960s. Lagging the Americans by a few years, the Soviet Union launched a number of Kosmos satellites in the 1960s that had weather and remote-sensing missions. However, it was the launch of Landsat 1 in the 1970s that heralded the start of EO as a discipline that could support other fields of research.

Whilst the continuing operation of Landsat has been in the balance a number of times due to funding questions, the programme has been in continuous operation since 1972 with Landsat 7 and Landsat 8 currently in orbit. Landsat data have been invaluable for long-term hydrological study, from wetland dynamics in the USA (Jin et al. 2017) to channel changes in Bangladesh (Dewan et al. 2017).

Hydrological remote sensing has been advanced further through the use of satellite-borne radar instruments, including the joint NASA-JAXA Tropical Rainfall Measuring Mission (TRMM) and more recent Global Precipitation Measurement (GPM) instrument. These sensors provide information on rainfall patterns across large areas that give valuable data in ungauged locations, notwithstanding significant uncertainties associated with the interpretation of data from an emerging technology. Space-borne microwave instruments are also in use to measure soil moisture, including European Space Agency's recent and ongoing SMOS and Sentinel missions, the principal limitations being the sensitivity only to surface soil moisture conditions and the mismatch between the horizontal scales of integration compared with *in situ* measurements available for ground truth.

From 2002 to 2017 the two Gravity Recovery and Climate Experiment (GRACE) spacecraft, GRACE-1 and GRACE-2 (a.k.a. Tom and Jerry) orbited the Earth, measuring changes in their relative velocities and separation caused by the Earth's gravitational field. GRACE has been used for both hydrological and hydrogeological monitoring (Owor et al. 2009; Shamsudduha et al. 2012; Bernknopf et al. 2018; Rodell et al. 2018), and continues with the launch into orbit of two GRACE Follow-On satellites (GRACE FO; https://gracefo.jpl.nasa.gov/mission/overview). Whilst the GRACE outputs offer useful large-scale assessments of water storage, it is essential that they be interpreted alongside locally-informed water management information, such as in cases where a small reduction in water volume in an aquifer may result in a large drop in water level (Alley and Konikow 2015).

Finally, the Surface Water Ocean Topography (SWOT) mission, scheduled for launch in 2021, is an EO mission planned specifically to use satellite altimetry to estimate river (and ocean) water elevations, measure changes in the levels of freshwater bodies, and to estimate the river discharge for the largest rivers (> 50–100 m width).

7.2.4 Land-based and Airborne Techniques

Whilst space-based EO has revolutionized hydrological monitoring, and provides scientists and planners with vast quantities of data at the click of a mouse, other remote-sensing techniques bridge the gap between direct measurements – such as river and rain gauges – and those generated from many kilometres overhead by satellites. At basin level, for example, accurate measurement of flow and discharge rates for individual river reaches is vital, for which ADCPs are widely used. Using the Doppler shift induced in the acoustic signal by the moving water column determines river velocity, and by combining this with measurement of the channel cross-section by tracking the bottom as it moves across a stream, discharge is calculated (Simpson and Oltmann 1993); it can also be used to measure sediment transport (Kostaschuk et al. 2005).

In hydrogeology, geophysical techniques such as vertical electrical sounding (VES) and electrical resistivity tomography (ERT) are used for measuring groundwater resources, siting boreholes and determining changes in groundwater level. By measuring the resistivity of the ground beneath them, these techniques help build a picture of the soil and rock layers, whether they are likely to be water-bearing and whether that water is likely to be saline or fresh. Time-domain electromagnetic (TDEM) methods operate using the principles of electromagnetic induction. Mills et al. (1988) described an example of a study using TDEM in a hydrological context, and gave an excellent explanation of the TDEM technique itself. The electromagnetic coupling upon which TDEM is based means that the electromagnetic field can be induced in the ground without the transmitter loop and receiving loop being in physical contact with it. Crucially, this allows TDEM equipment to be mounted beneath an airborne platform – a configuration not possible for VES and ERT.

Another airborne technique with hydrological application is Light Detection and Ranging (LiDAR). This uses laser pulses and returns to generate 3D representations of the landscape under surveillance, often from an aircraft. This can generate DEMs of around 1 m resolution as opposed to the resolution of around 10 m achieved using standard photogrammetric methods (Murphy et al. 2008). This is of particular benefit when monitoring or modelling the hydrology of low-relief scenarios (Jones et al. 2008) when a resolution of 10 m would be inadequate. At larger scales, radar altimetry from the Shuttle Radar Topography Mission (SRTM) has led to widely used digital elevation data products at 30–90 m resolution, which are used in support of many water-related applications (Lehner et al. 2008). More recently the satellite-borne Tandem-X sensor offers globally consistent elevation data with finer horizontal resolution and an overall absolute height accuracy of 3.5 m (Rizzoli et al. 2017).

Unmanned and autonomous vehicles are now being used in hydrological and hydrogeological monitoring, and their use is set to increase. The UK Environment Agency has developed a remote-controlled boat to make ADCP measurements, allowing more data to be more accurately generated with fewer surveyors involved (http://equipit.hrwallingford.com/products/arc-boat/the-arc-boat). UAV-based LiDAR is well-established, with LiDAR equipment light enough to be mounted on consumer-grade drones. As the lift capacity per unit cost of drones/UAVs increases and the weight of geophysics equipment comes down, the use of UAVs for geophysical survey is likely to become commonplace. These could be of particular benefit in hard-to-gauge settings or when flows go out of bank.

Whilst water quality monitoring and modelling is covered in Chapter 4, the dissolved chemicals in water provide another means of monitoring water and its movement, as do field-measurable parameters such as temperature and pH. Chemicals already present in water due to either natural or anthropogenic processes, or those added specifically for the task, can act as tracers, although the use of chemical and isotopic tracers is a large field of study in itself (Clark and Fritz 1997; Kendall and McDonnell 1998). Stable isotope ratios are used to track groundwater flow paths, the mixing of surface waters and the orographic controls on precipitation, and tracers can be used to estimate recharge rates and the residence time of groundwater (Gaye and Edmunds 1996).

7.2.5   Non-traditional Hydrological Monitoring Systems

As mobile communications become ubiquitous and the cost of data processing and data transmission reduces, the opportunity to automate and add telemetry to traditional monitoring methods is increasing. This has created two distinct opportunities for non-traditional methods of hydrological monitoring, and environmental monitoring more generally. These can be classed broadly as 'crowd-sourcing' and 'accidental infrastructure', although there is overlap between the two. The Trans African Hydro Meteorological Observatory (TAHMO) is a network of hydro-meteorological measurement stations being developed and deployed in sub-Saharan Africa (van de Giesen et al. 2014). TAHMO also links to schools to simultaneously provide secure and managed sites for their infrastructure and contribute to environmental education. The idea of 'accidental infrastructure' is that there are other forms of infrastructure that, at minimal extra cost, can provide additional environmental monitoring information. This idea is predicated on the fact than the costs of setting up, then managing and maintaining a system – especially ones in remote areas – are much higher than the costs of additional data acquisition or transmission. One example is the use of mobile phone infrastructure to monitor rainfall. Noting that water attenuates radio signals, Messer et al. (2006) measured signal attenuation between cellular communication masts during a rain event in Israel and demonstrated that attenuation from mobile phone masts could be used to monitor rainfall. Larger-scale studies were subsequently conducted in the Netherlands and Burkina Faso (Overeem et al. 2013; Doumounia et al. 2014). With commercial operators expanding mobile phone networks into areas where rain gauges are sparse, this has the potential to make a significant contribution. In a similar vein, taking advantage of advances in the remote monitoring of handpumps (Thomson et al. 2012) and their ubiquity in rural areas, handpumps have the potential to provide additional data about groundwater levels and rainfall (Colchester et al. 2017; Thomson et al. 2019).

7.3   Future Challenges

The traditional approach to monitoring is to rely on a single source, picking the most accurate and reliable sensor available, and curating this data-stream to ensure its availability and integrity. As more and more data become available, and can be shared more cheaply and rapidly, data fusion has become the norm. Combining different data-streams can generate clear benefits: Lam et al. (2017) provide an example of combining two high-quality datasets to generate better rating curves. What may be truly trans-formative about advances in the speed and cost of transmitting and processing data is the fusion of data that would not traditionally be combined, for example, combining a high-integrity, well-curated data source with one that is intermittent and error-prone, and doing this in near real-time. Pavelsky et al. (2014) described the complementary nature of SWOT data and gauge data. Data fusion techniques can dynamically exploit this complementarity to generate outputs of greater value than the sum of their parts. Khaleghi et al. (2013) outlined common problems encountered when fusing data from multiple sources and the methods used for tackling them. Referring to environmental data, Osborne et al. (2012) addressed the challenge of dealing with intermittent and delayed data – one which is critical when designing real-time systems that will rely on unreliable or periodic data sources, whilst Dutta and Morshed (2013) stated that 'differ-ent sources of environmental sensor or model based data could be linked to comple-ment and cross validate each other automatically to increase the reliability of the sensor network'.

Furthermore, data assimilation, the process of optimally combining an imperfect model and uncertain observations to correct model states, has been widely applied for reducing hydrological uncertainty (Liu et al. 2012). The model correction can be done using iterative maximum likelihood estimation, in which sequential updating of the state takes place, or by using variational approaches (e.g. Seo et al. 2009) in which a cost-minimization problem is solved over an assimilation window. The former category includes the ensemble Kalman filter (Reichle et al. 2008) and particle filter (DeChant and Moradkhani 2012), both of which interactively correct model states based on the likelihood estimation.

Despite the development of advanced techniques for quantifying and reducing model uncertainty, there are still significant challenges. First, observed variables used for model validation and calibration are themselves not free of uncertainties. This is due to errors in station sites, data collection, human intervention, and other sources that should always be taken into account (see Wilby et al. 2017). Second, there is often a scale mismatch between the model parameters (spatial aggregate) and measurements (point measurement), which makes it unrealistic for the model to match (Beven 2012).

7.4 Conclusion

Many of the pressing questions in water policy and management rely on estimates of future water resource availability and on predictions of hydrological extremes such as floods and droughts. These strands of information are derived from and underpinned by hydrological models that themselves rely, directly or indirectly, on measurements of stores and fluxes in the water cycle. Each of these activities is exposed to challenges that arise from the wide range of scales over which we observe variability in fluxes and het-erogeneity in hydrological states. Whilst the proliferation of hydrological models in the late twentieth and early twenty-first century is remarkable, we suggest that the next phase of development will be concerned with the combination of new process-based theoretical understanding combined with better area-wide constraints on hydrological fluxes and their variability. These developments are necessary to support model-based assessments of future change and to underpin the case for investment in water resource monitoring in the future. Taken together, the combination of enhanced global modelling and monitoring promise to transform our ability to assess water resource availability under a changing climate and to predict natural hazards.

References

Abbott, M.B., Bathurst, J.C., Cunge, J.A. et al. (1986a). An introduction to the European hydrological system — *Systeme Hydrologique Europeen,* "SHE", 1: History and philosophy of a physically-based distributed modelling system. *Journal of Hydrology* 87 (1–2): 45–59.

Abbott, M.B., Bathurst, J.C., Cunge, J.A. et al. (1986b). An introduction to the European hydrological system — *Systeme Hydrologique Europeen,* "SHE", 2: Structure of a physically-based distributed modelling system. *Journal of Hydrology* 87 (1–2): 61–77.

Alfieri, L., Burek, P., Dutra, E. et al. (2013). GloFAS – global ensemble streamflow forecasting and flood early warning. *Hydrology and Earth System Sciences* 17: 1161–1175. http://dx.doi.org/10.5194/hess-17-1161-2013.

Alfieri, L., Cohen, S., Galantowicz, J. et al. (2018). A global network for operational flood risk reduction. *Environmental Science and Policy* 84: 149–158. https://doi.org/10.1016/j.envsci.2018.03.014.

Alley, W.M. and Konikow, L.F. (2015). Bringing GRACE down to earth. *Groundwater* 53 (6): 826–829.

Arnold, J.G., Srinivasan, R., Muttiah, R.S., and Williams, J.R. (1998). Large area hydrologic modeling and assessment. Part I: model development. *Journal of the American Water Resources Association* 34 (1): 73–89.

Baldocchi, D., Falge, E., Lianhong, G. et al. (2001). FLUXNET: a new tool to study the temporal and spatial variability of ecosystem-scale carbon dioxide, water vapor, and energy flux densities. *Bulletin of the American Meteorological Society* 82 (11): 2415–2434.

Balsamo, G., Viterbo, P., Beljaars, A. et al. (2009). A revised hydrology for the ECMWF model, verification from field site to terrestrial water storage and impact in the integrated forecast system. *Journal of Hydrometeorology* 10: 623–643. https://doi.org/10.1175/2008JHM1068.1.

Barnett, T.P. and Pierce, D.W. (2009). Sustainable water deliveries form the Colorado River in a changing climate. *Proceedings of the National Academy of Sciences* 1106 (18): 7334–7338.

Beck, H.E., van Dijk, A.I.J.M., Levizzani, V. et al. (2017). MSWEP: 3-hourly 0.25° global gridded precipitation (1979–2015) by merging gauge, satellite, and reanalysis data. *Hydrology and Earth System Sciences* 21: 589–615. https://doi.org/10.5194/hess-21-589-2017.

Bernknopf, R., Brookshire, D.S., Kuwayama, Y. et al. (2018). The value of remotely sensed information: the case of a GRACE-enhanced drought severity index. *Weather, Climate, and Society* 10 (1): 187–203.

Best, M.J., Pryor, M., Clark, D.B. et al. (2011). The joint UK land environment simulator (JULES), model description – Part 1: Energy and water fluxes. *Geoscientific Model Development* 4 (3): 677–699.

Beven, K. (2012). *Rainfall-Runoff Modelling: The Primer*, 2e. Chichester: Wiley-Blackwell.

Beven, K. and Binley, A. (1992). The future of distributed models: model calibration and uncertainty prediction. *Hydrological Processes* 6 (3): 279–298.

Beven, K.J. and Cloke, H.L. (2012). Comment on "Hyperresolution global land surface modeling: Meeting a grand challenge for monitoring Earth's terrestrial water" by Eric F. Wood et al. *Water Resources Research* 48 (1): W01801. https://doi.org/10.1029/2011WR010982.

Beven, K. and Freer, J. (2001). Equifinality, data assimilation, and uncertainty estimation in mechanistic modelling of complex environmental systems using the GLUE methodology. *Journal of Hydrology* 249 (1–4): 11–29.

Beven, K.J. and Kirby, M.J. (1979). A physically based variable contributing area model of basin hydrology. *Hydrological Science Journal* 24: 43–69.

Bierkens, M.F.P. (2015). Global hydrology 2015: state, trends, and directions. *Water Resources Research* 51: 4923–4947. https://doi.org/10.1002/2015WR017173.

Bierkens, M.F.P., Bell, V.A., Burek, P. et al. (2015). Hyper-resolution global hydrological modelling: what is next? Everywhere and locally relevant. *Hydrological Processes* 29 (2): 310–320. https://doi.org/10.1002/hyp.10391.

Biswas, A.K. (1970). *History of Hydrology*. Amsterdam: North-Holland Publishing Company.

Blöschl, G. and Sivapalan, M. (1995). Scale issues in hydrological modelling: a review. *Hydrological Processes* 9: 251–290.

Brown, J.D., He, M., Regonda, S. et al. (2014). Verification of temperature, precipitation, and streamflow forecasts from the NOAA/NWS Hydrologic Ensemble Forecast Service (HEFS): 2. Streamflow verification. *Journal of Hydrology* 519 (D): 2847–2868. https://doi.org/10.1016/j.jhydrol.2014.05.030.

Burnash, R.J.C., Ferral, R.L., and McGuire, R.A. (1973). A Generalized Streamflow Simulation System – Conceptual Modeling for Digital Computers. Silver Springs: US Department of Commerce, National Weather Service; Sacramento: State of California Department of Water Resources.

Clark, M.P., Bierkens, M.F.P., Samaniego, L. et al. (2017). The evolution of process-based hydrologic models: historical challenges and the collective quest for physical realism. *Hydrology and Earth System Sciences* 21: 3427–3440. https://doi.org/10.5194/hess-21-3427-2017.

Clark, I.D. and Fritz, P. (1997). *Environmental Isotopes in Hydrogeology*. Boca Raton, FL: CRC Press/Lewis Publishers.

Clark, M.P., Kavetski, D., and Fenicia, F. (2011). Pursuing the method of multiple working hypotheses for hydrological modelling. *Water Resources Research* 47: W09301. https://doi.org/10.1029/2010WR009827.

Colchester, F.E., Marais, H.G., Thomson, P. et al. (2017). Accidental infrastructure for groundwater monitoring in Africa. *Environmental Modelling & Software* 91: 241–250.

Crétaux, J.-F., Jelinkski, W., Calmant, S. et al. (2011). SOLS: a lake database to monitor in near real time water level and storage variations from remote sensing data. *Advances in Space Research* 47 (9): 1497–1507. https://doi.org/10.1016/j.asr.2011.01.004.

Dawson, C.W. and Wilby, R.L. (2001). Hydrological modelling using artificial neural networks. *Progress in Physical Geography* 25 (1): 80–108.

De Roo, A.P.J., Wesseling, C.G., and Van Deursen, W.P.A. (2000). Physically based river basin modelling within a GIS: the LISFLOOD model. *Hydrological Processes* 14 (11–12): 1981–1992.

DeChant, C.M. and Moradkhani, H. (2012). Examining the effectiveness and robustness of sequential data assimilation methods for quantification of uncertainty in hydrologic forecasting. *Water Resources Research* 48: W04518. https://doi.org/10.1029/2011WR011011.

Dewan, A., Corner, R., Saleem, A. et al. (2017). Assessing channel changes of the Ganges-Padma River system in Bangladesh using Landsat and hydrological data. *Geomorphology* 276: 257–279.

Döll, P., Kaspar, F., and Lehner, B. (2003). A global hydrological model for deriving water availability indicators: model tuning and validation. *Journal of Hydrology* 270 (1–2): 105–134.

Döll, P. and Siebert, S. (2000). A digital global map of irrigated areas. *ICID Journal* 49 (2): 55–66.

Doumounia, A., Gosset, M., Cazanave, F. et al. (2014). Rainfall monitoring based on microwave links from cellular telecommunication networks: first results from a West African test bed. *Geophysical Research Letters* 41 (16): 6016–6022.

Dutta, R. and Morshed, A. (2013). Performance evaluation of South Esk hydrological sensor web: unsupervised machine learning and semantic linked data approach. *IEEE Sensors Journal* 13 (10): 3806–3815.

Eliasson, J. (2013). Hydrological science and its connection to religion in ancient Egypt under the pharaohs. *Advances in Historical Studies* 2 (3): 150–155.

Emerton, R.E., Stephens, E.M., Pappenberger, F. et al. (2016). Continental and global scale flood forecasting systems. *WIREs Water* 3 (3): 391–418. https://doi.org/10.1002/wat2.1137.

Famiglietti, J.S., Cazenave, A., Eicker, A. et al. (2015). Satellites provide the big picture. *Science* 349 (6249): 684–685.

Fekete, B.M., Robarts, R.D., Kumagai, M. et al. (2015). Time for in situ renaissance. *Science* 349 (6249): 685–686.

Franchini, M. (1996). Use of a genetic algorithm combined with a local search method for automatic calibration of conceptual rainfall runoff models. *Hydrological Sciences Journal* 41 (1): 21–39.

Freeze, R.A. and Harlan, R. (1969). Blueprint for a physically-based, digitally-simulated hydrologic response model. *Journal of Hydrology* 9 (3): 237–258. https://doi.org/10.1016/0022-1694(69)90020-1.

Gao, H., Birkett, C., and Lettenmeir, D.P. (2012). Global monitoring of large reservoir storage from satellite remote sensing. *Water Resources Research* 48: W09504.

Gaye, C.B. and Edmunds, W.M. (1996). Groundwater recharge estimation using chloride, stable isotopes and tritium profiles in the sands of northwestern Senegal. *Environmental Geology* 27: 246–251.

Gleeson, T., VanderSteen, J., Sophocleous, M.A. et al. (2010). Groundwater sustainability strategies. *Nature Geoscience* 3: 378–379. https://doi.org/10.1038/ngeo881.

Gochis, D.J., Barlage, M., Dugger, A. et al. (2018). *The WRF-Hydro modeling system technical description (Version 5.0)*. NCAR Technical Note. https://ral.ucar.edu/sites/default/files/public/WRFHydroV5TechnicalDescription.pdf.

GRDC (2018). Global Runoff Data Base – Status, Development, Use. https://www.bafg.de/GRDC/EN/01_GRDC/13_dtbse/database_node.html;jsessionid=DE9228235851CD15D9265EE2DD1CF39B.live21303 (accessed 23 April 2018).

Guowei, L. (2001). Hydrology in ancient time in China. *International Symposium on the Origins and History of Hydrology*, Dijon, 9–11 May, Université de Bourgogne.

Haddeland, I., Heinke, J., Biemans, H. et al. (2014). Global water resources affected by human interventions and climate change. *Proceedings of the National Academy of Science* 111 (9): 3251–3256.

Hanasaki, N., Kanae, S., and Oki, T. (2006). A reservoir operation scheme for global river routing models. *Journal of Hydrology* 327: 22–41.

Hanasaki, N., Kanae, S., Oki, T. et al. (2008a). An integrated model for the assessment of global water resources – Part 1: Model description and input meteorological forcing. *Hydrology and Earth Systems Sciences* 12: 1007–1025.

Hanasaki, N., Kanae, S., Oki, T. et al. (2008b). An integrated model for the assessment of global water resources – Part 2: Applications and assessments. *Hydrology and Earth Systems Sciences* 12: 1027–1037.

Hao, Z., AghaKouchak, A., Nakhjiri, N., and Farahmand, A. (2014). Global integrated drought monitoring and prediction system. *Scientific Data* 1: 140001. http://dx.doi.org/10.1038/sdata.2014.1.

Hirabayashi, Y., Mahendran, R., Koirala, S. et al. (2013). Global flood risk under climate change. *Nature Climate Change* 3: 816–821. http://dx.doi.org/10.1038/nclimate1911.

Hrachowitz, M. and Clark, M.P. (2017). HESS opinions: the complementary merits of competing modelling philosophies in hydrology. *Hydrology and Earth System Sciences* 21: 3953–3973. https://doi.org/10.5194/hess-21-3953-2017.

Jin, H., Huang, C., Lang, M.W. et al. (2017). Monitoring of wetland inundation dynamics in the Delmarva Peninsula using Landsat time-series imagery from 1985 to 2011. *Remote Sensing of Environment* 190: 26–41.

Jones, K.L., Poole, G.C., O'Daniel, S.J. et al. (2008). Surface hydrology of low-relief landscapes: assessing surface water flow impedance using LIDAR-derived digital elevation models. *Remote Sensing of Environment* 112 (11): 4148–4158.

Kendall, C. and McDonnell, J.J. (1998). *Isotope Tracers in Catchment Hydrology*. Amsterdam: Elsevier.

Khaleghi, B., Khamis, A., Karray, F., and Razavi, S. (2013). Multisensor data fusion: a review of the state-of-the-art. *Information Fusion* 14 (1): 28–44.

Kite, G. (2001). Modelling the Mekong: hydrological simulation for environmental impact studies. *Journal of Hydrology* 253: 1–13.

Konar, M., Evans, T.P., Levy, M. et al. (2016). Water resources sustainability in a globalizing world: who uses the water? *Hydrological Processes* 30 (18): 3330–3336. https://doi.org/10.1002/hyp.10843.

Konar, M., Hussein, Z., Hanasaki, N. et al. (2013). Virtual water trade flows and savings under climate change. *Hydrology and Earth System Sciences* 17: 3219–3234. https://doi.org/10.5194/hess-17-3219-2013.

Kostaschuk, R., Best, J., Villard, P. et al. (2005). Measuring flow velocity and sediment transport with an acoustic Doppler current profiler. *Geomorphology* 68 (1–2): 25–37.

Koster, R.D., Dirmeyer, P.A., Guo, Z. et al. (2004). Regions of strong coupling between soil moisture and precipitation. *Science* 305 (5687): 1138–1140. https://doi.org/10.1126/science.1100217.

Lam, N., Kean, J.W., and Lyon, S.W. (2017). Modeling streamflow from coupled airborne laser scanning and acoustic Doppler current profiler data. *Hydrology Research* 48 (4): 981–996.

Lehner, B., Verdin, K., and Jarvis, A. (2008). New global hydrography derived from spaceborne elevation data. *EOS, Transactions AGU* 89 (10): 93–94.

Liang, X., Lettenmaier, D.P., Wood, E.F., and Burges, S.J. (1994). A simple hydrologically-based model of land surface water and energy fluxes for GSMs. *Journal of Geophysical Research* 99 (D7): 14415–14428.

Lindström, G., Johansson, B., Persson, M. et al. (1997). Development and test of the distributed HBV-96 hydrological model. *Journal of Hydrology* 201: 272–288.

Lindström, G., Pers, C., Rosberg, J. et al. (2010). Development and testing of the HYPE (Hydrological Predictions for the Environment) water quality model for different spatial scales. *Hydrology Research* 41 (3–4): 295–319. https://doi.org/10.2166/nh.2010.007.

Liu, Y., Weerts, A., Clark, M. et al. (2012). Advancing data assimilation in operational hydrologic forecasting: progresses, challenges, and emerging opportunities. *Hydrology and Earth System Sciences* 16 (10): 3863–3887. https://doi.org/10.5194/hess-16-3863-2012.

Marston, L., Ao, Y., Konar, M. et al. (2018). High-resolution water footprints of production of the United States. *Water Resources Research* 54 (3): 2288–2316. https://doi.org/10.1002/2017WR021923.

McDonald, R.I., Weber, K., Padowski, J. et al. (2014). Water on an urban planet: urbanization and the reach of urban water infrastructure. *Global Environmental Change* 27: 96–105. https://doi.org/10.1016/j.gloenvcha.2014.04.022.

Messer, H., Zinevich, A., and Alpert, P. (2006). Environmental monitoring by wireless communication networks. *Science* 312 (5774): 713.

Mills, T., Hoekstra, P., Blohm, M., and Evans, L. (1988). Time domain electromagnetic soundings for mapping sea-water intrusion in Monterey County, California. *Groundwater* 26 (6): 771–782.

Miralles, D.G., Holmes, T.R.H., De Jeu, R.A.M. et al. (2011). Global land-surface evaporation estimated from satellite-based observations. *Hydrology and Earth System Science* 15: 453–469.

Montanari, A., Rosso, R., and Taqqu, M.S. (1997). Fractionally differenced ARIMA models applied to hydrologic time series. *Water Resources Research* 33 (5): 1035–1044.

Murphy, P.N.C., Ogilvie, J., Meng, F.-R., and Arp, P. (2008). Stream network modelling using lidar and photogrammetric digital elevation models: a comparison and field verification. *Hydrological Processes* 22 (12): 1747–1754.

Ng, C.N. and Young, P.C. (1990). Recursive estimation and forecasting of non-stationary time series. *Journal of Forecasting* 9 (2): 173–204. https://doi.org/10.1002/for.3980090208.

NIH (1990). *Hydrology in Ancient India*. Roorkee: National Institute of Hydrology.

Oki, T. and Kanae, S. (2006). Global hydrological cycles and world water resources. *Science* 313 (5790): 1068–1072. https://doi.org/10.1126/science.1128845.

Osborne, M.A., Roberts, S.J., Rogers, A., and Jennings, N.R. (2012). Real-time information processing of environmental sensor network data using bayesian gaussian processes. *ACM Transactions on Sensor Networks* 9 (1): 1–32.

Overeem, A., Leijnse, H., and Uijlenhoet, R. (2013). Country-wide rainfall maps from cellular communication networks. *Proceedings of the National Academy of Sciences* 110 (8): 2741–2745.

Owor, M., Taylor, R., Tindimugaya, C., and Mwesigwa, D. (2009). Rainfall intensity and groundwater recharge: empirical evidence from the Upper Nile Basin. *Environmental Research Letters* 4 (3): 35009.

Pavelsky, T.M., Durand, M.T., Andreadis, K.M. et al. (2014). Assessing the potential global extent of SWOT river discharge observations. *Journal of Hydrology* 519: 1516–1525.

Pechlivanidis, I.G., Jackson, B.M., Mcintyre, N.R., and Wheater, H.S. (2011). Catchment scale hydrological modelling: a review of model types, calibration approaches and uncertainty analysis methods in the context of recent developments in technology and applications. *Global Nest Journal* 13 (3): 193–214.

Pokhrel, Y.N., Hanasaki, N., Yeh, P.J.-F. et al. (2012). Model estimates of sea-level change due to anthropogenic impacts on terrestrial water storage. *Nature Geoscience* 5: 389–392.

Refsgaard, J.C., Storm, B., and Clausen, T. (2010). Système Hydrologique Europeén (SHE): review and perspectives after 30 years development in distributed physically-based hydrological modelling. *Hydrology Research* 41 (5): 355–377.

Reichle, R.H., Crow, W.T., and Keppenne, C.L. (2008). An adaptive ensemble Kalman filter for soil moisture data assimilation. *Water Resources Research* 44: W03423.

Rizzoli, P., Martone, M., Gonzalez, C. et al. (2017). Generation and performance assessment of the global TanDEM-X Digital Elevation Model. *ISPRS Journal of Photogrammetry and Remote Sensing* 132: 119–139.

Rodell, M., Famiglietti, J.S., Wiese, D.N. et al. (2018). Emerging trends in global freshwater availability. *Nature* 557 (7707): 651–659.

Rogers, D.P. and Tsirkunov, V.V. (2013). *Weather and Climate Resilience: Effective Preparedness Through National Meteorological and Hydrological Services*. Washington, DC: World Bank Publications.

Roobavannan, M., van Emmerik, T.H.M., Elshafei, Y. et al. (2018). Norms and values in socio-hydrological models. *Hydrology and Earth System Sciences* 22: 1337–1349.

Samaniego, L., Kumar, R., and Attinger, S. (2010). Multiscale parameter regionalization of a grid-based hydrologic model at the mesoscale. *Water Resources Research* 46: W05523. https://doi.org/10.1029/2008WR007327.

Schewe, J., Heinke, J., Gerten, D. et al. (2014). Multimodel assessment of water scarcity under climate change. *Proceedings of the National Academy of Science* 111 (9): 3245–3250. https://doi.org/10.1073/pnas.1222460110.

Seneviratne, S.I., Corti, T., Davin, E.L. et al. (2010). Investigating soil moisture–climate interactions in a changing climate: a review. *Earth-Science Reviews* 99 (3–4): 125–161.

Seo, D.J., Cajina, L., Corby, R., and Howieson, T. (2009). Automatic state updating for operational streamflow forecasting via variational data assimilation. *Journal of Hydrology* 367: 255–275.

Shamsudduha, M., Taylor, R.G., and Longuevergne, L. (2012). Monitoring groundwater storage changes in the highly seasonal humid tropics: validation of GRACE measurements in the Bengal Basin. *Water Resources Research* 48 (2) https://doi.org/10.1029/2011WR010993.

Sheffield, J., Wood, E.F., Chaney, N. et al. (2014). A drought monitoring and forecasting system for sub-Sahara African water resources and food security. *Bulletin of the American Meteorological Society* 95 (6): 861–882. https://doi.org/10.1175/BAMS-D-12-00124.1.

Simpson, M.R. and Oltmann, R.N. (1993). *Discharge-measurement system using an acoustic Doppler current profiler with applications to large rivers and estuaries*. Water Supply Paper 2395. San Francisco: US GPO.

Sivapalan, M., Konar, M., Srinivasan, V. et al. (2014). Socio-hydrology: use-inspired water sustainability science for the Anthropocene. *Earth's Future* 2 (4): 225–230. https://doi.org/10.1002/2013EF000164.

Sivapalan, M., Savenije, H.H.G., and Blöschl, G. (2012). Sociohydrology: a new science of people and water. *Hydrological Processes* 26 (8): 1270–1276. https://doi.org/10.1002/hyp.8426.

Slingo, J. and Palmer, T. (2011). Uncertainty in weather and climate prediction. *Philosophical Transactions of the Royal Society A: Mathematical, Physical and Engineering Sciences* 369: 4751–4767. https://doi.org/10.1098/rsta.2011.0161.

Sorooshian, S. and Gupta, V.K. (1995). Model calibration. In: *Computer Models of Watershed Hydrology* (ed. V.P. Singh), 23–67. Colorado: Water Resources Publications.

Stedinger, J.R., Vogel, R.M., Lee, S.U., and Batchelder, R. (2008). Appraisal of the generalized likelihood uncertainty estimation (GLUE) method. *Water Resources Research* 44: W00B06. https://doi.org/10.1029/2008WR006822.

Strangeways, I. (2003). *Measuring the Natural Environment*, 2e. Cambridge: Cambridge University Press.

Tetzlaff, D., Carey, S.K., McNamara, J.P. et al. (2017). The essential value of long-term experimental data for hydrology and water management. *Water Resources Research* 53: 2598–2604. https://doi.org/10.1002/2017WR020838.

Thielen, J., Bartholmes, J., Ramos, M.-H., and de Roo, A. (2009). The European flood alert system – Part 1: Concept and development. *Hydrology and Earth System Sciences* 13: 125–140.

Thomson, P., Bradley, D., Katilu, A. et al. (2019). Rainfall and groundwater use in rural Kenya. *Science of the Total Environment* 649: 722–730.

Thomson, P., Hope, R., and Foster, T. (2012). GSM-enabled remote monitoring of rural handpumps: a proof-of-concept study. *Journal of Hydroinformatics* 14 (4): 829.

Todini, E. (1996). The ARNO rainfall-runoff model. *Journal of Hydrology* 175 (1–4): 339–382.

van Beek, L.P.H., Wada, Y., and Bierkens, M.F.P. (2011). Global monthly water stress: 1. Water balance and water availability. *Water Resources Research* 47 (7): W07517. https://doi.org/10.1029/2010WR009791.

van de Giesen, N., Hut, R., and Selker, J. (2014). The trans-African hydro-meteorological observatory (TAHMO). *Wiley Interdisciplinary Reviews: Water* 1 (4): 341–348.

Vörösmarty, C.J., Green, P., Salisbury, J., and Lammers, R.B. (2000). Global water resources: vulnerability from climate change and population growth. *Science* 289 (5477): 284–288.

Wada, Y., Bierkens, M.F.P., de Roo, A. et al. (2017). Human–water interface in hydrological modeling: current status and future directions. *Hydrology and Earth System Sciences* 21: 4169–4193. https://doi.org/10.5194/hess-21-4169-2017.

Wada, Y., van Beek, L.P.H., Sperna Weiland, F.C. et al. (2012). Past and future contribution of global groundwater depletion to sea-level rise. *Geophysical Research Letters* 39 (L09402) https://doi.org/10.1029/2012GL051230.

Wada, Y., Wisser, D., and Bierkens, M.F.P. (2014). Global modeling of withdrawal, allocation and consumptive use of surface water and groundwater resources. *Earth System Dynamics* 5 (1): 15.

Wagner, W., Lemoine, G., and Rott, H. (1999). A method for estimating soil moisture from ERS scatterometer and soil data. *Remote Sensing of Environment* 70 (2): 191–207.

Ward, P.J., Jongman, B., Sperna Weiland, F.C. et al. (2013). Assessing flood risk at the global scale: model setup, results and sensitivity. *Environmental Research Letters* 8 (4): 1–10.

Wheeler, K.G., Basheer, M., Mekonnen, Z.T. et al. (2016). Cooperative filling approaches for the Grand Ethiopian Renaissance Dam. *Water International* 41 (4): 611–634. https://doi.org/10.1080/02508060.2016.1177698.

Whitehead, P.G., Wilson, E.J., and Butterfield, D. (1998). A semi-distributed integrated nitrogen model for multiple source in catchments INCA: Part I – Model structure and process equations. *Science of the Total Environment* 210-211: 547–558.

Wilby, R., Clifford, N., De Luca, P. et al. (2017). The "dirty dozen" of freshwater science: detecting then reconciling hydrological data biases and errors. *WIREs Water* 4 (3): e1209. https://doi.org/10.1002/wat2.1209.

Wu, H., Adler, R.F., Tian, Y. et al. (2014). Real-time global flood estimation using satellite-based precipitation and a coupled land surface and routing model. *Water Resources Research* 50: 2693–2717. http://dx.doi.org/10.1002/2013WR014710.

Xu, C.Y., Xiong, L., and Singh, V.P. (2017). Black-box hydrological models. In: *Handbook of Hydrometeorological Ensemble Forecasting* (eds. Q. Duan, F. Pappenberger, J. Thielen, et al.). Berlin, Heidelberg: Springer https://doi.org/10.1007/978-3-642-40457-3_21-1.

Yuan, X. and Wood, E.F. (2013). Multimodel seasonal forecasting of global drought onset. *Geophysical Research Letters* 40 (8): 4900–4905. https://doi.org/10.1002/grl.50949.

Zappa, M., Fundel, F., and Jaun, S. (2013). A "peak-flow box" approach for supporting interpretation and evaluation of operational ensemble flood forecasts. *Hydrological Processes* 27: 117–131. https://doi.org/10.1002/hyp.9521.

Zhao, R.J. (1992). The Xinanjiang model applied in China. *Journal of Hydrology* 135 (1–4): 371–381. https://doi.org/10.1016/0022-1694(92)90096-E.

Part II

Policy

Water policy has been described as a set of rules that guides decision-making about water over time and space. Issues range from sharing water resources and supplying drinking water services to reducing the risks posed by floods and waterborne diseases. Common across these diverse issues is the use of water policy to set societal goals for managing water to achieve sustainable development. Some policies go further than setting goals by prescribing a course of action, or set of solutions, drawing on a mix of regulatory instruments, economic incentives, information and infrastructure. In this context, water policy can be defined as a set of decisions that guide water management.

Water policy provides the decision architecture to shape and inform water management. An understanding of hydrology, climatology, ecology, engineering and public health offers the technical basis for policy-making, but data and capacity gaps often prevent the uptake of evidence in water policy decisions. Policy-making also raises governance issues involving who gets to decide water policy goals and how. Once policies are adopted, implementation moves into the realm of water management. In practice, these distinctions are blurred. There is an increasing drive to define the essential ingredients, or principles, for good water policy, which has led to tensions between the development of universal blueprints and the call for taking context seriously and being adaptive to change.

Existing models of the policy process struggle to capture the messy ways in which water policy is developed, contested and implemented. Water policy is often developed in response to crises which can be immediate in nature or chronic in persistence. Droughts, floods or outbreaks of waterborne disease raise water on the policy agenda, spurring action and investment. Water policy is inherently political, touching on the cultural, economic and ecological values of water. Controversies over who wins and loses fuel the politics of water. Water policy is developed across multiple scales from the local to global; the push for national water policy and global goals reflects the push for policy coherence. Water policy has been guided by successive 'paradigms' or dominant ways of thinking about water challenges and solutions. These paradigms explain the widespread search for transferable policy approaches, including the push for decentralization, integrated water resources management, river basin governance, water markets and the human right to water, among others. In this context, there is increasing need

Water Science, Policy, and Management: A Global Challenge, First Edition. Edited by Simon J. Dadson, Dustin E. Garrick, Edmund C. Penning-Rowsell, Jim W. Hall, Rob Hope, and Jocelyne Hughes.
© 2020 John Wiley & Sons Ltd. Published 2020 by John Wiley & Sons Ltd.

to analyse the design and effectiveness of different approaches to water policy to help decision-makers to identify the approaches suitable for different geographical and political economic conditions.

The following chapters chart water policy trends across a rapidly evolving set of challenges in a variety of geographies. Water resource planning and allocation has been a feature of society since the ancient civilizations of the Nile, Indus and Mesopotamia. Chapter 8 charts this evolution and addresses the emerging issues linked to groundwater development, rapid urbanization and the implications of solar and other energy technologies. Historically, regional water resource planning has largely been decoupled from the often highly localized tasks of delivering water services across urban and rural water systems. Chapters 9 and 10 explore the past and future of water services and water law, examining how water service providers are striving to meet the increasingly ambitious goals of safely-managed drinking water and how this is represented in debates over human rights and property rights to water. Hydroclimatic variability and extremes create increasing challenges for sustainable development of water resources, creating stress tests to address flood risk and manage shortages. Chapters 11, 12 and 13 address these threats by concentrating on the management of floods, wastewater, and production of knowledge in managing extremes. Multiple policy problems converge to create wicked problems that are hard to define with solutions that are costly to reverse.

The section does not claim to be comprehensive in its coverage of policy issues around the world, nor determine a new manifesto for the future. The contribution of this section is more modest, but points to three overarching policy issues for the future. First, path dependency in decision-making has an often disproportionate force and longevity that can block changes based on implementation experience and shifting risks and values. Second, as a counter-balance, policy is best thought of as always provisional. Environmental, economic, social and political processes are dynamic and unpredictable. Third, policy change is nonlinear and political. The assumption that policy is informed by evidence is widely disputed. Policies function across levels between legislative practice, institutional design and operational practices. They are rarely neatly aligned and merit close observation and evaluation. This latter observation points to useful lines of future research to understand, design and evaluate policy processes in the decades ahead.

8

Reallocating Water

Dustin E. Garrick[1], Alice Chautard[1], and Jonathan Rawlins[2]

[1] Smith School of Enterprise and the Environment, University of Oxford, UK
[2] OneWorld Sustainable Investments, Cape Town, South Africa

8.1 Water Crises as Allocation Challenges

Population growth, urbanization and increasing demand for food and energy are intensi-fying pressure on water resources, contributing to imbalances of supply and demand and growing perceptions of a 'global water crisis' (Gleick 2018; World Economic Forum 2018). Observers have lamented that the 'nature of the global freshwater crisis remains poorly defined and characterised' and propose five symptoms of the crisis: groundwater deple-tion, ecological destruction, drought-driven conflicts, unmet subsistence needs, and resource capture by elite (Srinivasan et al. 2012). Almost all of these challenges involve competition for water, which unfolds on many fronts – between people and the environ-ment, between cities and agriculture, and across neighbouring countries, amongst others. In short, water crises often stem from challenges of water allocation.

Water allocation involves rules and incentives that determine who gets water, how much, where and when. Like other natural resources, the allocation of water can occur through different modes of governance including markets, communities and the state – and varying blends of the three (Meinzen-Dick 2007). Regardless of the approach, allocation involves a set of rules and norms determining who can access, use, withdraw, manage and trade water (Schlager and Ostrom 1992). Allocation decisions unfold at multiple levels from the household and farm to the region and river basin. Water alloca-tion also evolves over time in response to changing hydrology, infrastructure, socioeco-nomic and political conditions. Some of the most pressing allocation challenges involve making adjustments as patterns of supply and demand change, ranging from acute issues (e.g. seasonal dryspells) to enduring changes (e.g. urbanization).

We begin with the benchmark for 'good' water allocation. Policy-makers and com-munities are trying to achieve a range of objectives in water allocation, from meeting basic needs and fostering economic development to preserving or restoring cultural and ecological flows. Because water allocation involves reconciling disputes over the value of water in its competing uses, the process is political and laden with values (Hellegers and Leflaive 2015). As a consequence, the process and outcomes of water allocation are

Water Science, Policy, and Management: A Global Challenge, First Edition. Edited by Simon J. Dadson, Dustin E. Garrick, Edmund C. Penning-Rowsell, Jim W. Hall, Rob Hope, and Jocelyne Hughes.
© 2020 John Wiley & Sons Ltd. Published 2020 by John Wiley & Sons Ltd.

judged according to multiple criteria – including efficiency and equity – which cannot always be maximized simultaneously (Dinar et al. 1997). An economically efficient allocation maximizes the net benefits to society generated by patterns of water use, ensuring that the combination of water uses with the highest net economic value are prioritized over lower valued uses. However, economic efficiency does not tackle the key questions regarding who captures the benefits or decides on their distribution. Therefore, the politics and practices of water allocation are invariably shaped by equity concerns regarding the process for distributing the benefits of water use between interests and actors, and over time. Disparities in access and use of water across gender, cultural and socioeconomic differences reflect the entrenched inequalities and power imbalances that lead to social and political exclusion in water allocation institutions (Wilder and Ingram 2018).

Water allocation is a central element of sustainable water resource management and the achievement of the SDGs, including Target 6.4 and the full suite of development objectives tied to water. There is also growing recognition by scientists and practitioners that extractions and development are approaching or already exceeding sustainable diversion limits in many parts of the world (Grafton et al. 2013). Rivers and groundwater systems have been developed and allocated before the full range of benefits of rivers are fully recognized and integrated into planning and allocation decisions. Even when sustainability considerations or environmental flows are integrated into allocation plans and rules, values and water uses evolve as societies and economies change, requiring adjustment, collective action and conflict resolution. Water allocation therefore involves the twin challenges of creating rules and incentives to distribute water whilst dealing with the vested interests of those who stand to lose due to changes from the status quo (Heinmiller 2009). It also requires storage and distribution infrastructure which determine the physical and technical possibilities for moving water between potential users.

Allocation challenges have led to growing experiences and experiments with water planning and water rights reform, blending social norms, policies, laws and incentives, which are informed by the available scientific understanding and water accounting that determine water availability and variability. Implementation and enforcement depend on political will, financing, and the underpinning cultural imperative. Water allocation therefore sits at the intersection of water science, policy and management, although we focus primarily here on the policy, political and institutional aspects.

This chapter examines the different paradigms and pathways for water allocation and reallocation. In the second section, we examine why water reallocation is difficult, elaborating a typology of barriers spanning cultural, technological and political-economic factors. The third section examines the pathways to water reallocation, introducing the concept of a ladder of interventions based on responses proportionate to the pressures and capacity for reform. The final section briefly explores the frontiers of water allocation, identifying the crosscutting challenges and emerging trends in water allocation in the context of rapid urbanization, climate change and sustainable development.

8.2 Navigating Reallocation

Responses to water scarcity involve three, interrelated categories (Molle 2003): supply augmentation, including storage and distribution infrastructure; demand management, including conservation; and allocation and reallocation, which includes pricing, permitting, court and community decisions to address competing demands for available supplies. In practice, these elements are inseparable; for example, infrastructure decisions are tantamount to allocation decisions, determining how and how much water is stored and where and when it can be delivered (Meinzen-Dick and Ringler 2008; Hooper and Lankford 2018).

Economic theory distinguishes an initial allocation from reallocation because the two processes may involve different modes of governance. An 'initial' allocation establishes water rights, whilst subsequent changes are governed by rules that determine when and how water can be redistributed. An initial allocation implies a formal process by the state or community to define rights or licences to access, withdraw, use, manage and trade water, and is often accompanied by an assessment of the renewable supplies under different infrastructure systems. Rarely is there a blank slate. However, informal water claims almost always pre-date efforts by governments and communities to allocate water formally. Therefore we argue that allocation almost always involves reallocation, either from informal users, the environment, or both.

In the rest of this chapter, we focus on water reallocation, namely a change to the volume, timing, location or quality of water delivery, intended to deliver specific goals and or achieve particular outcomes.[1] Responding to water allocation challenges requires institutional mechanisms to reallocate water according to equity and efficiency criteria, and negotiate the associated trade-offs, within the framework of extant political institutions.

There are many types of reallocation, which vary across three main dimensions (Marston and Cai 2016): (i) involuntary (administrative) or voluntary (negotiated, market) processes; (ii) intra-sectoral versus inter-sectoral transfers; and (iii) temporary or permanent agreements. Voluntary reallocation mechanisms include market-based transactions and negotiated settlements; involuntary reallocation mechanisms may involve administrative procedures or court decisions without consultation with affected water users. Involuntary reallocation may also occur implicitly, as land use changes or infrastructure is installed. Both voluntary and involuntary mechanisms may compensate those who reduce their water use. Typically, involuntary, permanent and uncompensated reallocation mechanisms across sectors or basin boundaries trigger the greatest political resistance (Molle and Berkoff 2009; Marston and Cai 2016).

Regardless of the reallocation process, the process is subject to multiple barriers even when there is a broad social agreement about the need to respond to water misallocation, such as when patterns of water use are inefficient, inequitable or unsustainable. Barriers to reallocation arise from water's cultural significance, economic importance, physical and technical complexity, and associated governance challenges. Local conditions determine the precise mixture of barriers in a given place and time. Here, we contend that the barriers to reallocation should be addressed head-on, with the first step being to identify what they are, and how they might be overcome.

The barriers to water reallocation are rooted in water's social, economic and physical characteristics. Water is difficult to bound and measure (e.g. compared with land) due to its mobility, variability and storage challenges (Libecap 2005). Water is a common pool resource, public good and private good in its competing uses (Garrick et al. 2017).

---

1 This section is adapted heavily from a background paper for the World Bank (Damania et al. 2017) and is reproduced here with permission. The authors would like to thank Quentin Grafton and James Horne for providing feedback on this section, which formed a section of Grafton, Garrick and Horne (2017).

Table 8.1 Barriers to water reallocation.

Socio-cultural dimensions	Natural and technological dimensions	Political economy dimensions
• Lack of shared norms and social capital • Divergent mental models, or worldviews • Weak communication and social exclusion • Large group size and/or social heterogeneity • Legacy of inequitable water use	• System boundaries and limits are poorly defined • Water supply is unreliable, including due to hydroclimatic variability • Inadequate or inappropriate infrastructure • Stranded infrastructure • Insufficient modelling, metering and monitoring	• Vested interests • Poorly defined property rights • Third party effects • High transaction costs • Limited administrative capacity • Institutional fragmentation • Inter-governmental coordination challenges

These challenges can be understood along two dimensions of water resources: their excludability (ability to restrict access) and rivalry (competition between users). As a common pool resource, excluding new users is difficult, and water consumption is rivalrous. For example, when a farmer irrigates her field, a portion of that water is evapotranspired and unavailable to other users in the short term. At the same time, water is a public good, involving water storage and watershed functions that also involve costly exclusion, but are non-rivalrous, meaning that if they are available for one, they are available to all within a given community, or public, which can lead to free-riding and underinvestment in storage and conservation. In some cases, the exclusion challenges have been addressed, as when water rights are separated from land and tradable for a well-defined parcel of water that can be bought and sold in a market; bottled water comprises one example of water as a pure private good, that comes closest to a traditional economic commodity. These unique social and economic characteristics cause complex and contentious trade-offs, particularly between consumptive and *in situ* uses. As a consequence, water requires a higher level of collective action than for many other resources (Hanemann 2006).

The barriers to improved water allocation can be divided into three interrelated categories: (i) socio-cultural dimensions, (ii) natural and technological dimensions and (iii) political economy dimensions. Some of the key factors in each of these categories are listed in Table 8.1 and elaborated below.

8.3 Socio-cultural Dimensions

Cultural barriers to an efficient and equitable water reallocation stem from different views of fairness and justice in water allocation. Water has multiple social, economic and ecological values, requiring trade-offs. Blomquist (2011) noted at least 14 'beneficial uses' for water rights in the Western US, ranging from *in situ* uses, such as instream flows, to water for irrigation. Society must balance priorities for basic human needs with rural and urban development imperatives requiring water for irrigation, hydropower, industrial and municipal, as well as spiritual and cultural values (Jackson 2006; Jackson et al. 2015).

Social norms refer to a 'behavioral pattern within a group, supported by a shared understanding of acceptable actions and sustained through social interactions within that group' (Nyborg et al. 2016, p. 42). Social norms related to fairness and cooperation affect social capital, and the associated trust and norms of reciprocity that enable or constrain water reallocation as social preferences change.

The lack of shared social norms can impede water reallocation. A cross-cultural study of norms in Fiji, Ecuador, Paraguay, New Zealand and the US, for example, found that all five countries rejected *some* aspects of water marketization due to strong norms of social justice (Wutich et al. 2013). Of the five countries in the study, only the advanced economies of the US and New Zealand considered the separation of water rights from land acceptable. By comparison, all three lower income countries expressed strong support for a human right to water, including the distribution of basic water needs for free. Nikolakis et al. (2013) illustrate that water markets and norms of social justice are not mutually exclusive in Northern Australia, where a majority of Indigenous respondents support water markets, but only if their existing land and water rights are not separated. These norms of social justice affect the difficulty and, perhaps, the willingness to consider a range of possible reallocation options.

Whilst norms are typically slow to change, abrupt shifts are possible when policy and technological changes converge. When such opportunities for beneficial change arise, such as from key events and triggers related to droughts, political changes and broader investments to cultivate the shared norms and political will, they need to be taken advantage of to move towards more efficient and equitable water allocation.

Efforts to reallocate water depend upon the capacity of diverse stakeholders to resolve conflicts and pursue agreements about how to allocate and reallocate water, particularly when addressing key equity dimensions related to poverty alleviation and inequality. Water users can have divergent mental models, which can impede conflict resolution. Mental models involve 'representations of an individual's or group's internally held understanding of the external world' (Hoffman et al. 2014, p.13016). Contrasting mental models constrain solutions to water allocation challenges, as experienced in the Crocodile River catchment of South Africa (Stone-Jovicich et al. 2011). Many stakeholders view illegal water takings as a problem and agree about the consequences of low river flows. Nevertheless, consensus amongst irrigators and conservationists remains elusive, both in relation to the causes of the problem and in setting priorities for future water use. Thus, the ability to coproduce knowledge about the water system and its dynamics is seen as a critical means of reconciling divergent mental models, provided that the process is transparent and considered fair.

Water reallocation becomes increasingly difficult as water moves greater distances across sectors, scales and jurisdictions (Meinzen-Dick 2014). For example, larger social groups may face additional barriers to water reallocation due to larger numbers and the associated group heterogeneity. Evidence from longstanding irrigation systems (Tang 1992), including more recent studies in the Philippines (Araral 2009), suggest it is harder to establish trust and norms of reciprocity. Contradictory evidence, however, suggests that the influence of group size is more nuanced and involves a potential trade-off between economies of scale as group size increases and the higher transaction costs associated with larger groups (Dinar et al. 1997). Typically, group size is positively associated with social heterogeneity, which increases the challenges posed by social, income, and gender inequality.

Efforts to redress historical inequality have led to commitments to meet basic water needs for human development, including explicit recognition of the human right to water in the Democratic Republic of Congo, Ecuador, Kenya, Nicaragua, South Africa, Uganda and Uruguay (Salman 2014). The progressive realization of equity in water allocation more broadly remains difficult, as evidenced by the challenges in delivering basic human water needs (25 l/person/day) under South Africa's 1998 National Water Act (Kidd 2016). But meeting basic needs involves questions of water supply and sanitation infrastructure, leaving water allocation reform to address competition for the water after these needs are met. Inequality of access or use includes resource 'capture' or when the allocation of water is skewed to benefit a small minority of the population, and has also been identified as a barrier to reallocation in the Jaguaribe (Brazil), Olifants (South Africa) and Sadah (Yemen) basins (Srinivasan et al. 2012).

8.3.1 Navigating the Changing Culture of Water in Spain

With a semi-arid climate and scarce precipitation levels (400 mm/year) (CHS (Confederacion Hydrografica del Segura) 1998), the Segura River Basin (SRB) in southeastern Spain is considered to be one of Europe's driest basins. The region also suffers from regular droughts and constitutes one of Europe's most severe cases of desertification (Oñate and Peco 2005).

Water allocation institutions have evolved over a thousand years to facilitate cooperation and conflict resolution amidst socio-cultural, economic and environmental changes. At the beginning of the twentieth century, traditional irrigated areas in the SRB encompassed 65 000 ha. In 1967, it reached 117 230 ha when the Tajo-Segura transfer (TST) was approved. By 2009 it had climbed to 230 000 ha (Ibor et al. 2011). Besides, the actual water volume transferred through the TST has been on average 15% lower than originally planned. Irrigated areas thus developed with considerably less water than expected. This deficit was in part met by pumping groundwater. The number of wells increased from 7829, before the arrival of the Tajo waters, to 20 350 in 1995 – most of them neither controlled nor registered (Calvo García-Tornel 2006).

Farmers who share a water outlet (e.g. river weir or well) self-organize into irrigation communities to manage common rights and use water (Ortega Reig 2015). Despite severe water scarcity, collective water management in southeastern Spain has survived for over a millennium. The resilience of these systems has been largely premised upon their robust and participatory institutions for devising and enforcing rules to share water and resolve conflicts when the resource becomes scarce (Maass and Anderson 1978). Fundamental changes in the way water was collected, transported and used across the twentieth century, however, have entailed a process of change in institutions managing irrigation.

A key change has been the modernization of irrigated areas; in many communities, gravity irrigation has been replaced by pipelines, laterals and drippers for pressured distribution. This has led to the disappearance of traditional rules for allocating water. For instance, practices like the *turno* (a set rotation order) were substituted by electrovalves automatically actioned by irrigation community officials. On the one hand, these works have contributed to improving the transparency, efficiency, reliability and fairness of water allocations. On the other, technological modernization, compounded by the increase in supply from the TST, means that, during 'non-drought periods', water is more readily available, and many communities now provide water 'on demand' to their farmers during these years. Farmers thus face fewer incentives to adapt to chronic water scarcity, increasing their vulnerability when droughts do occur.

Given path dependency and persistent barriers, water allocation reform will require greater attention to the behavioural, hydrological and political economic factors that impede change. This will require greater efforts to strengthen accounting and incentives in water allocation, whilst leveraging and achieving consistency with water policy at the national level and for the EU Water Framework Directive.

8.4   Natural and Technological Dimensions

Water allocation requires an understanding of the local conditions, including the hydrology and infrastructure of a given water resource system. A water resource system combines the natural and human dimensions of rivers, lakes, aquifers and other water systems.

Four key factors determine the effectiveness of water reallocation and the potential for more efficient and equitable processes and outcomes. First, any effort to reallocate water must determine the water supplies available for consumption and *in situ* uses from a given hydrological system and its infrastructure. The failure to recognize hydrological limits to water extraction is a persistent challenge. This challenge is exacerbated by the lack of alignment between political and hydrological boundaries. Inter-basin transfers and intra-basin distribution systems further determine the technical feasibility of reallocation to achieve efficient water allocation, whilst equitable allocations depend on access to infrastructure. Hydrological connections between surface and groundwater, and through the downstream return flows of upstream water use, increase the complexity of any effort to reallocate water. In turn, this makes a clear and shared understanding of the boundaries of a system critical.

Second, efforts to define hydrological limits require effective metering, monitoring and modelling. Inadequate information is a fundamental barrier that increases the transaction costs of reallocation and may exacerbate equity concerns due to information asymmetries. For example, uneven access to information contributes to information asymmetries and can undermine accounting efforts used for transparent and inclusive planning and allocation decisions. Even when technological innovations lower the cost of monitoring, as in the case of remote sensing and telemetry, lingering scientific uncertainty and political resistance may limit the legitimacy and uptake of the information. For example, water users may oppose efforts to meter water use for fear of losing water, or being charged more to recover costs, as illustrated by the Indian experience where farmers in Jaipur protested efforts to meter their water use for fear of pricing and other measures that would restrict their access (Birkenholtz 2016).

Third, 'hydrological stationarity', or the lack of it, is recognized as a challenge for water managers, posing constraints and opportunities for reallocation (Milly et al. 2008; see also Chapters 2 and 14, this volume). 'Difficult hydrology' associated with low runoff and unpredictable seasonal or inter-annual variability, which in many areas are projected to worsen with climatic changes, poses particular challenges by creating uncertainty about the availability of water and its timing for key uses. Irrigation and associated storage reservoirs, distribution systems and allocation institutions are one key response to variability and the lack of predictability. For example, the link between poverty rates in India and irrigation highlights the impact of freshwater variability on economies dominated by agriculture (Grey and Sadoff 2007). The conjunctive allocation and use of groundwater and surface water supplies is another way to overcome or mitigate scarcity or unpredictable water supplies by using groundwater storage, including active banking of surface water in aquifers, to buffer supply variability (Blomquist et al. 2004).

Fourth, inadequate or inappropriate storage (natural and artificial) and distribution infrastructure pose barriers to reallocation. Infrastructure decisions are tantamount to water allocation decisions because they can lock in patterns of water extraction and use (Bruns and Meinzen-Dick 2005). Increasing hydrological variability and uncertainty are also throwing into question the adequacy and appropriateness of existing infrastructure. The ability to store and distribute water also determines the technical feasibility of moving water from its current uses to satisfy emerging demands or address historic inequalities in water allocation. The distinction between storage and distribution is critical in this context, as storage does not guarantee distribution and access.

Inadequate aquifer storage and insufficient or inappropriate water infrastructure (reservoir storage and distribution systems) contributes to water shortages, unsustainable water allocation patterns and vulnerability to shocks (Srinivasan et al. 2012). Limited aquifer storage and groundwater overdraft is a major challenge (e.g. the Hai River in China and Lerma-Chapala in Mexico).

Surface water storage infrastructure has a more complex effect on water allocation, causing problems when there is too little, too much or the wrong type of water storage. Inadequate infrastructure is a problem in the Volta River, for example, where irrigated agriculture is limited; conversely, excessive or uncoordinated infrastructure create allocation challenges in both centralized (Aral Sea) and more decentralized (groundwater development in the Krishna Basin) infrastructure schemes (Srinivasan et al. 2012). Thus, a key consideration is whether water infrastructure balances the competing needs for stability and flexibility in water allocation, enabling adjustments to shifting availability and preferences.

Despite the barriers to reallocation posed by hydrological constraints and infrastructure systems, efforts to define and map the hydrological system, and establish hydrological limits on water consumption (including environmental flows), can establish the enabling conditions for subsequent legal and institutional reforms and new incentives and norms. For example, an audit of water use in 1995 was a key step at the beginning of the water reform process in Australia. Strategic basin assessments and participatory river basin modelling projects have also emerged elsewhere as key means of addressing the challenges of metering and monitoring water allocation, and enhancing their legitimacy.

In transboundary rivers with barriers to data-sharing across political borders, hydroeconomic modelling, including remotely sensed data on irrigation and groundwater use, have informed basin planning for infrastructure development and water allocation. A strategic basin assessment led by the World Bank in the Ganges using such tools concluded that effective groundwater management should be prioritized over the development of upstream reservoirs (Sadoff et al. 2013). The simulation of water supply scenarios can also address socio-cultural resistance to change, as illustrated by the development of shortage sharing rules and new allocation mechanisms in the Colorado River of the US and Mexico (Garrick et al. 2008).

8.5 Political Economy Dimensions

The political economy of water and challenges to the reallocation of water arise from a range of interrelated institutional impediments: poorly defined water rights, third party effects, stakeholder exclusion, high transition and transaction costs, and ineffective institutional structures constraining water reallocation (Marston and Cai 2016). Vested interests, institutional fragmentation and coordination challenges pose further challenges. Almost all water resources cross political borders in some form across users' associations, states and/or countries, requiring coordination between governments and across tiers of governance (Garrick 2015).

Poorly defined water rights limit the incentives for water reallocation. Whilst water is almost always formally owned by the government, rights to *use* water are defined through a range of methods, including court adjudications and administrative procedures, as well as a host of community-level and informal approaches, with varying levels of security, flexibility and enforcement (see Chapter 10). The costs of formally defining water rights and subsequently monitoring and enforcing them can be substantial, and often prohibitively so, in countries at all income levels. For example, court adjudications over water rights in the Western US have lasted decades and involved millions of dollars in lawyers' fees (Feller 2007). In sub-Saharan Africa, complex and costly permitting processes associated with colonial water rights reforms have limited smallholder access to water permits (van Koppen and Schreiner 2018).

The separation of water rights from land ownership is viewed as an important means of enabling reallocation, but this is not always feasible due to the cultural values or legal constraints. Even in the most ambitious efforts to separate land and water rights, such as those in Australia, the process has required significant resources and coordination across jurisdictions (Bjornlund and O'Callaghan 2004; Young and McColl 2009). Efforts to reform property rights to water are also complicated by the hierarchy of property rights from the individual water-user level to water districts, states, countries and transboundary agreements. Thus, changes at one level can have a 'knock-on' effect at lower or higher levels (Easter and McCann 2010). Reallocation of water out of irrigation districts also involves significant barriers, which tend to magnify as water moves longer distances across sectors, political borders or basin boundaries due to the hydrological and/or infrastructural constraints and associated third-party effects. Efforts to reform water rights systems have largely depended on negotiation amongst the parties impacted.

Concerns over *third party effects* are rooted in issues of equity: barriers that are most difficult to address when water moves across sectors and jurisdictions (Robison and Kenney 2012). Third party effects impede reallocation due to the (i) perceived negative impacts on community values, culture and livelihoods and (ii) threat of socioeconomic and population decline due to decreases in land value, tax base and supporting industries (Marston and Cai 2016). The hydrological and ecological impacts of changes in water use present another challenge, such as the effects of channel lining on groundwater recharge in the Yellow River (Sun et al. 2013).

There are several examples of third party effects triggering resistance and posing barriers to future reallocation. The transfer of water from agriculture in the Owens Valley, California, to support urban populations in Los Angeles is perhaps the most oft-cited example of third-party effects due to the perceived consequences of the transfer on the agricultural community, regional economy and Mono Lake ecosystem. Recent analysis demonstrated that both sides – the rural region and city – gained economically from the transfer, albeit unevenly. For example, the Owens Valley (source region) saw its land values increase 11-fold in the 30-year period from 1900 to 1930 overlapping with the negotiations, whilst a neighbouring county saw its land values only double during this period. Further comparisons demonstrated that the farmers who sold their priorities were better off than staying in agriculture, particularly accounting for the effects of the Great Depression (Libecap 2009). Nevertheless, the Owens Valley transfer demonstrates how the perceptions of unfair and inequitable water allocation can impede subsequent reallocation efforts (Libecap 2009). In a more recent and prominent example, an assessment of stakeholder impacts associated with the south-to-north China transfer identified localized adverse impacts immediately upstream and downstream of the Danjiangkou Reservoir at the area of origin (Pohlner 2016).

Third-party effects have led to: (i) 'no harm' standards in water reallocation; (ii) efforts to compensate the affected communities; and (iii) regional development projects, including infrastructure intended to boost agricultural productivity in the affected region. Amongst the key barriers to implementing the 'no harm rule' and managing third-party effects is the lack of information about historical data on consumptive water use, patterns of return flows, and the economic, social and ecological effects of changes in water use. Any effort to reallocate water must, therefore, account for the size, duration and severity of third-party effects, particularly the negative impacts (Grafton et al. 2012). Estimating these effects is often fraught, reflecting poor-quality baseline data, and consultative processes that are time-intensive and have the ability to undermine a political will to proceed, even if the evidence is compelling.

High transition and transaction costs pose a barrier to reallocation even when there is broad agreement about the need and direction of the proposed response. Transaction costs have been described as the economic equivalent of friction, and refer to the resources required to define, manage and transfer property rights (Garrick 2015). Transaction costs involve efforts to search for willing buyers or sellers, negotiate over prices, secure administrative approvals and monitoring, and enforce the ensuing trade, all of which can be biased towards preserving the status quo. Even when the transaction costs of water reallocation are relatively modest, they can negate the potential gains from water reallocation, as illustrated by modelling in Italy and Spain (Pujol et al. 2006). As water scarcity intensifies, institutional reforms to water rights and their administration can facilitate water reallocation and reduce transaction costs.

A key distinction exists between *transaction costs* of reallocation under the prevailing rules of the game versus the *transition costs* associated with efforts to change the rules of the game themselves, that is, to move from one institutional status quo to another to create more efficient or equitable institutional mechanisms for water reallocation (Challen 2000). Transaction costs are symptomatic of the underlying socio-cultural factors, hydrological and infrastructure constraints and institutional impediments previously identified. Garrick et al. (2013) noted that transaction costs can comprise up to 70% of the total costs of market-based transactions, with the proportion of total costs highest for transactions reallocating water from agriculture to environmental purposes.

Institutional fragmentation and inter-governmental coordination challenges add another layer of difficulty to all of the other barriers by raising the costs and complexity of reallocation. Institutional fragmentation refers to the division or overlap in governance functions across multiple agencies or units, such as when ministries of energy, agriculture, water and infrastructure all play a role in carrying out water planning or other related water governance functions. In the USA, for example, over 20 federal agencies share responsibility for water resource management (Gleick 2010). Even Mexico's relatively more centralized water governance frameworks include seven ministries in the technical council of its National Water Commission (OECD 2013).

Inter-governmental coordination refers to the challenge of coordinating governance functions across independent or semi-independent jurisdictions. Political borders pose barriers to reallocation when countries or subnational governments attempt to block the net export of water from their jurisdiction. Controversies over upstream–downstream allocation decisions for the Nile River illustrate the challenges of coordinating across sovereign nation-states of Egypt, Ethiopia and Sudan. Similar challenges arise in federated states between subnational governments, as illustrated by the tensions triggered by upstream groundwater pumping in the Rio Grande of the US and Mexico or the inter-state allocation conflicts arising in the Cauvery Basin of India.

The heightened costs of collective action across political borders have prompted efforts to establish intergovernmental agreements. Treaties, compacts and partnership agreements are used to share the costs, benefits and risks of water allocation and to establish associated river basin organizations, dispute resolution and monitoring efforts to lower the costs of addressing misallocation (De Stefano et al. 2012; Garrick and De Stefano 2016).

8.5.1 Barriers to Reallocation from Agriculture to the City of Cape Town

The City of Cape Town (CCT) is the second most populous city in South Africa and the economic hub and capital of the Western Cape Province. From 2014 to 2017 the Western Cape experienced the most severe meteorological drought in recorded history. Coupled with increasing rates of in-migration and a burgeoning population, these drivers culminated in a socioeconomic drought and water crisis of global significance that has seen increased levels of conflict and cooperation between agricultural and urban water users. Importantly, the crisis has exposed political, regulatory and institutional deficiencies, and yielded new experiments to address these gaps and barriers.

In response to the severe drought, disproportionate curtailments of 45% and 60% were placed on urban and agricultural users respectively. These restrictions have indirectly reallocated water to CCT as the average daily demand for water in relation to total system yield from agriculture decreased by ~6% and increased by ~5.5% for CCT.

A multitude of barriers exist that have hindered effective drought planning, management and response. Notably from an institutional and political perspective, the Western Cape is the only province in South Africa run by the official opposition party. Despite the National Department of Water and Sanitation (DWS) controlling all bulk water infrastructure in the country, CCT financed and owns three of the six major dams supplying the city and surrounding areas. Moreover, the DWS suffers from crippling technical and regulatory incapacities and corruption, which has arguably been the reason for their unwillingness to devolve allocation powers to more localized institutions such as Catchment Management Agencies (SAWC 2017). These inter-governmental coordination challenges severely limited reallocation as a drought response measure.

The need for allocative equity is simultaneously one of the biggest drivers and barriers to effective reallocation of water. The need to redress historical injustices in water allocations has driven policy change, such as the 2008 Water Allocation Reform strategy (Movik 2010; Merrey 2011). However, water allocation reforms have been hampered by the highly unequal education and wealth distributions (van Koppen and Schreiner 2014). Thus, previously disadvantaged individuals often lack the technical and financial capacity to utilize any new allocation of water, amongst other third-party impacts. Additionally, uncertainty over land tenure amongst commercial farmers has resulted in poor stakeholder participation in water reallocation processes.

Despite the prioritization of urban water use over agriculture and the gifting of $10\,\mathrm{Mm}^3$ to the CCT from private farm dams, reallocation is not seen as a viable way to manage water shortages or accommodate new growth. Notwithstanding the diverse water mix that the CCT is trying to develop in order to improve climate resilience going forward, there have been no significant system-wide institutional changes undertaken to address the underlying socio-political problems driving conflict for water across the urban–rural divide.

8.6 A Ladder of Interventions?

Intensified competition for water increases the need to overcome the barriers to reallocation outlined above. Regardless of the specific policy tools appropriate for local conditions, increasing ambition in water allocation reform often involves a sequence of interventions that entail progressively more difficult sets of barriers. Success depends on the ability to address institutional and governance barriers that range from those that are relatively easy to manage to those that are much harder (Hellegers and LeFlaive 2015). In this context, efforts to improve water allocation may involve a ladder of interventions – a set of increasingly complex policy and institutional responses, as pressures and capacity grow.

The appropriate starting point will depend on the pressures, capacity and barriers involved in a particular location and time. Population and economic growth may trigger increasing competition amongst rural water users and between sectors, although governance capacity may remain limited until wealth and political will increase. In such settings, a water availability study or monitoring innovations may be an appropriate entry point (e.g. Hirpa et al. 2018). Studies and the underpinning monitoring systems and modelling offer a means of building capacity and reconciling different mental models about the status of the water resource and key demands. The development of shared knowledge will be constrained by data limitations and institutional fragmentation, which are particularly challenging in lower income countries.

Growing urbanization and industrialization brings increasing pressure for reform. Added pressure and capacity create the need and potential for more complex water allocation reforms to address inter-sectoral misallocation, including competition between different rural uses (mining and agriculture) and across rural, urban and environmental uses. Water rights reforms, pricing mechanisms or related measures may be needed to address local supply–demand imbalances and the shifting social preferences that come with poverty alleviation and growth. Socio-economic inequality is a prime barrier to reform or its enforcement, particularly when rural livelihoods depend on informal rights to water threatened by economic activity in new industrial sectors. Local pressures may also involve uncertainty about where to draw the boundaries around the system, as competition for water creates interdependencies across rural settlements, agricultural regions and cities. Vested interests and legacy issues in rural regions may combine with poorly defined water rights and weak administrative capacity to pose formidable barriers to water reallocation.

River basin closure – a condition in which downstream water needs are unmet – or groundwater overdraft may accompany regional development as emerging water needs add to existing water uses for agricultural and rural settlements. Growing rural–urban linkages can bring new pressure for reallocation between sectors, although there is no guarantee that urbanization and industrial growth will bring sufficient management capacity to address the necessary trade-offs, particularly in transboundary river basins and aquifers. Basin or sub-basin allocation reforms confront some of the most vexing and intractable barriers due to the large group sizes and the associated social, cultural and economic heterogeneity. The larger geographical territory also raises questions about the adequacy, appropriateness and coordination of water infrastructure. In these contexts, basin planning and conflict management may prove necessary due to the third-party effects, creating substantial inter-governmental coordination challenges.

8.7 Frontiers of Water Allocation

The experiences from the global review, and the case studies in the Americas, Africa and Europe, highlight a diverse range of contexts, drivers, barriers and responses in water allocation. They illustrate how allocation disputes evolve with time and shifting values. In both case studies, agricultural development, and the water allocation institutions for irrigated agriculture, have struggled to adapt to urban demand and the growing recognition of ecological flows. The escalating demographic, socioeconomic and environmental pressures will only magnify water allocation challenges under future trajectories of population growth and resource use. These trends will reinforce the calls for water allocation reform and require accelerated progress.

The synthesis of the academic research, policy studies and cases of water allocation reform in this chapter is insufficient for broad generalizations or simple paths to reform. Any analysis of water allocation has stressed that context and history matter – this is plainly evident and accepted. However, there is still much we can learn from the global experience about the types of barriers and the ladder of interventions for improving water allocation. The review suggests there are distinct and recurrent types of barriers that explain why water allocation and reallocation is difficult and so heavily contested – socio-cultural barriers, natural and technological constraints, and political economic factors. Understanding these barriers offers a first step for better tailoring the sequence, or 'ladder', of interventions for a proportionate response to intensified competition for water.

Notwithstanding the importance of context, the global experience suggests several areas for water allocation reform and priorities to guide institutional analysis and development. First, there are opportunities to use existing institutional arrangements better by understanding and addressing the barriers to water allocation, and developing the information and incentives to support institutional capacity and implementation. Institutional reforms may still prove necessary when existing policies and laws have major gaps, or the implementation barriers prove intractable, but several countries have adopted strong policy frameworks for water allocation without investing in the institutional strengthening and coordination to deliver. Breakthroughs in behaviour change, water accounting and benefit-sharing illustrate the potential to address the cultural, technological and political economic barriers to water reallocation, respectively.

Second, global challenges present unprecedented competition for water, but also novel opportunities and capacities. Urbanization and climate change capture the major drivers of changes in demand and supply. Meeting SDGs for sustainable cities and climate change will present the need – and opportunity – for reallocation. This transformation will threaten the informal sector, poor and most vulnerable, and the associated norms and local practices for water allocation, requiring new models of collective action that better coordinate and, where appropriate, integrate formal and informal approaches. The subsidiarity principle suggests that local authority and capacities should be fostered rather than crowded out.

Third, crisis management in the context of droughts, migration or other supply and demand spikes can create reactive and brittle allocation institutions. From Cape Town to California, droughts have created unprecedented openings for water allocation reform, but also windows for bad ideas. Long-term planning and prioritization of sustainability and adaptability are constrained by the political economic factors introduced above. Investments in water accounting, understanding water risk, strengthening water rights and developing incentives can deliver significant returns. Achieving progress in these areas will require greater capacity to diagnose the barriers impeding change, and harness the drivers and innovations to build momentum for reform.

References

Araral, E. (2009). What explains collective action in the commons? Theory and evidence from the Philippines. *World Development* 37: 687–697.

Birkenholtz, T. (2016). Dispossessing irrigators: water grabbing, supply-side growth and farmer resistance in India. *Geoforum* 69: 94–105.

Bjornlund, H. and O'Callaghan, B. (2004). Property implications of the separation of land and water rights. *Pacific Rim Property Research Journal* 10: 54–78.

Blomquist, W. (2011). A political analysis of property rights. In: *Property in Land and Other Resources* (eds. D.H. Cole and E. Ostrom), 369–384. Cambridge, MA: Lincoln Institute of Land Policy.

Blomquist, W., Schlager, E., and Heikkila, T. (2004). *Common Waters, Diverging Streams. Linking Institutions to Water Management in Arizona, California, and Colorado.* Washington, DC: Resources for the Future.

Bruns, B.R. and Meinzen-Dick, R. (2005). Frameworks for water rights: an overview of institutional options. In: *Water Rights Reform: Lessons for Institutional Design* (eds. B.R.

Bruns, C. Ringler and R. Meinzen-Dick), 3–26. Washington, DC: International Food Policy Research Institute (IFPRI).

Calvo García-Tornel, F. (2006). Sureste español: Regadío, tecnologías hidráulicas y cambios territoriales. *Scripta Nova* 218 (04) http://www.ub.edu/geocrit/sn/sn-218-04.htm.

Challen, R. (2000). *Institutions, Transaction Costs, and Environmental Policy: Institutional Reform for Water Resources*. Cheltenham, UK: Edward Elgar Publications.

CHS (Confederacion Hydrografica del Segura) (1998). *Plan Hidrológico de la Cuenca del Segura*. Murcia: Confederación Hidrográfica del Segura.

Damania, R., Desbureaux, S., Hyland, M. et al. (2017). *Uncharted Waters*. Washington, DC: World Bank.

De Stefano, L., Duncan, J., Dinar, S. et al. (2012). Climate change and the institutional resilience of international river basins. *Journal of Peace Research* 49: 193–209.

Dinar, A., Rosegrant, M.W., and Meinzen-Dick, R.S. (1997). *Water Allocation Mechanisms: Principles and Examples*. Policy, Research Working Paper No. WPS 1779. Washington, DC: World Bank.

Easter, K.W. and McCann, L.M. (2010). Nested institutions and the need to improve international water institutions. *Water Policy* 12: 500–516.

Feller, J.M. (2007). The adjudication that ate Arizona water law. *Arizona Law Review* 49: 405.

Garrick, D.E. (2015). *Water Allocation in Rivers under Pressure: Water Trading, Transaction Costs and Transboundary Governance in the Western US and Australia*. Cheltenham, UK: Edward Elgar Publishing.

Garrick, D.E. and De Stefano, L. (2016). Adaptive capacity in federal rivers: coordination challenges and institutional responses. *Current Opinion in Environmental Sustainability* 21: 78–85.

Garrick, D.E., Hall, J.W., Dobson, A. et al. (2017). Valuing water for sustainable development. *Science* 358 (6366): 1003–1005.

Garrick, D., Jacobs, K., and Garfin, G. (2008). Models, assumptions, and stakeholders: planning for water supply variability in the Colorado River Basin. *Journal of the American Water Resources Association* 44: 381–398. https://doi.org/10.1111/j.1752-1688.2007.00154.x.

Garrick, D., Whitten, S., and Coggan, A. (2013). Understanding the evolution and performance of water markets and allocation policy: a transaction costs analysis framework. *Ecological Economics* 88: 195–205.

Gleick, P.H. (2010). Roadmap for sustainable water resources in southwestern North America. *Proceedings of the National Academy of Sciences* 107 (50): 21300–21305. https://doi.org/10.1073/pnas.1005473107.

Gleick, P.H. (2018). Transitions to freshwater sustainability. *Proceedings of the National Academy of Sciences* 115 (36): 8863–8871. https://doi.org/10.1073/pnas.1808893115.

Grafton, R. Q., Garrick, D. E., and Horne, J. (2017). Water Misallocation: Governance Challenges and Responses. https://www.researchgate.net/profile/R_Grafton/publication/331297130_Water_Misallocation_Governance_Challenges_and_Responses/links/5c70c265a6fdcc4715942db3/Water-Misallocation-Governance-Challenges-and-Responses.pdf.

Grafton, R.Q., Libecap, G.D., Edwards, E.C., and Landry, C. (2012). Comparative assessment of water markets: insights from the Murray-Darling basin of Australia and the Western USA. *Water Policy* 14 (2): 175–193.

Grafton, R.Q., Pittock, J., Davis, R. et al. (2013). Global insights into water resources, climate change and governance. *Nature Climate Change* 3: 315–321.

Grey, D. and Sadoff, C.W. (2007). Sink or swim? Water security for growth and development. *Water Policy* 9: 545.

Hanemann, W.M. (2006). The economic conception of water. In: *Water Crisis: Myth or Reality* (eds. P.P. Rogers, M.R. Llamas and L.M. Cortina), 61–91. London: CRC Press.

Heinmiller, B.T. (2009). Path dependency and collective action in common pool governance. *International Journal of the Commons* 3 (1): 131–147.

Hellegers, P. and LeFlaive, X. (2015). Water allocation reform: what makes it so difficult? *Water International* 40 (2): 273–285.

Hirpa, F.A., Dyer, E., Hope, R. et al. (2018). Finding sustainable water futures in data-sparse regions under climate change: insights from the Turkwel River basin, Kenya. *Journal of Hydrology: Regional Studies* 19: 124–135.

Hoffman, M., Lubell, M., and Hillis, V. (2014). Linking knowledge and action through mental models of sustainable agriculture. *Proceedings of the National Academy of Sciences* 111 (36): 13016–13021.

Hooper, V. and Lankford, B.A. (2018). Unintended water allocation: gaining share from the ungoverned spaces of land and water transformations. In: *The Oxford Handbook of Water Politics and Policy* (eds. K. Conca and E. Weinthal), 248. Oxford: Oxford University Press.

Ibor, C.S., Mollá, M.G., Reus, L.A., and Genovés, J.C. (2011). Reaching the limits of water resources mobilization: irrigation development in the Segura river basin, Spain. *Water Alternatives* 4 (3): 259–278.

Jackson, S. (2006). Compartmentalising culture: the articulation and consideration of indigenous values in water resource management. *Australian Geographer* 37: 19–31.

Jackson, S., Pollino, C., Maclean, K. et al. (2015). Meeting indigenous peoples' objectives in environmental flow assessments: case studies from an Australian multi-jurisdictional water sharing initiative. *Journal of Hydrology* 522: 141–151.

Kidd, M. (2016). Compulsory licensing under South Africa's National Water Act. *Water International* 41 (6): 916–927.

Libecap, G.D. (2005). The problem of water. *Workshop on New Institutional Economics and Environmental Issues*, INRA-ENESAD CESAER. http://www.u.arizona.edu/~libecapg/downloads/TheProblemOfWater.pdf.

Libecap, G.D. (2009). Chinatown revisited: Owens Valley and Los Angeles – bargaining costs and fairness perceptions of the first major water rights exchange. *Journal of Law, Economics, and Organization* 25: 311–338.

Maass, A. and Anderson, R.L. (1978). *And the Desert Shall Rejoice: Conflict, Growth and Justice in Arid Environments*. Cambridge, MA: MIT Press.

Marston, L. and Cai, X. (2016). An overview of water reallocation and the barriers to its implementation. *WIREs Water* 3: 658–677. https://doi.org/10.1002/wat2.1159.

Meinzen-Dick, R. (2007). Beyond panaceas in water institutions. *Proceedings of the National Academy of Sciences* 104 (39): 15200–15205.

Meinzen-Dick, R. (2014). Property rights and sustainable irrigation: A developing country perspective. *Agricultural Water Management* 145: 23–31.

Meinzen-Dick, R. and Ringler, C. (2008). Water reallocation: drivers, challenges, threats, and solutions for the poor. *Journal of Human Development* 9 (1): 47–64.

Merrey, D.J. (2011). Time to reform South African water reforms: a review of "transforming water management in South Africa" (ed. B. Schreiner and R. Hassan). *Water Alternatives* 4 (2): 252–255.

Milly, P.C.D., Betancourt, J., Falkenmark, M. et al. (2008). Climate change: stationarity is dead: whither water management? *Science* 319: 573–574.

Molle, F. (2003). *Development trajectories of river basins: a conceptual framework*. IWMI Research Report H033886. International Water Management Institute.

Molle, F. and Berkoff, J. (2009). Cities vs. agriculture: a review of intersectoral water reallocation. *Natural Resources Forum* 33 (1): 6–18.

Movik, S. (2010). Allocation discourses: South African water rights reform. *Water Policy* 13 (2): 161–177.

Nikolakis, W.D., Grafton, R.Q., and To, H (2013). Indigenous values and water markets: survey insights from northern Australia. *Journal of Hydrology* 500: 12–20.

Nyborg, K., Anderies, J.M., Dannenberg, A. et al. (2016). Social norms as solutions. *Science* 354 (6308): 42–43.

OECD (2013). *Making Water Reform Happen in Mexico. OECD Studies on Water*. Paris: OECD Publishing https://doi.org/10.1787/9789264187894-en.

Oñate, J.J. and Peco, B. (2005). Policy impact on desertification: stakeholders' perception in Southeast Spain. *Land Use Policy* 22 (2): 103–114.

Ortega Reig, M.V. (2015). *Collective management of irrigation in eastern Spain. Integration of new technologies and water resources*. PhD thesis. Universitat Politècnica de València.

Pohlner, H. (2016). Institutional change and the political economy of water megaprojects: China's south-north water transfer. *Global Environmental Change* 38: 205–216.

Pujol, J., Raggi, M., and Viaggi, D. (2006). The potential impact of markets for irrigation water in Italy and Spain: a comparison of two study areas. *Australian Journal of Agricultural and Resource Economics* 50 (3): 361–380.

Robison, J.A. and Kenney, D.S. (2012). Equity and the Colorado River compact. *Environmental Law* 42 (4): 1157–1208.

Sadoff, C., Harshadeep, N.R., Blackmore, D. et al. (2013). Ten fundamental questions for water resources development in the Ganges: myths and realities. *Water Policy* 15 (S1): 147–164.

Salman, S.M. (2014). The human right to water and sanitation: is the obligation deliverable? *Water International* 39: 969–982.

SAWC (2017). *Report on the State of the Department of Water and Sanitation*. South African Water Caucus. www.fse.org.za/Downloads/SAWCStateofDWSReport.pdf (accessed 26 July 2017).

Schlager, E. and Ostrom, E. (1992). Property-rights regimes and natural resources: a conceptual analysis. *Land Economics* 68 (3): 249–262.

Srinivasan, V., Lambin, E., Gorelick, S. et al. (2012). The nature and causes of the global water crisis: syndromes from a meta-analysis of coupled human-water studies. *Water Resources Research* 48 (10) https://doi.org/10.1029/2011WR011087.

Stone-Jovicich, S.S., Lynam, T., Leitch, A., and Jones, N.A. (2011). Using consensus analysis to assess mental models about water use and management in the Crocodile River catchment, South Africa. *Ecology and Society* 16 (1).

Sun, X., Speed, R., and Shen, D. (2013). *Water Resources Management in the People's Republic of China*. Routledge.

Tang, S.Y. (1992). *Institutions and Collective Action: Self-Governance in Irrigation*. California: ICS Press.

van Koppen, B. and Schreiner, B. (2014). Moving beyond integrated water resource management: developmental water management in South Africa. *International Journal of Water Resources Development* 30 (3): 543–558.

van Koppen, B. and Schreiner, B. (2018). *A hybrid approach to decolonize formal water law in Africa*. IWMI Research Report 173. International Water Management Institute.

Wilder, M. and Ingram, H. (2018). Knowing equity when we see it. In: *The Oxford Handbook of Water Politics and Policy* (eds. K. Conca and E. Weinthal). Oxford: Oxford University Press.

World Economic Forum (WEF) (2018). Global Risks Report. Davos, Switzerland: WEF. https://www.weforum.org/reports/the-global-risks-report-2018.

Wutich, A., Brewis, A., York, A.M., and Stotts, R. (2013). Rules, norms, and injustice: a cross-cultural study of perceptions of justice in water institutions. *Society and Natural Resources* 26: 795–809.

Young, M.D. and McColl, J.C. (2009). Double trouble: the importance of accounting for and defining water entitlements consistent with hydrological realities. *Australian Journal of Agricultural and Resource Economics* 53: 19–35.

9

Rural Water Policy in Africa and Asia

Rob Hope[1,2], Tim Foster[3], Johanna Koehler[2], and Patrick Thomson[1,2]

[1] School of Geography and the Environment, University of Oxford, UK
[2] Smith School of Enterprise and the Environment, University of Oxford, UK
[3] University of Technology Sydney, Australia

9.1 Fifty Years of Rural Water Policy in Africa and Asia

Providing safe and reliable drinking water in rural Africa and Asia has been an enduring and elusive policy challenge (White et al. 1972; Falkenmark 1982). Scattered and remote settlements with low population density and often low and variable income, or subsistence livelihoods, have presented engineering, financial and institutional challenges to the model of piped network systems that have dominated in most urban contexts. With often higher levels of deprivation in rural areas by education, health or welfare measures, there has been a moral imperative to find inclusive and sustainable policy solutions at least cost. Whilst drinking water coverage has generally improved, only one in five countries below 95% coverage in 2015 is currently on track to achieve universal basic drinking water services by 2030 (UNICEF/WHO 2017). With the advent of the Sustainable Development Goal era (2015–2030), the ambition to provide safely managed drinking water on-site, free of contamination and when demanded, for everyone, is without precedent (UNICEF/WHO 2017). In this context, we propose three periods of rural water policy in Africa and Asia between 1980 and 2030 to (i) identify four pillars of rural water policy design, (ii) consider how they have adapted over time, and (iii) propose priorities for progress to achieve at least basic drinking water services for all by 2030.

The 1980s launched the first International Drinking Water Supply and Sanitation Decade (Falkenmark 1982), focusing in particular on regions where economic development was limited, rural poverty widespread and national governments were emerging after decades of colonial rule. We argue that this policy period of 'Community Access' largely continued for two decades and established a common set of pillars across equality, institutional, finance, and accountability domains until around 2000. The next period sees shifts in policy pillars away from 'Community Access' (1980–2000) to one of 'Rights and Results' (2000–2020). Scanning the horizon, we chart the next decade to the conclusion of the Sustainable Development Goals in 2030, characterizing it as one of 'Regulated Services' (2020–2030). We argue that the ambition of the policy period

Water Science, Policy, and Management: A Global Challenge, First Edition. Edited by Simon J. Dadson, Dustin E. Garrick, Edmund C. Penning-Rowsell, Jim W. Hall, Rob Hope, and Jocelyne Hughes.
© 2020 John Wiley & Sons Ltd. Published 2020 by John Wiley & Sons Ltd.

ahead should be to target the long-heralded but elusive goal of 'basic drinking water services' for all, which would have dramatic implications for an estimated 633 million people in rural Africa and Asia without basic drinking water in 2015[1] (UNICEF/ WHO 2018).

Achieving universal rural water services presents unique challenges due to resource excludability and rivalry dynamics, cultural and political diversity, social and political inequalities, information asymmetry and financial fragility, hydroclimatic and technological change, and often high levels of poverty. Understanding why and when shifts and reform in water policy occur in response to, or despite, these multiple drivers and constraints over time is poorly documented. This matters in terms of past and present global development goals which have driven billions of dollars of investment, often with uneven and disappointing outcomes. The contribution of this chapter is to examine these issues in three areas. First, we propose four guiding pillars which characterize three periods of rural water policy covering half a century. Second, we examine the pillars in the context of shifting policy discourses over time in rural Africa and Asia. Third, we reflect on the prospects to change practices to achieve the Sustainable Development Goal of universal and equitable drinking water services for rural populations in Africa and Asia by 2030.

9.2 Pillars of Rural Water Policy

Water policy is always provisional and context-dependent (Briscoe 2014), whether in the building of a nation (Blackbourn 2006), managing a river basin (Barry 1997), as a socio-cultural response to water scarcity (Wade 1988; Mosse 2003), or as part of the political economy of development aid (Mosse 2004; Lankester 2013). Pillars that guide water policy emerge and are championed irregularly, often by particular institutes, multilateral organizations, or as outputs from signature conferences. One would point to a handful of globally influential agreements and commitments relevant to rural water policy:

- The United Nations' Water Conference and the Mar del Plata Water Action Plan (1977) highlighting the issue of 'community water supplies' to promote reliable water supplies in rural and urban communities. Specifically, 'all peoples … have the right to access drinking water of quantities and of a quality sufficient for their basic needs' (Falkenmark 1977, p. 225). This led to the International Decade of Drinking Water Supply and Sanitation between 1980 and 1990 (Falkenmark 1982).
- The Dublin Statement on Water and Sustainable Development (ICWE 1992) with four guiding principles: (i) effective and holistic management of water resources, (ii) subsidiarity with public participation at the lowest appropriate level, (iii) empower women in provision, management and safeguarding of water, and (iv) water as an economic good with affordable pricing.

---

1 The UN/WHO Joint Monitoring Programme estimates that 343 million rural people in sub-Saharan Africa (57% of the rural population), 177 million rural people in Central and South Asia (14% of rural population), 81 million rural people in East and Southeast Asia (8%), and 32 million rural people in Western Asia and Northern Africa (17%) were without basic water services in 2015 (UNICEF/WHO 2018).

- The United Nations' Millennium Development Goal (2000), Target 7C, to halve by 2015 the proportion of people without sustainable access to safe drinking water and basic sanitation.
- The Bonn Charter (2004) to achieve 'good safe drinking water that has the trust of consumers'. Identifying water risks through the supply and distribution system with a focus on water safety plans, measurement and standards.
- In 2010, the United Nations' General Assembly, through resolution 64/292, recognized the human right to water and sanitation, and acknowledged that clean drinking water and sanitation are integral to the realisation of all rights. The right to drinking water is realised by non-discriminatory access to sufficient, safe, physically accessible and affordable services.
- The United Nations' Sustainable Development Goal (2015), Target 6.1, by 2030, to achieve universal and equitable access to safe and affordable drinking water for all. The other water targets feature the needs of women and girls, and those in vulnerable situations, for sanitation (6.2) and the role of local communities in water and sanitation management (6B).

Whilst there is an inevitable challenge in consolidating many salient perspectives and principles into a few core pillars, we propose the following for this analysis: (i) equality, (ii) institutions,[2] (iii) finance, and (iv) accountability. The selection criteria reflect core aspects of water policy as it has evolved and been applied from the 1980s through to the 2030 horizon (see Table 9.1).

Equality reflects the moral imperative to meet basic needs and protect human rights along with the political and economic self-interest of a common platform for growth and prosperity. The South African poet, Antije Krog, is quoted in the 1998 National Water Act: 'with water we will wash away the sins of the past', indicating the political salience of delivering a (free) basic service to *all* citizens in rejection of the racially biased policies of the apartheid era. The pillar has evolved over time, highlighting vulnerable or excluded groups, with increased focus on women and girls, acknowledging the gendered inequalities and avoidable costs in water collection for health, dignity and development. The articulation and emphasis around equity has shifted in focus from

Table 9.1 A typology of rural water policy pillars and periods from 1980 to 2030.

Pillars	Community Access 1980–2000	Rights and Results 2000–2020	Regulated Services 2020–2030
Equality	Access	Rights	Regulation
Institutions	Subsidiarity	Decentralization	Pluralism
Finance	Subsidy	Blended	Domestic
Accountability	Expenditure	Spot-checks	Monitoring

Source: Authors.

---

2 Institutions are defined in line with North's (1990, p. 3) notion of 'humanly devised constraints that shape human interaction... [and] structure incentives in human exchange, whether political, social, or economic'.

'basic needs' in the 1970s to 'human rights' with a focus on non-discrimination, and we argue that the evolution of policy responses is moving towards 'regulation' with the goal of leaving no one behind (UNICEF/WHO 2017). Naturally, these categories are not mutually exclusive, but highlight changing political emphasis and governance structures which progressively internalize equity provision as it moves from a global dialogue to national and subnational responsibility.

Institutions are the human-devised rules that constrain behaviour, share information and provide incentives for collective action (North 1990). Water institutions have devised a remarkable range of adaptive, innovative, and sometimes divisive, norms and rules over time from the despotic behaviour of early irrigation societies (Wittfogel 1957) to the cultural inclusivity of sharing drinking water despite scarcity through religious edicts and cultural norms (see Chapter 10). Here, we consider the issues of scale, responsibility and authority in delivering drinking water services. In the context of newly-independent states, the economic and political logic of local management in remote, low density and scattered rural communities was pragmatic (Mamdani 1996). An unwritten social contract required national and global governments to mobilize and subsidize finance for water infrastructure to build a 'basic service', with the communities responsible for ongoing operation and maintenance. With maturing states and increasing resources and capacity, the mantra of decentralization of services has become dominant in Africa and Asia (Conyers 2007; Crawford and Hartmann 2008), and the geography of accountability has become closer, linking political cycles and financial flows to visible improvements in the daily lives of rural voters. Increasingly policy and practice are challenging the false dichotomy between private sector or public delivery with emerging models of pluralism – referred to in different contexts of multiple legal orders and questions of efficiency and equity, including value-pluralism (Ingram et al. 1986; Meinzen-Dick and Nkonya 2007; Cleaver 2012; Wilder and Ingram 2018) – and which are here defined as sharing water risks and institutional responsibilities between the state, the market and communities (Koehler et al. 2018).

The finance pillar recognizes the enduring and evolving challenge of funding sustainable rural water services. In 1982, Malin Falkenmark noted that if the challenge for the first half of the international decade of drinking water and sanitation was building infrastructure, the second half would be dominated by the maintenance challenge. The increasing interest in hybrid finance (see Chapter 15) to promote sustainability identifies three primary flows of finance, from tariffs (customers), taxes (government), and transfers (donors), which acknowledges reliance on transfers alone from international aid is no longer viable (OECD 2009). With an average, annual global cost of USD 114 billion per year to meet the safely managed drinking water and sanitation targets (Hutton and Varughese 2016), there is urgency to identify alternative financing models with an increasing role for domestic finance, which notionally should be more sustainable in non-fragile states. The current shift in the debate away from water being considered a free good is critical for making progress towards financial sustainability at the local level.

Finally, the accountability pillar captures the sharpening political scrutiny and advances in information systems to monitor the equity and performance of water service delivery. Early focus was on measuring inputs, by tracking expenditure within an overseas development assistance (ODA) paradigm, and provided a benchmark for improving the design, local manufacture and construction of infrastructure. Sustainability was promoted by investing in community management with limited alternatives for system monitoring. Increasing (although largely anecdotal) evidence of weak sustainability and extensive downtime of handpumps questioned devolving all the responsibilities and risks to communities (Harvey and Reed 2004). In response to growing concerns on infrastructure functionality after construction, there has been a growth in 'spot-checks' at national and subnational levels (Foster 2013), although issues of collection frequency, data quality and response mechanisms to failures have questioned the benefit to address the underlying challenges (Thomson et al. 2012a, b).

The emergence of mobile networks across rural Africa and Asia and the widespread adoption of mobile phones have allowed basic data to be collected from rapid surveys in a relatively low-cost fashion, and ushered in the possibility of moving away from using infrequent spot-checks to assess infrastructure functionality and towards the continuous monitoring of service provision. As well as increasing accountability and enabling regulation of rural water services, a 'surveillance-response' approach (Thomson et al. 2012b) allows information on the functionality of water infrastructure to be rapidly incorporated into day-to-day maintenance and repair planning, thereby reducing downtimes and improving levels of service. The wider uptake of smart sensors in urban piped systems using mobile or satellite data transmission is now being matched in the rural space with novel sensor technologies for rural infrastructure beginning to be developed and deployed (Thomson et al. 2012a; Nagel et al. 2015; Thomas 2016; Welle et al. 2016). This automated flow of low-cost data on hourly or daily handpump usage has helped reduce downtimes of a month or more to less than three days, and also provided insights into institutional and financial design, with observed data indicating the variable demand for rural water (Thomson et al. 2019). The next generation of machine learning applications are processing data to predict shallow groundwater depth (Colchester et al. 2017) and failure events (Greeff et al. 2018) to increase the value of this 'accidental infrastructure'.

9.3 Community Access, 1980–2000

The 'Community Access' period is roughly identified between 1980 and 2000 and is characterized by access, subsidiarity, subsidy and expenditure. The period charts the first global intervention for rural water supply during the International Decade of Drinking Water Supply and Sanitation, concluding with the start of the Millennium Development Goals. For many African and Asian countries it marked the emergence and development of independent states after decades of colonial rule. Newly-won independence presented significant political, economic and social challenges, with often two-thirds or more of the population in scattered, rural communities depending on subsistence livelihoods with limited health, education, energy or water services.

This period acknowledged the enormous scale of the institutional, financial, and technological challenge of providing basic water services, and the priority was to increase coverage through low-cost approaches that did not rely on smoothly functioning government bureaucracies. For example, a multi-country analysis of domestic water supplies in rural Kenya, Tanzania and Uganda in the late 1960s found that average daily water use per person was around 10l collected from sources around half a kilometre from the household in 2.5 trips per day (White et al. 1972). The water sources were mainly surface water (streams, springs, wells or reservoirs) and would be largely classified as unimproved sources due to concerns and uncertainty around water quality, distance and reliability. Globally, in 1980, 32% of an estimated 1.49 billion people living in rural areas had access to an improved water supply (Therkildsen 1988:49). A decade earlier, the proportion of rural people with access was 13%. Perhaps motivated both by the scale of the current deficit and an almost tripling of access in the previous decade, the International Decade for Drinking Water and Sanitation proposed a target of universal drinking water access by 1990. Of 94 countries polled by the WHO, 13% committed to universal rural water access, with the majority (70%) opting for coverage of between 50% and 95%. Of the 26 African countries polled, four out of five opted for the lower target. Early concerns with meeting the universal goal cited insufficient investment and over-reliance on external funds, limited institutional capacity, and the omission of operation and maintenance in costing and management (Biswas 1981).

Looking at recent estimates of rural water access in 'least developed countries', the level of coverage has increased for basic rural water access from 43% in 2000 to 52% in 2015 (UNICEF/WHO 2017). If we consider the sub-Saharan Africa region, progress has been faster, from 29% to 43%, compared with Central and Southern Asia rising from 78% to 86%. It is sobering to note that a rural African was less likely to have basic water access in 2000 (29%) than in 1980 (32%), though caveats apply on the comparability and quality of measurement data. Equality in access to drinking water provided a policy and development imperative for action. Events such as the devastating drought in the Indian states of Bihar and Uttar Pradesh in 1967 provided a stimulus for more effective and equitable rural water policy (Mudgal 1997). UNICEF played a pivotal role in responding to the drought and supporting the Government of India in the development of community handpumps with a standardized design (Indian Mark II and III) and deployment of 2.3 million handpumps by 1996 with an estimated coverage of 500 million people (Mudgal 1997, p. 17). The National Drinking Water Mission established in 1986 provided government focus and financing for progress in increasing access with community ownership and management. Similar initiatives and programmes emerged across Africa and Asia, often with non-governmental organizations (NGOs) as conduits between government policy and community implementation.

The dominant institutional choice of this period was community management (Arlosoroff et al. 1987; Briscoe and de Ferranti 1988). This reflects the subsidiarity principle to allocate management to the lowest appropriate level (Marshall 2008). With a global commitment to increase access, the institutional capacity to manage such an unprecedented programme of infrastructure deployment required local community participation and management. It was assumed that an increase in community participation could solve the problem of sustainability associated with previous top-down approaches. Kleemeier (2000) argued that this emphasis on community participation and management in rural water supply from the late 1970s onwards was part of a general trend towards a bottom-up, basic needs strategy of development, adopted in response to the perceived failure of top-down plans. Such an approach was also consistent with various actors' diverse experiences, views and ideologies: practitioners with experience of failed government projects and neo-liberal funders' views on reducing the role of government could be reconciled with the ideas of community empowerment prevalent, and sometimes romanticized, within western NGOs (Harvey and Reed 2006) and the World Health Organization.

Considering 'development projects as policy experiments' (Rondinelli 1991), an innovative programme of work evolved in East Africa to provide technical and institutional support to shape and formalize this institutional approach (Arlosoroff et al. 1987; Briscoe and de Ferranti 1988; Sara and Katz 1998; Narayan 1995). With a strong commitment to community participation and ownership, project design required communities to take on operational, maintenance and associated financial responsibilities. In Africa, as in Asia, the technical design of the dominant handpump technologies (Afridev, India Mark II, No. 6) allowed for low-cost repairs but a strong degree of collective action amongst diverse communities.

In Pakistan, analysis of the collective action outcomes for the maintenance of 132 Aga Khan Rural Support Programmes (AKRSP) in Balistan concluded 'good projects' can work in 'bad communities', provided there is a focus on improved project design (Khwaja 2001). However, an intriguing finding is a U-shaped distribution of infrastructure maintenance such that communities with very high or very low social equality achieve better maintenance outcomes than the messy majority. A multi-decadal analysis of handpump sustainability in coastal Kenya from the 1980s echoes the Pakistan work in how the legacy of poor project design (here, groundwater with low water quality) leads to less sustainable payment behaviours for community handpumps three decades later (Foster and Hope 2016). Wider studies have reflected on the provisional and insufficient nature of community management as the policy priorities have shifted from access to services over time (Harvey and Reed 2004; Banerjee and Morella 2011; Chowns 2015; Whaley and Cleaver 2017). Echoing these collective action shortcomings is evidence from Cambodia that shows the operational performance of communally managed handpumps is significantly outstripped by privately owned handpumps managed at the household level (Foster et al. 2018b).

This period was characterized by an often implicit government or donor subsidy for capital infrastructure, with communities bearing future operational and maintenance costs (Fonseca and Pories 2017). The estimated costs of achieving universal water and sanitation by 1990 ranged from $300 to 600 billion (1978 US dollars) subject to the use of standard technology, but with no account for future maintenance costs (Biswas 1981, p. 161). This is an order of magnitude more than the previous decade, with the assumption that national governments would be willing and able to cover the majority of the ongoing costs. In Tanzania, Therkildsen (1988) documented donor behaviour in the 1980s as 'Watering White Elephants' with significant investments but prescriptive 'ends' to achieving rural water access without adequate flexibility or institutional strengthening with local partners, at all levels. Projects failed as the planning logic of meeting an externally derived target failed to understand local context and dynamic challenges of sustainable systems (Mosse 2004).

In effect, financing through subsidy led to accountability by monitoring expenditure rather than outcomes. With inadequate technical provision and a limited focus on monitoring the delivery of services, projects and programmes are deemed to be 'successful' if resources are allocated and liquidated as planned. Assumptions are made on an arbitrary number of 'beneficiaries' per waterpoint with the local community competently managing the system into the future, divesting government and external actors of any future responsibility for service delivery. The symbols, language and framing of the project or programme is controlled by a planning calculus where the role, engagement and voice of local people is often abstracted or assumed in the public representation of a well-meaning intervention skilfully completed. The assumption that collective action in diverse rural communities will be able to manage financial risks is increasingly

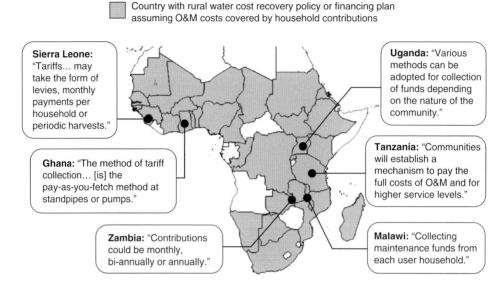

Country with rural water cost recovery policy or financing plan assuming O&M costs covered by household contributions

Sierra Leone: "Tariffs… may take the form of levies, monthly payments per household or periodic harvests."

Ghana: "The method of tariff collection… [is] the pay-as-you-fetch method at standpipes or pumps."

Zambia: "Contributions could be monthly, bi-annually or annually."

Uganda: "Various methods can be adopted for collection of funds depending on the nature of the community."

Tanzania: "Communities will establish a mechanism to pay the full costs of O&M and for higher service levels."

Malawi: "Collecting maintenance funds from each user household."

Figure 9.1 The legacy of community-based management in Africa. Based on information presented in Banerjee and Morella (2011) and WHO and UN-Water (2014). Banerjee and Morella (2011) listed countries with a rural water cost recovery strategy. WHO and UN-Water (2014) listed countries with a 'financing plan [which] defines if operating and basic maintenance is to be covered by tariffs or household contributions'. Quotes taken from the following *sources:* Malawi Ministry of Irrigation and Water Development (2010), Tanzania Ministry of Water and Livestock Development (2002), Zambia Ministry of Local Government and Housing (2007), Uganda Ministry of Water and Environment (2011), Sierra Leone Ministry of Water Resources (2013), Ghana Community Water & Sanitation Agency (2011). (*See color representation of this figure in color plate section*).

questioned by the high non-functionality rates documented in many countries (Harvey and Reed 2004; Banks and Furey 2016). However, the policy frameworks in much of Africa still assume communities will bear these costs and/or manage the process (Figure 9.1).

9.4 Rights and Results, 2000–2020

The next period we identify is 'Rights and Results' where the global targets of the Millennium Development Goal era (2000–2015) overlap with the United Nations' ratification of the Human Right to Water and Sanitation in 2010. This marks a policy departure from a 'basic needs' focus to growing donor support and national adoption of 'decentralization' as a development approach composed of a set of public policies that transfer responsibilities, resources, or authority from higher to lower levels of government (Falleti 2005), often with the goal to support delivery of basic services, particularly to the poor. In one sense, the decentralization process provides a more transparent institutional extension to the subsidiarity approach to clarify who is accountable. Growing concerns on water supply sustainability led to the design and adoption of spot-checks to overcome the sector failings earlier illustrated in Tanzania in the 1980s. The implementation of spot-checks seeks to physically confirm whether a waterpoint was functioning on the day of a visit and reflects a broader shift in sector finance from expenditure and a project mentality to thinking about services and blended finance.

The global inequalities and deprivations in basic drinking water supplies found political voice and significant progress in the 'Community Access' period, but the results by the end of the 1990s illustrated a relatively modest advance, compared with expectations, with the global proportion of people with improved water supplies increasing from 32% to 43%. The Joint Monitoring Programme of WHO and UNICEF played an important role in monitoring these numbers, and hence they triggered implicit competition between political leaders for good results compared with neighbours. Whilst these numbers can be questioned, the daily water collection ritual for mainly women and girls in rural Africa and Asia echoed the biblical 'drawers of water' (Joshua, 9:21) even at the turn of a new millennium with momentum advancing the human right to water (Langford and Russell 2017). In 2002, General Comment No. 15 on the Human Right to Water of the Committee on Economic, Social and Cultural Rights provided recognition by the UN General Assembly. In 2010, both the human rights to water and sanitation were given explicit recognition by the UN General Assembly and the Human Rights Council (see Chapter 10).

However, a rights-based approach did not gain a political consensus (at the state level) and revealed long-established fault-lines between legal positivism and economic consequentialism (Seymour and Pincus 2008). These fault-lines reflect fundamentally different perspectives between individual rights and consumer choices, the tension over which has moral or political precedent, and the degree to which there can be common ground (Hope 2017). For example, an alternative legal perspective questions the nature and type of injustices most suited to a legal right and the associated limits of rights (Osiatyński 2009). This reflects upon the historical abuse of rights, largely by states and the powerful, and the insight that 'the exercise of freedom and protection of rights is proportional to the strength and institutional organization of civil society' (Osiatyński 2009, p. 71). This reflection places greater emphasis on non-state actors than the first UN Special Rapporteur on Human Right to Water and Sanitation who acknowledges the limits of rights but grounds the approach in the state and its behaviour: 'Obviously, a human rights perspective is not a solution by itself, but these rights should influence and inspire national laws, policies and strategies and broader international commitments and policies, as well as the work of international organisations' (Langford and Russell 2017: xvi).

One consideration in a rural context is that individual rights may not align with consumer choice in delivering water to the poor. Donor choice as well as a 'demand-responsive approach' following consumer preferences dominated the 'Community Access' period where an economic rationale focused on consumer preferences and priorities to shape institutional design and development outcomes (Sara and Katz 1998). However, potential efficiency gains have to be reconciled with distributional inequalities, institutional variability and the politics of collective decision-making in rural communities, which are far from ensuring equitable outcomes (Mamdani 1996; Cleaver 1999; Khwaja 2001; Koehler et al. 2015).

Decentralization has emerged as a policy response in Africa and Asia to transfer authority, administration and finance for more effective and equitable delivery of basic services, including drinking water (Crook 2003; Robinson 2007; Crawford and Hartmann 2008). It partly addresses the concerns above on the role of intermediaries, or meso-level institutions, in ensuring policy is delivered at the right scale. The leap from national government to community management has long raised a principal-agent problem where information flows on performance delivery are slow, incomplete or entirely absent (Hope et al. 2012). Subnational and local levels of government, such as Gram Panchayats, county governments, Kebeles, or Union Pourashavas, would fit this criterion in India, Kenya, Ethiopia and Bangladesh respectively. As noted, it is an extension or clarification of the subsidiarity principle with a more explicit and legal obligation for delivery of services, particularly where a constitutional mandate exists.

Finance is a critical component of any decentralized water system. With increasing evidence that rural water systems were not functioning as designed from the 1970s (Biswas 1981) through the 1990s (Harvey and Reed 2004; RWSN 2009), the need for a revised financing model to complement or support community payments became evident (Kleemeier 2000; Whittington et al. 2009; Banerjee and Morella 2011; Foster and Hope 2017). The application of the 3T model – tariffs, taxes, and transfers (OECD 2009) – created interest in rural water finance, with models emerging in West Africa and India. With the support of the World Bank, Manobi in Senegal developed a performance-based, rural piped programme which has translated to other countries (Narkevic and Kleemeier 2010; Ndaw 2015). In India, WaterAid incubated a handpump maintenance programme in Uttar Pradesh, leveraging government subsidies with user payments for more effective maintenance systems (Hope and Rouse 2013). In both cases, there was government or donor financial support to complement user payments, and key performance indicators of sustainability by financial or operational criteria. In Kenya, a Water Services Maintenance Trust Fund model has also been piloted in two counties which characterizes the logic and structure of a performance-based approach (Figure 9.2).

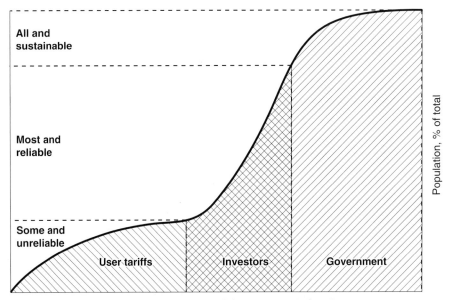

Finance for maintaining water infrastructure, % of total

Figure 9.2 A schematic for pooling financial risk through a blended finance model, the Kenya Water Services Maintenance Trust Fund. *Source:* authors.

Improving accountability to justify financial allocations created interest and donor demand for regular spot-checks as a simple auditing method. Enumerators would be appointed to make random visits to validate whether a handpump or piped system is functioning in a subsequent year(s) and compute a sustainability score based on a mix of institutional, social, technical and financial indicators (Godfrey et al. 2014). Thomson et al. (2012b) discussed the hope and hype of these and other monitoring systems, identifying caveats in assumptions and responses, particularly where the quality of data may not be known and the danger of assuming it is of high quality when it may not be. Spot-checks have become the basis for waterpoint mapping, with enumerators completing mobile surveys using tablets or smartphones. For performance-based finance programmes, such data are becoming routine in performance monitoring to guide sector performance and shape future investments.

9.5 Regulated Services, 2020–2030

The final and horizon-scanning period is the decade leading to the completion of the Sustainable Development Goals in 2030. We characterize this as 'Regulated Services', with changing priorities for the four pillars reflecting the evolution of the rural water supply debate since the 1980s. In terms of equality, it shifts from access and rights to placing greater emphasis on regulation of water services. Independent, enforceable and sustainably financed regulation of drinking water services is a critical dimension in institutional separation between water policy and service delivery (Rouse 2013). Regulation increasingly recognizes pluralism in service delivery approaches, with a clearer role for the state, market and communities beyond bricolage[3] processes (Cleaver 2012) that unfold consciously and unconsciously. Furthermore, pluralism goes beyond dichotomies and controversies between the state and market (Ostrom 2010) in favour of 'clumsy' contextual solutions (Thompson 2013) where state, market and communities define more collaborative models, allocating risks and responsibilities equitably and effectively at the right scale (Koehler et al. 2018). We advance the increasing role of domestic finance in recognition of the limited, uncertain and political flows of development assistance, and discuss emerging models and platforms. The advances in new sensor technologies and remote monitoring, which may increasingly shape the operational landscape as part of the 'surveillance-response' paradigm, will, at effectively zero extra cost, generate data-streams that can increase regulatory effectiveness and open up new channels of performance-based finance (Thomson and Koehler 2016).

Universal provision of drinking water is enhanced with regulatory oversight. If a human right to water is legislated at the national level, or a policy of universal, safe drinking water is enacted at a subnational level, this can only be meaningful with adequate and independent regulatory provision. It is an essential but often absent part of the water institutional landscape in rural Africa and Asia. In Kenya, there is evidence that where election margins are tighter, politicians may demonstrate greater

---

3 Cleaver (2012) defines bricolage as a process in which people consciously and unconsciously draw on existing social formulae (styles of thinking, models of cause and effect, social norms and sanctioned social roles and relationships) to patch or piece together institutions in response to changing situations.

responsibility in promoting drinking water improvements in their county water ministries (Koehler 2018). In such cases regulatory oversight is especially important to avoid widening regional disparities and potentially false data. The political economy of policy design may be little motivated to include and fund an independent authority to monitor and enforce policy delivery. However, for water consumers and citizens this is a critical dimension to protect their rights and to avoid abuses (see Chapter 17).

In the context of the Sustainable Development Goal of progressing from a 'basic' to 'safely managed' water, a number of regulatory issues seem salient. First, the benchmark of safely managed drinking water is on-site, on-demand and without contamination (UNICEF/WHO 2017). It is difficult to see how such private delivery of water in the rural contexts of Africa or Asia is financially or institutionally feasible by 2030. Without dismissing this goal, it would seem pragmatic to first achieve basic water supplies for everyone. The history of economic development points to a common pattern of increasing piped water access in urban areas, which leads to an inflexion point in increasing piped water in rural areas (Figure 9.3). The figure illustrates that rural piped supplies are limited (<20%) until around 50% of the urban population has piped supplies. Further, many African countries are making greater and disproportionate progress in urban piped access (above the dotted line) compared with Asia which has a more even pattern.

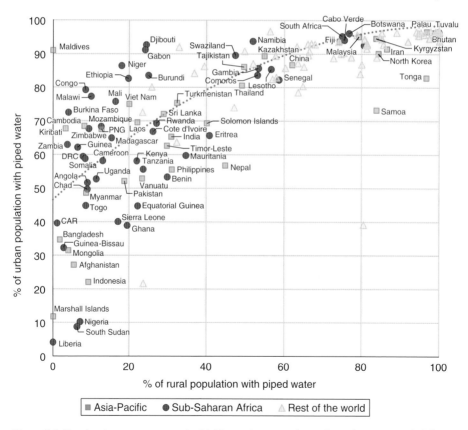

Figure 9.3 Piped water coverage as main drinking water source in rural vs. urban areas: sub-Saharan Africa, Asia, and the rest of the world in 2015. *Source:* Authors' analysis of data from UNICEF/WHO (2018). (*See color representation of this figure in color plate section*).

Given the historical trends already discussed, the achievement of universal basic delivery in rural Africa and Asia by 2030 would be an unprecedented public policy success.

Regulation of basic water access in rural areas is non-trivial, which partly explains why there has been limited progress over decades. Whilst there have been major advances in remotely monitoring daily reliability and estimated quantities of water from handpumps (Thomson et al. 2012a; Nagel et al. 2015) and small piped systems (Ndaw 2015), similar progress has not been observed for monitoring water quality in Africa or in Asia (Peletz et al. 2016). The uncertainty in determining reliable estimates of water quality (Nowicki et al. 2018) from rural water sampling using established and emerging methods identifies an unresolved scientific challenge before operational and political issues complicate matters further. This is amplified by the well-rehearsed issue of sampling at the water source or at point-of-use (Wright et al. 2004). In turn, this speaks to the legal obligation and associated liability within a rights-based framework where lack of clarity understandably makes governments reluctant to commit to the unknown. In most cases, governments have interest in monitoring publicly-provided waterpoints aligning with the SDG water quality targets on bacterial contamination (*Escherichia coli*) and certain chemicals (e.g. arsenic). Progress to resolve the methodological problems of identification and measurement at scale is a central challenge for rural water regulation in communities, schools, clinics, and places of work.

Monitoring by surveillance offers systematic, objective and continuous data to measure, monitor and improve rural water performance. The incremental advances in 'spot-checks' to increase accountability do not address the potential biases, errors and redundancy of a human check one day every year or more (<1% of daily usage). In-situ monitoring devices build from established systems in most urban contexts where standard or smart meters provide necessary data for utility management and planning, promoting accountability in policy, planning and finance cases (Figure 9.4).

First, the policy case is that public investments in surveillance of rural water infrastructure can objectively monitor the daily quantity of water supplied and system

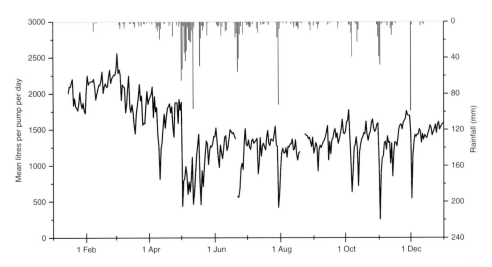

Figure 9.4 Handpump usage and rainfall in coastal Kenya, 2014. *Source:* Thomson et al. (2019). Reproduced with permission of Elsevier.

reliability at scale. Obligation 37(c) of General Comment 15 (UN Economic and Social Council 2003) refers to the obligation 'to monitor the extent of the realization, or the non-realization, of the right to water'. Whilst General Comment 15 is not a binding document, and obligations under the Economic and Social Council Covenant are qualified by the 'progressive realization' clause, they are increasingly being incorporated into national legislation and policy (Thomson et al. 2012b). For countries to know if they are fulfilling their policy commitments, they must understand the levels of water services they are actually providing. It is known that rural waterpoints serve people, livestock and sometimes small-scale irrigation. This matters in periods of drought where demand escalates as surface water alternatives evaporate. If a system is unreliable at such points of peak demand, the social and economic implications lead to negative impacts, often hitting the most vulnerable hardest. For example, the cost of humanitarian responses for addressing water infrastructure failure in droughts run into hundreds of millions of US dollars, which could be reduced through investing in reliably maintained water systems (Godfrey and Hailemichael 2017).

Second, the planning case builds the latter point on understanding risks in system design. Where demand is known to be higher, there is the need to guide future infrastructure investments or decommissioning of redundant waterpoints. Third, the financial case from surveillance offers a radical change from building infrastructure to providing services with quantifiable and transparent metrics which could release longer-term finance at lower unit cost through efficiency gains. The investment case would shift to long-term delivery of services validated by a portfolio of performance metrics.

The financial support for surveillance would increasingly shift to domestic funding. By this, we mean the major source of long-term finance would be from national or local government and users. Where these domestic commitments are made, performance-based flows of external finance would pool with domestic finance in special project vehicles such as trust funds and the like. Critical to this type of arrangement is the surveillance system, to provide credible, timely and objective data for investors, but equally for the regulatory authority. Whilst surveillance would create an additional cost, the theoretical case suggests that the reduction in inefficiencies in failed or abandoned water systems would offset the support for smart monitoring systems. For example, two rural maintenance programmes in the Central African Republic and Kenya estimate monitoring at less than 5% of local operational costs in delivery models for handpump and small piped schemes for over 200 000 people in communities, schools and clinics (Lane 2018).

The emergence and growth of local, social enterprises to support the monitoring and maintenance of rural water infrastructure in Senegal (Manobi), Burkina Faso (Hydro Vergnet), Tanzania (Msabi), Kenya (FundiFix), Cambodia and Vietnam (piped schemes owned and operated by micro-enterprises) aligns a public policy imperative with an economic opportunity. This growing market response (see e.g. Cain 2018) to a policy and regulatory gap presents new institutional opportunities and risks to rethink the dynamic between social regulation and collective representation (Koehler et al. 2018). The community access period was defined by a linear relationship between the state and the community with the limitations discussed earlier. Here we see a new institutional space being created between the state, market and communities where risks and responsibilities may be more effectively identified and allocated. The pluralism of these risk dynamics recognizes that states apply rules to reduce risk, the market sees risk as opportunity, and communities share risk through collective action. Each of the domains has an important contribution but, to date, they have not been formally understood as a collective and systematic response. Early evidence from Kenya suggests that this provides a promising but unproven approach to improving rural water sustainability for communities, schools and clinics in a coordinated approach at scale (Koehler et al. 2018).

Regulation is a necessary condition to provide independent oversight of the market structure and behaviour of such arrangements. The extent, capacity and powers of regulation can vary from passive monitoring to active sector development. Common functions of economic and water quality regulation require associated investments in monitoring systems and legal provision to take proportionate action where provision of services fails across categories of coverage, reliability, physical access, quality or affordability. Government failure by policy design or legal provision, or market failure by inequitable service models, cannot be remedied by weak regulation. The influence and impacts of regulation are proportional to its institutional investment and legal mandate. Decades of under-investment, or no investment, in rural water regulation reflects a wider malaise with a historical bias to investing in infrastructure over institutions, characterized by a political priority for access over services. Determining the sequence, nature and timing of infrastructure and institutional investments is a complex but critical determinant for achieving and maintaining sustainable water systems (Grey and Sadoff 2006). As such, we advance the idea of 'regulated services' as a logical extension in the continuum of progress towards rural water sustainability, with the caveat that weak regulation as a political fix is inadequately underlining the importance of domestic finance through stable government support and investment.

9.6   Limits to Progress

Our discussion of four policy pillars over three historical periods has demonstrated how elusive the goal of universal water service delivery has proven to be, despite significant economic, social and political developments in Africa and Asia. So what are the limits to progress, and to what extent can they be addressed? Two broad findings emerge, with the acknowledgement of the selective nature of the pillars chosen and the indicative typologies of 'community access', 'rights and results', and 'regulated services'.

First, institutions matter. This has long been recognized and, in many ways, has characterized the rural water sector through the idea of community-based management as a pragmatic response to weak or absent government in earlier periods. However, the bias towards access and building infrastructure without investing in appropriate institutions to independently monitor and maintain service delivery, inform planning and shape policy, has created a graveyard of well-meaning intentions across Africa and Asia. Institutional allocation of risks and responsibilities between the state, market and community has created a disproportionate burden on those least able to manage service delivery effectively. New ideas and emerging evidence on pluralist arrangements offer a less prescriptive and more inclusive approach, reflecting the growth of performance-based enterprises supporting communities and pooling operational and financial risks at scale. Institutional design should be at scale in managing and regulating rural water systems, infrastructure and stakeholders.

The political economy and legacy of uncoordinated investments in infrastructure casts doubt on the prospects of changing donor, government and contractor behaviours to promote greater accountability. Where the ends matter more than the means (Therkildsen 1988), sustainability is trumped by worthy but short-term goals. Shifting sector psychology from date-stamped targets to an institutional model that can be sustained over time is a defining challenge and opportunity. Putting a 'cart of well-meaning targets' before identifying an institutional 'horse' has not worked in the past and presents a warning of the likely political pressure to meet the advancing SDG target in 2030. We propose the idea of 'regulated services' to convene alternative ideas of pluralist arrangements, domestic finance and surveillance to design and test new institutional approaches where there is recognition that rural water sustainability is a function of institutional investment and political commitment at the appropriate operational level.

Second, rural information systems offer new opportunities to improve accountability and financial sustainability. The dramatic and recent growth of mobile and satellite network coverage, ownership of mobile phones, mobile payments, sensor technology, and associated data analytics present opportunities to strengthen institutional design (Foster et al. 2012; Hope et al. 2012; Nagel et al. 2015). We qualify these comments with the recognition that data can be costly and meaningless without an institutional purpose for collection, curation and link to action (Thomson et al. 2012b). But, for decades the rural water sector has operated in a virtual data vacuum, with decision-making processes largely opaque and results generally uncertain. Significant advances in remote monitoring of daily water demand (Nagel et al. 2015; Thomson et al. 2019) can provide objective metrics for monitoring, regulation and management systems.

In conclusion, the Sustainable Development Goal presents the third wave of interventions to improve rural water supply at scale which, if achieved without sustainable institutional models in the process, may be at the cost of the longer-term prize of maintaining sustainable drinking water for all. One response in the short term is to increase investments in designing and testing existing and emerging institutional models in rural Africa and Asia, to evaluate the trade-offs in performance across institutional, financial and operational dimensions. Building a global community of practice with a robust evidence base will allow policy and planning institutions to balance and negotiate short-term political goals with long-term sector sustainability.

Acknowledgements

This work is an output from the REACH programme (www.reachwater.org.uk) funded by UK Aid from the UK Department for International Development (DFID) for the benefit of developing countries (Aries Code 201880). However, the views expressed, and the information contained in it are not necessarily those of or endorsed by DFID, which can accept no responsibility for such views or information or for any reliance placed on them. Additional funding and support for this chapter came from the New Mobiles Citizens and Waterpoint Sustainability in Rural Africa project supported by ESRC/DFID (Grant number: ES/J018120/1); the University of Oxford Clarendon Fund; and the Worshipful Company of Water Conservators.

References

Arlosoroff, S., Tschannerl, G., Grey, D. et al. (1987). *Community Water Supply: The Handpump Option*. Washington, DC: World Bank.

Banerjee, S.G. and Morella, E. (2011). *Africa's Water and Sanitation Infrastructure: Access, Affordability, and Alternatives*. Washington, DC: World Bank.

Banks, B. and Furey, S.G. (2016). *What's Working, Where, and for How Long: A 2016 Water Point Update to the RWSN (2009) Statistics.* St. Gallen, Switzerland: GWC/Skat.

Barry, J.M. (1997). *Rising Tide: The Great Mississippi Flood of 1927 and How It Changed America.* New York: Simon and Schuster.

Biswas, A. (1981). Water for the third world. *Foreign Affairs* 60 (1): 148–166.

Blackbourn, D. (2006). *The Conquest of Nature: Water, Landscape and the Making of Modern Germany.* London: Pimlico.

Briscoe, J. (2014). The Harvard water federalism project – process and substance. *Water Policy* 16 (S1): 1–10.

Briscoe, J. and de Ferranti, D. (1988). *Water for Rural Communities: Helping People Help Themselves.* Washington, DC: World Bank.

Cain, A. (2018). Informal water markets and community management in peri-urban Luanda, Angola. *Water International* 43 (2): 205–216. https://doi.org/10.1080/02508060. 2018.1434958.

Chowns, E. (2015). Is community management an efficient and effective model of public service delivery? Lessons from the rural water supply sector in Malawi. *Public Administration and Development* 35 (4): 263–276.

Cleaver, F. (1999). Paradoxes of participation: questioning participatory approaches to development. *Journal of International Development* 11 (4): 597–612.

Cleaver, F. (2012). *Development Through Bricolage: Rethinking Institutions for Natural Resource Management.* London: Earthscan.

Colchester, F., Marais, H.G., Thomson, P. et al. (2017). Accidental infrastructure for groundwater monitoring in Africa. *Environmental Modelling and Software* 91: 241–250.

Conyers, D. (2007). Decentralisation and service delivery: lessons from sub-Saharan Africa. *IDS Bulletin* 38 (1): 18–32.

Crawford, G. and Hartmann, C. (2008). *Decentralisation in Africa: A Pathway out of Poverty and Conflict.* Amsterdam: Amsterdam University Press.

Crook, R.C. (2003). Decentralisation and poverty reduction in Africa: the politics of local–central relations. *Public Administration and Development* 23 (1): 77–88. https://doi.org/10.1002/pad.261.

Falkenmark, M. (1977). UN water conference: agreement on goals and action plan. *AMBIO* 6 (4): 222–227.

Falkenmark, M. (1982). *Rural Water Supply and Health. The Need for a New Strategy.* Uppsala: Scandinavian Institute of African Studies.

Falleti, T.G. (2005). A sequential theory of decentralization: Latin American cases in comparative perspective. *American Political Science Review* 99 (3): 327–346.

Fonseca, C. and Pories, L. (2017). *Financing WASH: how to increase funds for the sector while reducing inequalities.* Position Paper for the Sanitation and Water for All Finance Ministers Meeting. The Hague, Netherlands: IRC, water.org, Ministry of Foreign Affairs and Simavi.

Foster, T. (2013). Predictors of sustainability for community-managed handpumps in sub-Saharan Africa: evidence from Liberia, Sierra Leone, and Uganda. *Environmental Science and Technology* 47: 12037–12046.

Foster, T. and Hope, R. (2016). A multi-decadal and social-ecological systems analysis of community waterpoint payment behaviours in rural Kenya. *Journal of Rural Studies* 47: 85–96. https://doi.org/10.1016/j.jrurstud.2016.07.026.

Foster, T. and Hope, R. (2017). Evaluating waterpoint sustainability and access implications of revenue collection approaches in rural Kenya. *Water Resources Research* 53 (2): 1473–1490. https://doi.org/10.1002/2016WR019634.

Foster, T., Hope, R.A., Thomas, M. et al. (2012). Impacts and implications of mobile water payments in East Africa. *Water International* 36 (7): 788–804.

Foster, T., Willetts, J., Lane, M. et al. (2018a). Risk factors associated with rural water supply failure: a 30-year retrospective study of handpumps on the south coast of Kenya. *Science of the Total Environment* 626: 156–164. https://doi.org/10.1016/j.scitotenv.2017.12.302.

Foster, T., Shantz, A., Lala, S., and Willetts, J. (2018b). Factors associated with operational sustainability of rural water supplies in Cambodia. *Environmental Science: Water Research & Technology* 4 (10): 1577–1588.

Godfrey, S. and Hailemichael, G. (2017). Life cycle cost analysis of water supply infrastructure affected by low rainfall in Ethiopia. *Journal of Water Sanitation and Hygiene for Development* 7 (4): 601–610. https://doi.org/10.2166/washdev.2017.026.

Godfrey, S., Van der Velden, M., Muianga, A. et al. (2014). Sustainability check: five-year annual sustainability audits of the water supply and open defecation free status in the 'one million initiative', Mozambique. *Journal of Water Sanitation and Hygiene for Development* 4 (3): 471–483.

Greeff, H., Manandhar, A., Thomson, P. et al. (2018). Distributed inference condition monitoring system for rural infrastructure in the developing world. *IEEE Sensors Journal.* https://doi.org/10.1109/jsen.2018.2882866.

Grey, D. and Sadoff, C. (2006). Water for Growth and Development. Thematic Documents of the IV World Water Forum. Mexico City: Comision Nacional del Agua.

Harvey, P. and Reed, R. (2004). *Rural Water Supply in Africa: Building Blocks for Handpump Sustainability.* Loughborough: Water, Engineering and Development Centre, Loughborough University.

Harvey, P. and Reed, R. (2006). Community-managed water supplies in Africa: sustainable or dispensable? *Community Development Journal* 42 (3): 365–378.

Hope, R. (2017). A poor choice? Public policy, social choice and the human right to water. In: *The Human Right to Water – Theory, Policy and Practice* (eds. M. Langford and A. Russell), 601–621. Cambridge: Cambridge University Press.

Hope, R. and Rouse, M. (2013). Risks and responses to universal drinking water security. *Philosophical Transactions of the Royal Society* 371: 1–23. https://doi.org/10.1098/rsta.2012.0417.

Hope, R., Foster, T., and Thomson, P. (2012). Reducing risks to rural water security in Africa. *AMBIO* 41 (7): 773–776.

Hutton, G. and Varughese, M. (2016). *The Costs of Meeting the 2030 Sustainable Development Goal Targets on Drinking Water, Sanitation, and Hygiene.* Washington, DC: World Bank.

ICWE (1992). The Dublin Statement and Report of the Conference. *International Conference on Water and the Environment: Development Issues for the 21st Century,* Dublin, Ireland (26–31 January 1992). Dublin: ICWE.

Ingram, H., Scaff, L., and Silko, L. (1986). Replacing confusion with equity: alternatives for water policy in the Colorado River basin. In: *New Courses for the Colorado River: Major Issues for the Next Century* (eds. G.D. Weatherford and F.L. Brown), 177–199. Albuquerque: University of New Mexico Press.

Khwaja, A.I. (2001). *Can good projects work in bad communities? Collective Action in the Himalayas.* Working Paper Series rwp01-043. Harvard University, John F. Kennedy School of Government.

Kleemeier, E. (2000). The impact of participation on sustainability: an analysis of the Malawi rural piped scheme program. *World Development* 28 (5): 929–944.

Koehler, J. (2018). Exploring policy perceptions and responsibility of devolved decision-making for water service delivery in Kenya's 47 county governments. *Geoforum* 92: 68–80. https://doi.org/10.1016/j.geoforum.2018.02.018.

Koehler, J., Thomson, P., and Hope, R. (2015). Pump-priming payments for sustainable water Services in Rural Africa. *World Development* 74: 397–411. https://doi.org/10.1016/j.worlddev.2015.05.020.

Koehler, J., Rayner, S., Katuva, J. et al. (2018). A cultural theory of drinking water risks, values and institutional change. *Global Environmental Change* 50: 268–277. https://doi.org/10.1016/j.gloenvcha.2018.03.006.

Lane, A. (2018). *Financing Functionality: A Comparison of Emerging Maintenance Service Providers for Rural Water in Sub-Saharan Africa*. MSc thesis. School of Geography and the Environment, University of Oxford, UK.

Langford, M. and Russell, A. (2017). *The Human Right to Water – Theory, Policy and Practice*. Cambridge: Cambridge University Press.

Lankester, T. (2013). *The Politics and Economics of Britain's Foreign Aid – The Pergau Dam Affair*. London: Routledge.

Mamdani, M. (1996). *Citizen and Subject. Contemporary Africa and the Legacy of Late Colonialism*. Princeton: Princeton University Press.

Marshall, G. (2008). Nesting, subsidiarity, and community-based environmental governance beyond the local level. *International Journal of the Commons* 2 (1): 75–97.

Meinzen-Dick, R. and Nkonya, L. (2007). Understanding legal pluralism in water and land rights: lessons from Africa and Asia. In: *Community-Based Water Law and Resource Management Reform in Developing Countries* (eds. B. Van Koppen, M. Giordano and J. Butterworth), 12–27. Wallingford: CAB International.

Ministry of Irrigation and Water Development (MIWD) (2010). *Implementation Guidelines for Rural Water Supply and Sanitation*. Lilongwe, Malawi: Ministry of Irrigation and Water Development.

Ministry of Local Government and Housing (MLGH) (2007). *National Guidelines for Sustainable Operation and Maintenance of Hand Pumps in Rural Areas*. Lusaka, Zambia: Ministry of Local Government and Housing.

Ministry of Water and Environment (MWE) (2011). *National Framework for Operation and Maintenance of Rural Water Supplies in Uganda*. Kampala, Uganda: Ministry of Water and Environment.

Ministry of Water and Livestock Development (MWLD) (2002). *National Water Policy*. Dar es Salaam, Tanzania: Ministry of Water and Livestock Development.

Ministry of Water Resources (MWR) (2013). *Rural Water Supply and Small Towns Strategy Document*. Freetown, Sierra Leone: Ministry of Water Resources.

Mosse, D. (2003). *The Rule of Water. Statecraft, Ecology and Collective Action in South India*. New Delhi: Oxford University Press.

Mosse, D. (2004). *Cultivating Development. An Ethnography of Aid Policy and Practice*. London: Pluto Press.

Mudgal, A.K. (1997). *India Handpump Revolution – Challenge and Change*. HTN Working Paper: WP 01/97. St Gallen: Swiss Centre for Development Cooperation in Technology and Management.

Nagel, C., Beach, J., Iribagiza, C., and Thomas, E. (2015). Evaluating cellular instrumentation on rural handpumps to improve service delivery – a longitudinal study in rural Rwanda. *Environmental Science & Technology* 49 (24): 14292–14300. https://doi.org/10.1021/acs.est.5b04077.

Narayan, D. (1995). *The contribution of people's participation: evidence from 121 rural water supply projects*. Environmentally Sustainable Development Occasional Paper Series: No. 1. Washington, DC: World Bank.

Narkevic, J. and Kleemeier, E. (2010). *A global review of private operator experiences in rural areas. Private Operator Models for Community Water Supply*. Rural Water Supply Series Field Note (February 2010). Nairobi: World Bank Water and Sanitation Program.

Ndaw, M.F. (2015). *ICT services to improve performances of rural water private operators in West Africa*. Field Note, Water and Sanitation Program. Washington, DC: World Bank.

North, D.C. (1990). *Institutions, Institutional Change and Economic Performance*. Cambridge: Cambridge University Press.

Nowicki, S., Lapworth, J.L., Ward, J.S.T. et al. (2018). Tryptophan-like fluorescence as a measure of microbial contamination risk in groundwater. *Science of the Total Environment* 646 (2019): 782–791.

OECD (2009). *Managing Water for All: An OECD Perspective on Pricing and Financing*. Paris: OECD.

Osiatyński, W. (2009). *Human Rights and Their Limits*. Cambridge: Cambridge University Press.

Ostrom, E. (2010). Beyond markets and states: polycentric governance of complex economic systems. *American Economic Review* 100: 641–672.

Peletz, R., Kumpel, E., Bonham, M. et al. (2016). To what extent is drinking water tested in sub-Saharan Africa? A comparative analysis of regulated water quality monitoring. *International Journal of Environmental Research and Public Health* 13 (3): 275.

Robinson, M. (2007). Does decentralisation improve equity and efficiency in public service delivery provision? *IDS Bulletin* 38 (1): 7–17.

Rondinelli, D.A. (1991). Decentralizing water supply services in developing countries: factors affecting the success of community management. *Public Administration and Development* 11 (5): 415–430. https://doi.org/10.1002/pad.4230110502.

Rouse, M. (2013). *Institutional Governance and Regulation of Water Services. The Essential Elements*. London: IWA Publishing.

RWSN (2009). *Myths of the Rural Water Supply Sector. RWSN Perspective No. 4*. Gland, Switzerland: Rural Water Supply Network. http://www.rural-water-supply.net/en/resources/details/226.

Sara, J. and Katz, T. (1998). *Making Rural Water Supply Sustainable: Report on the Impact of Project Rules*. Washington, DC: UNDP/World Bank.

Seymour, D. and Pincus, J. (2008). Human rights and economics: the conceptual basis for complementarity. *Development Policy Review* 24 (4): 387–405.

Therkildsen, O. (1988). *Watering White Elephants? Lessons from Donor Funded Planning and Implementation of Rural Water Supplies in Tanzania*. Uppsala: Scandinavian Institute of African Studies.

Thomas, E. (2016). Introduction. In: *Broken Pumps and Promises: Incentivizing Impact in Environmental Health* (ed. E. Thomas), 1–4. Heidelberg: Springer.

Thompson, M. (2013). Clumsy solutions to environmental change: lessons from cultural theory. In: *A Changing Environment for Human Security* (eds. L. Sygna, K. O'Brien and J. Wold), 424–432. Abingdon: Routledge.

Thomson, P. and Koehler, J. (2016). Performance-oriented monitoring for the water SDG – Challenges, tensions and opportunities. *Aquatic Procedia* 6: 87–95. https://doi.org/10.1016/j.aqpro.2016.06.010.

Thomson, P., Hope, R., and Foster, T. (2012a). GSM-enabled remote monitoring of rural handpumps: a proof-of-concept study. *Journal of Hydroinformatics.* 14 (4): 29–39.

Thomson, P., Hope, R., and Foster, T. (2012b). Is silence golden? Of mobiles, monitoring, and rural water supplies. *Waterlines* 31 (4): 280–292.

Thomson, P., Bradley, D., Katilu, A. et al. (2019). Rainfall and groundwater use in rural Kenya. *Science of the Total Environment* 649: 722–730.

UN Economic and Social Council (2003). General Comment No. *15: The Right to Water.* Committee on Economic, Social and Cultural Rights. https://www2.ohchr.org/english/issues/water/docs/CESCR_GC_15.pdf (accessed 3 June 2019).

UNICEF/WHO (2017). *Progress on Drinking Water, Sanitation and Hygiene: 2017 Update and SDG Baselines.* Geneva: World Health Organization (WHO) and the United Nations Children's Fund (UNICEF).

UNICEF/WHO (2018). *Joint Monitoring Programme for Water Supply, Sanitation and Hygiene – Data.* https://washdata.org/data.

Wade, R. (1988). *Village Republics. Economic Conditions for Collective Action in South India.* Cambridge: Cambridge University Press.

Welle, K., Williams, J., and Pearce, J. (2016). ICTs help citizens voice concerns over water – or do they? *IDS Bulletin* 47 (1) https://doi.org/10.19088/1968-2016.105.

Whaley, L. and Cleaver, F. (2017). Can 'functionality' save the community management model of rural water supply? *Water Resources and Rural Development* 9: 56–66.

White, G.F., Bradley, D.J., and White, A.U. (1972). *Drawers of Water: Domestic Water Use in East Africa.* Chicago: University of Chicago Press.

Whittington, D., Davis, J., Prokopy, L. et al. (2009). How well is the demand-driven, community management model for rural water supply systems doing? Evidence from Bolivia, Peru and Ghana. *Water Policy* 11 (6): 696–718.

Wilder, M. and Ingram, H. (2018). Knowing equity when we see it: water equity in contemporary global contexts. In: *The Oxford Handbook of Water Politics and Policy* (eds. e.K. Conca and E. Weinthal), 49–75. Oxford: Oxford University Press.

Wittfogel, K.A. (1957). *Oriental Despotism: A Comparative Study of Total Power.* New Haven and London: Yale University Press.

World Health Organization and UN-Water (2014). *UN-Water global analysis and assessment of sanitation and drinking-water (GLAAS) 2014 report: investing in water and sanitation, increasing access, reducing inequalities.* Geneva: World Health Organization.

Wright, J., Gundry, S., and Conroy, R. (2004). Household drinking water in developing countries: a systematic review of microbiological contamination between source and point of use. *Tropical Medicine and International Health* 9 (1): 106–117.

10

The Human Right to Water

Rhett Larson[1], Kelsey Leonard[2], and Richard Rushforth[3]

[1] *Sandra Day O'Connor College of Law, Arizona State University, USA*
[2] *Department of Political Science, McMaster University, Canada*
[3] *School of Informatics, Computing, and Cyber Systems, Northern Arizona University, USA*

10.1   The Legal and Historical Background of the Human Right to Water

The rationales for establishing and enforcing a human right to water can be roughly placed into three categories (Larson 2013). The first rationale is that the recognition of a human right to water acts as a legal bulwark against social and economic inequality (Thompson 2011). The second rationale for a human right to water is perhaps best expressed by the political philosopher John Rawls. Put very simply, Rawls argued that governments should guarantee provision of certain 'primary goods' essential for the realisation of all other rights and responsibilities (Rawls 1971). Water is the quintessential Rawlsian primary good, and a human right to water appropriately puts first things first (Haeusermann 1998). The third rationale for the recognition of a human right to water is to advance governmental accountability. This policy aim relates to a fundamental legal doctrine in many parts of the world – the public trust doctrine (Sax 1970, p. 471). Under the public trust doctrine, the state holds title to water resources as a trustee, with a fiduciary obligation to manage that resource for the benefit of all citizens. The public trust doctrine, although applied differently in different countries, is a recognized legal doctrine around the world (Blumm and Guthrie 2012, p. 741). A human right to water may provide a mechanism whereby citizens can enforce the state's fiduciary obligation owed to them under the public trust doctrine.

These three broad rationales supporting a human right to water have existed as long as human civilization. Various conceptions of human rights have similarly existed throughout human history, from early formulations of natural law by the Greeks, to the proclamations of Persian emperor Cyrus the Great regarding freedom of religion and racial equality, to the Great Law of Peace of the Haudenosaunee Confederacy championing the rights of all beings through unity and peace (Jacobs 2000; Weissbrodt and de la Vega 2007, pp. 13–14). Given the persistence of both human rights and the rationales for a human right to water, it is unsurprising that some form of a human right to water

Water Science, Policy, and Management: A Global Challenge, First Edition. Edited by Simon J. Dadson, Dustin E. Garrick, Edmund C. Penning-Rowsell, Jim W. Hall, Rob Hope, and Jocelyne Hughes.
© 2020 John Wiley & Sons Ltd. Published 2020 by John Wiley & Sons Ltd.

has existed in many different cultures and legal traditions for centuries. Some of these legal approaches to the human right to water have been spread by conquest and colonization, and have taken root and developed differently around the world.

One of the earliest and most widespread conceptions of a human right to water is embedded in one of the most ancient and influential legal systems – Sharia law. The importance placed on water rights in Sharia law is not surprising, given the roots of this religious code in the arid Arabian peninsula (Salzman 2006, p. 100). Sharia law recognizes a right to irrigate, called *shirb*, and another called *shafa*, or the 'right of thirst', which establishes a universal right for all humans and their animals to quench their thirst (Salzman 2006, p. 100). These rights are implemented in different ways depending on varying geographies and on sectarian interpretations, but include certain generalizable principles, including a focus on equity in water distribution (Mallat 1995). And whilst Sharia law is codified in only a few national legal systems, it remains influential in many countries because it provides a foundation for the Ottoman water code, called the *Majalla*, which continues to inform water policy in areas formerly under Ottoman control (Ahmad 2000, p. 159). These Sharia water rights are sometimes interpreted as absolute prohibitions against charging for water services, as prohibiting charging for water but not for water delivery services, or, depending on distinctions between private and public water sources, in determining when, how, and how much to charge for water services (Morill and Simas 2009).

Similar to the role of the *Majalla* throughout the Ottoman Empire, the Roman Empire imported and expanded other early forms of the human right to water into the legal systems of many nations. In modern nations that are former Roman colonies, property interests in water are allocated based on riparian water rights with roots in Roman law (Dellapenna 2011, p. 61). Under a riparian water rights regime, landowners with property abutting a waterbody have the right to make 'reasonable use' of that water (Sax and Abrams 1986, pp. 154–157). 'Natural' uses are presumed reasonable, whilst 'artificial' water uses are subject to a court's evaluation of the relative interests of co-riparian landowners (Trelease 1954). A natural use of water based on riparian rights satisfies basic needs, and is typically limited to household-level drinking, sanitation, hygiene, and subsistence-level irrigation and livestock watering (Ausness 1986, p. 416). All other appropriations of water are considered artificial, and therefore subject to the reasonableness evaluation (Haddaway 1929, p. 453). The ancient Roman riparian regime thus includes a legal recognition that water uses for basic human needs are prioritized above other uses.

The different rights to *shirb* and *shafa* under Sharia law, and the distinction between artificial and natural uses, illustrate one of the challenges in formulating and implementing a human right to water, which is delineating its boundaries. When we speak of a human right to water, is it a reference only to drinking water, to domestic water uses in general (including hygiene and sanitation), or to a broader right that incorporates access to water resources for the development of food and energy? This debate is beyond the scope of a single chapter, but the human right to water addressed here references rights to domestic water uses.

The British Empire also similarly planted elements of a human right to water in its former colonies. Under British law, the king could grant charters (monopolies) to certain enterprises, but such charters included certain duties, including the common law duty to serve (Rossi 1998, pp. 1244–1248). That duty applied to water uses for mills and ferries in medieval England, but has come to apply to other public utilities as those services developed in later centuries throughout the former British Empire. The duty to serve functions as a *quid pro quo* of sorts for the state granting monopolies to public water utilities – in exchange for the monopoly, the enterprise assumes the duty to serve all customers without discrimination. The duty to serve that is held by a public utility represents a proto-human right to water, in that it guarantees an affordable price and access to all similarly situated people served by the utility.

There are other potential examples of historical approaches to a human right to water beyond those of the Ottoman, Roman and British empires. These three examples of imperial sowing of the seeds of the human right to water are not intended as an exhaustive list of early approaches to human rights to water, nor is referencing these approaches any kind of endorsement or defence of these empires nor of colonialism in general. But these imperial seeds did take root in many parts of the world and influenced the development of water law, and an effective evaluation of the current flowering of a modern conception of a human right to water requires an understanding of the legal concepts from whence it sprang. These three examples illustrate major challenges to the effective implementation of a sustainable and equitable human right to water. The example of *shafa* illustrates concerns for affordability. The example of the riparian right concept of 'natural uses' illustrates the concern for quantifying a minimum amount of water guaranteed by a human right. The example of the British common law duty to serve illustrates the need for equitable provision of water resources without discrimination. More fundamentally, perhaps, each of these historical approaches reflects a major limitation to the human right to water – it is inherently anthropocentric, and may limit cost recovery and cost internalization to facilitate economic and ecological sustainability. The human right to water is, unsurprisingly, concerned mainly with humans.

These examples of foundational elements of the modern human right to water also reflect a fundamental challenge inherent in the formulation and implementation of any human right – delineating the parameters of the government's mandate. Human rights are often described in one of two ways (Berlin 1969). On the one hand, rights could be characterized as negative rights. Such rights are protections against government interference without due process and a sufficiently compelling countervailing government interest. Such negative rights include, for example, freedom from religious, racial or ethnic discrimination embodied in many national constitutions and in the UN Covenant on Civil and Political Rights (the UNCCPR) (UN G.A. Res. 2200: United Nations 1967; Dworkin 1977).

On the other hand, human rights may be described as positive rights. Positive rights establish affirmative duties, requiring governments to take action to ensure them, and may include, for example, rights to education, healthcare or housing, or rights such as those guaranteed under the UN Covenant on Economic, Social, and Cultural Rights (the UNCESCR) (UN G.A. Res. 2200: United Nations 1967). The distinction between positive and negative rights, whilst descriptively helpful, is problematic, in part because even the protection of negative rights requires the guaranteed provision of enforcement mechanisms, such that many rights, like magnets, have a positive and negative pole (Cross 2001). The examples of natural use protections in Roman riparian rights, *shafa* rights under Sharia law, and the duty to serve under British common law, reflect this positive/negative rights polarity. Riparian rights are private property rights protected against arbitrary governmental intrusion, yet the preference given to natural uses imposes a duty on the state to ensure some minimum quantity for other riparian users. The right of *shafa* under Sharia law arguably requires water provision, yet also prohibits the state from imposing certain pricing structures on certain water sources. The duty to serve under British common law appears at first glance to be a paradigmatic example of a positive right – the duty to provide a service – yet in many ways is a true negative right because it prohibits state-sponsored discrimination.

These examples of historical legal doctrines attempting to respond to timeless water challenges have evolved into modern legal innovations more explicitly grounded in the rhetoric of human rights (McCaffrey 1992). For example, in 2002, the UN Human Rights Commission issued General Comment 15 to the UNCESCR, which interprets that covenant to implicitly include a right to water as a prerequisite to the realisation of other explicitly protected rights (United Nations Economic and Social Council, 2003: General Comment 15). The UN General Assembly adopted a resolution recognizing an international human right to water in 2010 (UN G.A. Res. 64/292: United Nations 2010). The 2010 UN Resolution declared that the 'right to safe and clean drinking water... [is] a human right that is essential for the full enjoyment of life and all human rights' (UN G.A. Res. 64/292: United Nations 2010).

At the domestic level, more than 40 countries have expressly recognized a right to water in their respective constitutions (Larson 2013). Nevertheless, the vast majority of countries do not guarantee water under their constitutions. And even in those countries with a constitutional right to water, the formulation of that right often leaves critical water management questions open for interpretation, including the amount, price and proximity to the point of use necessary to satisfy such a right.

Despite the strong rationales for, and long history of, a human right to water, the recognition and implementation of such a right can be problematic. A human right to water can be difficult to enforce. The international human right to water is largely formulated in General Comment 15, which is a non-binding interpretation of the UNCESCR. The 2010 UN General Assembly Resolution is similarly a non-binding political statement, and 41 countries abstained from signing that resolution (Sammis 2010). The UNCESCR itself lacks an Optional Protocol to allow for any claims to be brought for its violation. At the domestic level, the human right to water is similarly difficult to enforce. It is largely formulated as an aspirational right that nations must progressively achieve within the limits of available resources. If not clearly described by law, courts are left to make difficult determinations regarding the rate and scope of realisation of the right, and complex delineations of the state's obligations in terms of quantity, quality, cost and access (Larson 2013).

A human right to water may also be formulated or implemented in a way that is not economically or ecologically sustainable. If the right requires water to be underpriced, water providers may not recover costs, and thus cannot reinvest revenues to ensure that infrastructure is maintained and upgraded, professionals are well-trained, and the system is monitored for quality control (Zetland 2010). Furthermore, if the human right to water results in water being underpriced, then water consumers may not fully internalize the costs of their consumption, and lack sufficient incentives to conserve scarce water resources (Larson 2013).

Nevertheless, the limits and challenges of a human right to water can be overcome or mitigated if accompanied by other policy interventions. Such interventions may include enhanced institutional capacity of adjudicating bodies, direct water subsidies to indigent households, or tariff structures tied to water consumption. Perhaps most importantly for the equitable and sustainable implementation of the human right to water, such a right should be joined with effective measures to counter corruption in the water sector and greater protection of negative rights associated with water, including protections against discriminatory or arbitrary water provision or pricing and processes for stakeholder participation in water policy decision-making (Larson 2015).

With challenges and opportunities at both the international and domestic levels, the human right to water discourse appears to be at something of a tipping point – a precarious balance between the hope for equitable, affordable clean water provision and the concern of vague, unenforceable, or unsustainable government mandates.

10.2 Defining the Human Right to Water

In 1998, water scholar Peter Gleick wrote a seminal article in *Water Policy*, where he argued that access to a safe and adequate amount of water was a basic human right enshrined in the UN Declaration on Human Rights (Gleick 1998). However, the evolution of water policy discourse for a human right to water can be traced decades prior to Gleick's scholarship, and in fact continues to be refined within global water governance diplomacy. The first explicit articulation of a human right to water occurred at the 1977 Mar del Plata Water Conference, with additional iterations informed by UN instruments and conferences in the decades since (see Table 10.1). However, despite ongoing refinements and attempts for global mobilization for a human right to water, implementation is disparate. The disparate nature of implementation of the human right to water may result from the lack of a common understanding of the meaning of this 'basic human right'. In the dominant discourse of the human right to water, the articulation of the right is centred on the individual and their 'basic needs' to sustain life and livelihood (Gleick 1996; Hall et al. 2014). This definition does not account for the collective or the ecological, cultural, socioeconomic and political dependencies on water (Schmidt and Mitchell 2014, p. 59).

Whilst the push towards international recognition of the human right to water extends, arguably, back to the Mar Del Plata water conference in 1977 (Salman 2014), it was not until 2010 that the United Nations recognized the human right to water and sanitation with Resolution 64/292 (United Nations 2010). The Resolution recognized that both clean drinking water and sanitation are 'essential to the realization of all human rights' (UN G.A. Res. 64/292) and builds upon General Comment No. 15 adopted by the UN Committee on Economic, Social and Cultural Rights in 2002, which defined 'the right to water as the right of everyone to sufficient, safe, acceptable and physically accessible and affordable water for personal and domestic uses'. Each of the components to the human right to water – sufficient, safe, acceptable, physically accessible and affordable – are further defined by the United Nations.

For reference, the United Nations has defined the sufficient, safe, acceptable, physically accessible, and affordable aspects of the human right to water as follows (United Nations 2014):

- *Sufficient.* The water supply for each person must be sufficient and continuous for personal and domestic uses. These uses ordinarily include drinking, personal sanitation, washing of clothes, food preparation, personal, and household hygiene. According to

Table 10.1 Timeline of human right to water international policy development.

Year	Human Right to Water Instrument
1972	Stockholm United Nations Conference on the Human Environment[a]
1977	Water Conference, Mar del Plata[b]
1979	Convention on the Elimination of All Forms of Discrimination against Women (CEDAW)[c]
1981	International Drinking Water Supply and Sanitation Decade (1981–1990)[d]
1989	Convention on the Rights of the Child[e]
1990	The Montreal Charter on Drinking and Sanitation, Montreal, Canada[f]
1992	International Conference on Water and the Environment, Dublin, Ireland[g]
1992	Dublin Statement on Water and Sustainable Development[h]
1992	Rio Declaration – Agenda 21 'Programme of Action for Sustainable Development'[i]
1997	Convention on the Law of the Non-Navigational Uses of International Watercourses[j]
1999	Protocol on Water and Health to the 1992 Convention on the Protection and Use of Transboundary Watercourses and International Lakes[k]
2000	Millennium Development Goals for 2000–2015[l]
2002	UN Committee on Economic, Social, and Cultural Rights (CESCR) adopted General Comment No. 15 (GC15)[m]
2003	International Year of Freshwater[n]
2004	UN General Assembly adopted Resolution 58/217 to establish the International Decade for Action, 'Water for Life' (2005–2015)[o]
2007	United Nations Declaration on the Rights of Indigenous Peoples[p]
2010	Human right to water was declared in the UN Resolution 64/292[q]
2015	Sustainable Development Goals for 2015–2030[r]

Source: Authors' adapted table from multiple sources available in footnotes for references (all accessed 15 October 2018), and United Nations (2018) 'The Human Right to Water and Sanitation Milestones' (http://www.un.org/waterforlifedecade/pdf/human_right_to_water_and_sanitation_milestones.pdf)
[a] https://sustainabledevelopment.un.org/milestones/humanenvironment
[b] https://www.who.int/water_sanitation_health/unconfwater.pdf
[c] http://www.un.org/womenwatch/daw/cedaw
[d] http://apps.who.int/iris/handle/10665/141107
[e] https://www.ohchr.org/en/professionalinterest/pages/crc.aspx
[f] http://www.sie-see.org/en/article/the-montreal-charter-on-drinking-water-and-sanitation
[g] https://www.ircwash.org/sites/default/files/71-ICWE92-9739.pdf
[h] http://www.un-documents.net/h2o-dub.htm
[i] https://sustainabledevelopment.un.org/outcomedocuments/agenda21
[j] http://legal.un.org/avl/ha/clnuiw/clnuiw.html
[k] https://www.wipo.int/edocs/trtdocs/en/unece-ln/trt_unece_ln.pdf
[l] https://research.un.org/en/docs/dev/2000-2015
[m] https://www2.ohchr.org/english/issues/water/docs/CESCR_GC_15.pdf
[n] http://www.un-documents.net/a55r196.htm.
[o] http://www.un-documents.net/a58r217.htm
[p] https://www.un.org/development/desa/indigenouspeoples/declaration-on-the-rights-of-indigenous-peoples.html
[q] http://www.un.org/ga/search/view_doc.asp?symbol=A/RES/64/292
[r] https://sustainabledevelopment.un.org.

the World Health Organization (WHO), between 50 and 100 l of water per person per day are needed to ensure that most basic needs are met and few health concerns arise.

- *Safe.* The water required for each personal or domestic use must be safe, therefore free from microorganisms, chemical substances, and radiological hazards that constitute a threat to a person's health. Measures of drinking-water safety are usually defined by national and/or local standards for drinking-water quality. WHO Guidelines for drinking-water quality provide a basis for the development of national standards that, if properly implemented, will ensure the safety of drinking-water.
- *Acceptable.* Water should be of an acceptable colour, odour and taste for each personal or domestic use. [...] All water facilities and services must be culturally appropriate and sensitive to gender, lifecycle and privacy requirements.
- *Physically accessible.* Everyone has the right to a water and sanitation service that is physically accessible within, or in the immediate vicinity of the household, educational institution, workplace, or health institution. According to WHO, the water source has to be within 1000 m of the home and collection time should not exceed 30 minutes.
- *Affordable.* Water, and water facilities and services, must be affordable for all. The United Nations Development Programme (UNDP) suggests that water costs should not exceed 3% of household income.

Human rights are informed by the political and cultural understandings of the society that constructs them. In this way, the Universal Declaration on Human Rights ratified in 1948 quintessentially represents Eurocentric values emerging from a global polity war-torn and battered by the scourge of twentieth century genocide and nuclear proliferation. The Marshall Plan provided unheralded investment for many states to reconstruct or build new water systems. Given this social construction of the human rights framework, there are a few criticisms that social scientists raise with regards to the human right to water. The human right to water does not account for cultural or spiritual water needs. This discrepancy is especially important for the jurisdictional overlap in water governance with Indigenous Peoples. Notably, in 2007 the United Nations Declaration on the Rights of Indigenous Peoples (UNDRIP) was adopted and contains numerous explicit references to Indigenous Peoples' individual and collective rights to water (United Nations 2007). UNDRIP also serves as an instrument of international law recognizing Indigenous Peoples' rights to water alongside treaties, reserved rights and Indigenous law (Marshall and Neuman 2011). UNDRIP conflicts with other human right to water instruments by articulating economic, political and cultural rights to water and not just an adequate amount for individual consumption (Robison et al. 2018). The disparate understandings of the human right to water are further complicated by the variance in domestically codified water rights globally.

10.2.1 Difference Between Human Right to Water and Water Rights

There is a difference between the human right to water and a water right. Water rights systems globally vary based on interpretations of property law and processes for water allocation, and are informed by political and cultural differences. Thus, a 'water rights system is, fundamentally, one that identifies the total available resource and divides it, via the grant of water rights, amongst different users' (Speed 2009, p. 390). A water rights system enables certain water management and policy tools, including but not limited to: government-facilitated reallocations; centralized control of water abstractions; pricing mechanisms; and water trading markets (Speed 2009, pp. 390–391). A water rights system is often employed because it can (i) instil confidence in the diversity of users that water withdrawals are sustainable; (ii) provide security in stable allocation of resource for future water investments; and (iii) reduce conflicts arising from lack of allocation transparency and inefficient use (Speed 2009, p. 390). Water governance is further complicated by the disparate meanings at the local, national and global levels assigned to water rights, the human right to water, and property rights (Schmidt and Mitchell 2014).

Human rights are not water rights. Therefore, the discourse shift from a justiciable property right to soft law articulations of a 'human right to water' can be a political tool to deny legitimate water claims of marginalized populations, such as Indigenous Peoples and the urban poor (Bakker 2007). There are grave water injustices that have been committed over time around the world, many rooted in violent occupation and colonialism. The human right to water framework has been critiqued as aiding the neoliberal and water privatization agenda of the global elite and not adequate for promoting global water justice (Bakker 2007; Perera 2015, p. 199). Comparatively, some scholars have argued that the human right to water provides a remedy for ailing water rights systems that have disenfranchised marginalized communities (Hale 2007, p. 140). Water rights systems are an extension of the state's 'water sovereignty' and exist within the national and subnational political and legal frameworks (Marshall and Neuman 2011). As such, the human right to water does not have the same justiciability as a 'water right' unless codified within domestic law. This is one of the greatest challenges facing the implementation of the human right to water – the inability to codify and translate the right directly to water management and policy. Furthermore, the lack of justiciability domestically for the human right to water exacerbates a key problem within water rights systems, where individual water users cannot sue to ensure the accessibility, quality and affordability of water appropriated for public use by water utility services (Hale 2007, p. 142). Ultimately, implementation of the human right to water not only requires domestic justiciability through policy reform, but the necessary infrastructure and systems design to provide water accessibility.

10.3 Implementing the Human Right to Water

The UN definition of the human right to water creates an explicit set of design criteria for water agencies around the world. Viewed through this lens, the human right to water is no longer a passive definition; rather it is a proactive guide to the planning and construction of new water systems and a set of criteria to assess how far extant water systems go towards achieving this universal human right.

Historically, the predominant model for water resource engineering and development has been the 'hydraulic mission' style of developing both large, centralized water infrastructure projects – such as dams, aqueducts, irrigation canal networks and piped water systems – and the requisite bureaucracies for water storage and control, energy generation and provision, domestic provision and irrigation, with little to no consideration for the environment (Molle et al. 2009). These infrastructure projects help meet 'hard' water demands such as drinking water, irrigation, energy production and industrial uses (Gleick 2002, p. 373). However, for large, centralized water infrastructure projects to succeed, or at least be viable, population density is the most important factor as small, fluctuating populations present challenges for water financing and the generation of long-term, sustained revenues required to cover maintenance costs on these projects (Humphreys et al. 2018). Nonetheless, where these conditions have been met, large, centralized water infrastructure projects have 'mitigated the impacts of rainfall variability on water availability for food and basic livelihood' and increased per capita GDP (Brown and Lall 2006).

The United Nations developed the Millennium Development Goals (MDGs) as a unifying framework for global development and a vision to ameliorate poverty worldwide across many dimensions by the year 2015 (United Nations 2015). Of the eight MDGs, Target 3 of Goal 7 specifically addresses access to clean water: 'To halve the proportion of the universal population without sustainable access to clean and safe drinking water and basic sanitation by 2015'. In the intervening years between 2000 and 2015, 400 million people gained access to improved non-piped drinking water and 2.1 billion people gained access to improved piped drinking water (WHO and UNICEF 2017, p. 12). As of 2015, 6.5 billion people, or approximately 89% of the global population, had access to improved drinking water (WHO and UNICEF 2017, p. 10).

Progress towards the MDG of providing sustainable access to clean and safe drinking water between 2000 and 2015 was not uniform across the globe, making the Sustainable Development Goal (SDG) of universal and equitable access by 2030 a far greater challenge. Indeed, countries in Eastern Asia, Northern Africa, Western Asia, Latin America and the Caribbean have experienced marked increases in the availability of piped water between 1990 and 2015. Piped water is the predominant source of drinking water in these regions, with 74%, 86%, 89% and 89% piped drinking water coverage, respectively (WHO and UNICEF 2015). Further, in these regions, coverage by improved water sources does not fall below 6% of the population. The Caucasus and Central Asia have a similarly high coverage by piped water sources, but a precipitous increase in coverage was not observed between 1990 and 2015 (WHO and UNICEF 2015). Southeastern Asia and Southern Asia did experience a precipitous increase in improved water access during this time-frame, with the predominant drinking water source improvement in water sources that are not piped (WHO and UNICEF 2015). Sub-Saharan Africa and Oceania are the world's largest regions with the least access to improved water sources, with nearly two-thirds of sub-Saharan Africa and 52% of Oceania without coverage by improved water sources (WHO and UNICEF 2017, p. 3). Further, the only countries where less than half of their citizens have access to improved drinking water sources are located in these two world regions (WHO and UNICEF 2017, p. 3).

Both sub-Saharan Africa and Oceania present challenges to the 'hydraulic mission' style of water resources development, as they tend to have higher rural populations and lower population densities. In these regions, the prerequisites for sustaining large, centralized infrastructure – the ability to finance the project long-term, the population density and economic activity necessary to generate revenues for ongoing maintenance, and technical capital to provide the necessary ongoing maintenance – are unlikely to be met due to a variety of cultural (e.g. nomadic populations), economic (e.g. cost to build piped water supply prohibitively high), or geographical reasons (e.g. constrained by available water resources or island nation).

As we move towards the 2030 goal of UN Sustainable Development Goals (SDGs) with the target of 'achieve[ing] universal and equitable access to safe and affordable drinking water for all', how do we provide access to improved drinking water to those that do not already have access to it and ensure the human right to water?

In defining the human right to water, General Comment No. 15 lays out a set of explicit design criteria to assess how well a country, or individual water system, goes towards realising essential human rights. Whilst large, centralized water systems have fulfilled much of humanity's right to water, the areas of the world that still lack access to improved drinking water supplies also lack the predicates for large, centralized infrastructure systems. Therefore, what options remain for providing safe and clean drinking water? Decentralized water supply systems – such as community-scale water kiosks – can complement large, centralized water infrastructure projects by providing access to improved drinking water sources to small, rural populations where developing these systems is either prohibitively expensive or otherwise infeasible. Water infrastructure systems, whether centralized or decentralized, and the governance systems for provisioning water exert power through their design and implementation (Morrison et al. 2017) and obduracy (Hommels 2005). Therefore, challenges to providing universal and equitable access to safe drinking water centres on bridging the gap between what services are possible through engineering solutions, and what services can be provided equitably and affordably.

It is perhaps the nature of such a right that its full realisation will be an ongoing process, and that process is now taking place during a time in which water resource engineering and development has undergone a shift in how best to provide sufficient, safe, acceptable, physically accessible and affordable water in the face of major population changes – both growth and migration – and climate change.

10.4   Gap Between Policy Articulation and Implementation of the Human Right to Water

The human right to water is debated on its policy ideas and discourse, and has promoted progress in water access in many instances, but nevertheless struggles for broad and effective implementation (Mirosa and Harris 2012). Generally, the global community can agree that to live with human dignity requires access to a safe and adequate amount of water for life and livelihood, as articulated in General Comment 15 and the UN Resolution 64/292. However, what is less clear is who is responsible for the delivery of the water to individual beneficiaries of this human right. In 2002, General Comment 15 stated that the states have the responsibility to guarantee the human right to 'water sovereignty' to their citizenry, guided by the three principles of 'Respect, Protect and Fulfill' (General Comment 15). These principles require UN member states to, within their sovereign duties, ensure: (i) individuals or corporations do not infringe on the water rights of others; (ii) governments do not interfere unjustly with a person's access to water; and (iii) governments use all available resources to implement progressively the right to water (Langford 2005, pp. 277–279; see also Langford and Russell 2017). Criticism of this path for mobilization of the human right to water includes its framing within the UN system, which excludes 'non-state' actors such as Indigenous Peoples, non-recognized states (e.g. Taiwan), or the increasing issue of humanitarian contexts or fragile states where rule of law, accountability and legitimacy are not met. The local governance of water is more complex than the nation-state framework and often involves extensive multilevel and inter-jurisdictional actors for collaborative governance (Marshall and Neuman 2011). The disconnect amongst actors for implementation of the human right to water presents a significant problem.

Additionally, the human right to water generally does not include the rights of nature (i.e. rivers) which may have far-reaching ramifications for resolving future water conflicts, given climate change and increasing extreme climate events. The human right to water is inherently anthropocentric, technocratic and positivist, ascribing superiority to humans, and discounting non-Western political systems that embody a water ethic informed by reciprocity and relationality with other life on this planet. As water policy scholars Schmidt and Mitchell have highlighted, the total allocation of water to individual beneficiaries of the human right to water discounts 'ecological viability' (Schmidt and Mitchell 2014, p. 55).

The current understandings of the human right to water evolved over decades and, although currently only human-use centred, could evolve to be more 'holistic' and inclusive of ecosystem health (Armstrong 2009). The viability of human access to an adequate amount and quality of water cannot be separated from ecological needs. Furthermore, the mobility of humanity gives us the flexibility to adapt where natural systems cannot (Armstrong 2009, p. 144). In this way, the human right to water necessitates technological, legal and policy innovations to source alternative water supplies, when needed, to protect biodiversity and critical habitats. The waterscapes across the planet have been manipulated by the hydraulic mission model that in many instances has left them in struggling conditions. Therefore, universal and equitable access to safe and affordable drinking water will require 'striking a balance between human resource use and ecosystem protection' (Vörösmarty et al. 2010).

There are remaining questions as to the implementation of the human right to water, with the advent of legal personhood being granted to rivers in New Zealand and India (O'Donnell and Talbot-Jones 2018). Within a human right to water framework, how do we account for the natural world? There is a unique balance between meeting the needs of individuals for life and livelihood and our ecological responsibilities to the planet. Many nation-states have also included constitutional protections for the environment for the security of future generations (Daly 2012). The language of protections for 'future generations' also appears in General Comment 15, whereby the human right to water is defined within parameters of 'sustainability' but lacks clarity on how to guarantee the right for future generations (Ziganshina 2008, p. 127; Salman 2014, p. 974).

A greater challenge to implementation is the varying solution paths advocated by social scientists, lawyers and engineers. The legal, social science and engineering communities have different disciplinary understandings of the human right to water that are narrowly defined (Linton 2014, pp. 116–117). Often social scientists and legal scholars have inconsistent interpretations of the law and policy requirements for effective water governance and implementation of the human right to water. This divide is further compounded with the engineering community's often 'exclusively technical solutions to water crises' that conflict with social scientists' reliance on policy innovations (Gerlak and Mukhtarov 2013). Ultimately, addressing the gap in progressive articulations of the human right to water and the required on-the-ground implementation will need blended knowledge mobilizations that champion water technology, law and policy equally.

10.5 Key Policy Challenges Facing the Human Right to Water

Despite international articulations of a human right to water like General Comment 15 and the 2010 General Assembly Resolution, the world is still faced with ever-increasing water security challenges. These challenges arise from an array of factors that span the globe and affect all countries regardless of development status. There is also concern that the human right to water is generally practicable in times of water plenty, but that when there is water scarcity it cannot surmount the existing power dynamics of water inequities and would be incapable of resolving water conflicts (Wilder 2012). Moreover, there is no clear guidance on how the human right to water aligns with transboundary water governance. Many of the world's largest basins are shared by two or more nations that lack transboundary water agreements delineating a shared approach for water access and determination as to whether the parties agree to secure a human right to water. Delivering and managing large complex basins may be further complicated when coupled with the human right to water feeding confusion about how to guarantee this human right in the face of transboundary competition and conflict where sovereign welfare supersedes (Bakker 2007).

In countries with mature water infrastructure systems, realising the human right to water is a continuous process. Whilst we may focus on the ribbon-cutting ceremonies for new projects, all water infrastructure projects age and require ongoing investment, maintenance and adaptive management. In Flint, Michigan, city residents have been poisoned for years by their water supply and continue to live without access to safe and clean drinking water (Hawthorne 2018). Where old infrastructure is neglected and not managed with respect to new conditions, and there is political indifference to ensuring the provision of clean water, crises like Flint happen and the human right to water is violated in major developed economies.

Water planning must incorporate population dynamics and climate change. The city of Cape Town, South Africa, faced the prospect of running out of water for the first time this year, highlighting the potential for increased water governance crises exacerbating existing inequalities to water access for the urban poor, women, and other marginalized communities. Water scarcity and security are inextricably linked to failures in water policy, planning and management. In Canada, there are over 150 First Nation communities who are on Drinking Water Advisories, with some communities suffering without access to safe water for over two decades (Health Canada 2018). All countries must ensure the human right to water equally amongst all populations. In areas where conventional water resources are already stressed, population growth and climate change must factor into current and future water planning to ensure the human right to water is preserved. Conservation, both mandated and voluntary, are potential soft water paths (Gleick 2003) to securing new water, but in instances where conservation cannot meet the differences between projected supply and demand, alternative hard water paths must be explored.

The challenges faced by these communities are not isolated incidences; rather, they are endemic across political scales and water regimes. Some have looked to the human right to water as the next policy evolution to improve water governance globally, but key challenges remain. Firstly, there is no shared definition for the human right to water.

Second, there is danger in conflating the human right to water with a water right within a larger property rights framework. Consequently, there is a gap in the progressive policy articulation and on-the-ground implementation of the human right to water.

10.6 Conclusion

The history of the development of a human right to water is characterized by constant evolution and re-evaluation in the face of consistent challenges of sustainability and equity inherent in defining the quantity, quality, and affordability of water. Rigid minimum standards that cannot adapt to varying cultural and geographical contexts are as likely to fail as ambiguous articulations of the parameters of the right. The formulation, interpretation and implementation of the human right to water should be even more adaptive in the future, as the impact of climate change on water variability will require a collaborative and flexible approach to the formulation of a human right to water.

Acknowledgements

This chapter was in part supported by the National Science Foundation under Grant No. ACI-1639529 and USDA Grant No. 2017-08812.

References

Ahmad, A. (2000). Islamic water law as a comparative model for maintaining water quality. *Journal of Islamic Law and Culture* 5 (2): 159–186.

Armstrong, A.C. (2009). Viewpoint – further ideas towards a water ethic. *Water Alternatives* 2 (1): 138–147.

Ausness, R. (1986). Water rights, the public trust doctrine, and the protection of instream uses. *University of Illinois Law Review* 1986: 407–437.

Bakker, K. (2007). The "commons" versus the "commodity": alter-globalization, anti-privatization and the human right to water in the global south. *Antipode* 39 (3) https:// doi.org/10.1111/j.14667-8330.2007.00534.x.

Berlin, I. (1969). *Four Essays on Liberty*. Oxford: Oxford University Press.

Blumm, M.C. and Guthrie, R.D. (2012). Internationalizing the public trust doctrine: natural law and constitutional and statutory approaches to fulfilling the Saxion vision. *University of California at Davis Law Review* 45: 741–808.

Brown, C. and Lall, U. (2006). Water and economic development: the role of variability and a framework for resilience. *Natural Resources Forum* 30 (4): 306–317. https://doi.org/10.1111/j.1477-8947.2006.00118.x.

Cross, F. (2001). The error of positive rights. *UCLA Law Review* 48 (4): 857–924.

Daly, E. (2012). Constitutional protection for environmental rights: the benefits of environmental process. *International Journal of Peace Studies* 17 (2): 71–80.

Dellapenna, J.W. (2011). The evolution of riparianism in the United States. *Marquette Law Review* 53: 61.

Dworkin, R. (1977). *Taking Rights Seriously*. Massachusetts: Harvard University Press.

Gerlak, A. and Mukhtarov, F. (2013). River basin organizations in the global water discourse: an exploration of agency and strategy. *Global Governance* 19: 307–326.

Gleick, P.H. (1996). Basic water requirements for human activities: meeting basic needs. *Water International* 21 (2): 83–92.

Gleick, P.H. (1998). The human right to water. *Water Policy* 1 (5): 487–503. https://doi.org/10.1016/S1366-7017(99)00008-2.

Gleick, P.H. (2002). Water management: soft water paths. *Nature* 418 (6896): 373.

Gleick, P.H. (2003). Global freshwater resources: soft-path solutions for the 21st century. *Science* 302 (5650): 1524–1528.

Haddaway, A. (1929). Water rights – is a water right an easement? *Texas Law Review* 7: 453.

Haeusermann, J. (1998). *A Human Rights Approach to Development*. London: Department for International Development of the UK Government.

Hale, S. (2007). The significance of justiciability: legal rights, development, and the human right to water in the Philippines. *SAIS Review of International Affairs* 27 (2): 139–150.

Hall, R.P., Van Koppen, B., and Van Houweling, E. (2014). The human right to water: the importance of domestic and productive water rights. *Science and Engineering Ethics* 20 (4) https://doi.org/10.1007/s11948-013-9499-3.

Hawthorne, M. (2018). Flint Researchers Find Alarming Levels of Lead in Cicero, Berwyn Tap Water, Suggesting Thousands of Older Homes at Risk. *Chicago Tribune*, 10 August 2018. https://www.chicagotribune.com/news/ct-met-cicero-berwyn-lead-water-testing-20180809-story.html.

Health Canada (2018). *Drinking Water Advisories in First Nations Communities* [Data file and code book]. https://open.canada.ca/data/en/dataset/5f73fff7-2011-48b9-af52-ffb31e68539c.

Hommels, A. (2005). Studying obduracy in the city: toward a productive fusion between technology studies and urban studies. *Science, Technology, & Human Values* 30 (3): 323–351.

Humphreys, E., Van der Kerk, A., and Fonseca, C. (2018). Public finance for water infrastructure development and its practical challenges for small towns. *Water Policy* 20 (S1): 100–111.

Jacobs, B. (2000). *International Law/ Great Law of Peace*. University of Saskatchewan. https://www.collectionscanada.gc.ca/obj/s4/f2/dsk3/SSU/TC-SSU-07042007083651.pdf.

Langford, M. (2005). The United Nations concept of water as a human right: a new paradigm for old problems? *International Journal of Water Resources Development* 21 (2): 273–282. https://doi.org/10.1080/07900620500035887.

Langford, M. and Russell, A. (eds.) (2017). *The Human Right to Water: Theory, Practice and Prospects*. Cambridge: Cambridge University Press.

Larson, R. (2013). The new right in water. *Washington and Lee Law Review* 70 (4): 2181–2267.

Larson, R. (2015). Adapting human rights. *Duke Environmental Law & Policy Forum* 26 (1): 1–51.

Linton, J. (2014). Modern water and its discontents: a history of hydrosocial renewal. *Wiley Interdisciplinary Reviews: Water* 1 (1): 111–120. https://doi.org/10.1002/wat2.1009.

Mallat, C. (1995). The quest for water use principles. In: *Water in the Middle East: Legal, Political and Commercial Implications* (eds. C. Mallat and J.A. Allan), 72–85. London: I. B. Tauris.

Marshall, D. and Neuman, J. (2011). Seeking a shared understanding of the human right to water: collaborative use agreements in the Umatilla and Walla basins of the Pacific northwest. *Willamette Law Review* 43 (3): 403.

McCaffrey, S. (1992). A human right to water: domestic and international implications. *Georgetown International Environmental Law Review* 5 (1): 1–24.

Mirosa, O. and Harris, L.M. (2012). Human right to water: contemporary challenges and contours of a global debate. *Antipode* 44 (3): 932–949. https://doi.org/10.1111/j.1467-8330.2011.00929.x.

Molle, F., Mollinga, P.P., and Wester, P. (2009). Hydraulic bureaucracies and the hydraulic mission: flows of water, flows of power. *Water Alternatives* 2 (3): 328–349.

Morill, J. and Simas, J. (2009). Comparative analysis of water laws in MNA countries. In: *Water in the Arab World* (eds. N.V. Jagannathan, A.S. Mohamed and A. Kremer), 285–308. Washington, DC: International Bank of Reconstruction and Development/World Bank Middle East and North Africa (MNA) Region.

Morrison, T.H., Adger, W.N., Brown, K. et al. (2017). Mitigation and adaptation in polycentric systems: sources of power in the pursuit of collective goals. *Wiley Interdisciplinary Reviews: Climate Change* 8 (5): e476.

O'Donnell, E.L. and Talbot-Jones, J. (2018). Creating legal rights for rivers: lessons from Australia, New Zealand, and India. *Ecology and Society* 23 (1): 7.

Perera, V. (2015). Engaged universals and community economies: the (human) right to water in Colombia. *Antipode* 47 (1): 197–215. https://doi.org/10.1111/anti.12097.

Rawls, J. (1971). *A Theory of Justice*. Massachusetts: Harvard University Press.

Robison, J.A., Cosens, B.A., Jackson, S. et al. (2018). Indigenous water justice. *Lewis and Clark Law Review* 22 (3): 841–921.

Rossi, J. (1998). The common law "duty to serve" and protection of consumers in an age of competitive retail public utility restructuring. *Vanderbilt Law Review* 51: 1244–1248.

Salman, M.A. (2014). The human right to water and sanitation: is the obligation deliverable? *Water International* 39 (7): 969–982. https://doi.org/10.1080/02508060.2015.986616.

Salzman, J. (2006). Thirst: a short history of drinking water. *Yale Journal of Law and Humanities* 18 (3): 94–121.

Sammis, J. (2010). Explanation of Vote by John F. Sammis, US Deputy Representative to the Economic and Social Council, on Resolution A/64/L.63/Rev.1, the Human Right to Water. http://usun.state.gov/briefing/statements/2010/145279.htm.

Sax, J. (1970). The public trust doctrine in natural resource law: effective judicial intervention. *Michigan Law Review* 68: 471–565.

Sax, J. and Abrams, R. (1986). *Legal Control of Water Resources*. St. Paul, Minnesota: West Publishing Company.

Schmidt, J.J. and Mitchell, K.R. (2014). Property and the right to water: toward a non-liberal commons. *Review of Radical Political Economics* 46 (1): 54–69. https://doi.org/10.1177/0486613413488069.

Speed, R. (2009). A comparison of water rights systems in China and Australia. *International Journal of Water Resources Development* 25 (2) https://doi.org/10.1080/07900620902868901.

Thompson, B. (2011). Water as a public commodity. *Marquette Law Review* 95: 32–38.

Trelease, F. (1954). Coordination of riparian and appropriative rights to the use of water. *Texas Law Review* 33.

United Nations (1967). *International Covenant on Economic, Social and Cultural Rights*, G.A. Res. 2200, 21 U.N. GAOR Supp. 49, U.N. Doc. A/6316, entered into force on Jan. 3, 1976 [hereinafter UN G.A. Res. 2200].

United Nations (2007). *United Nations Declaration on the Rights of Indigenous Peoples*, G.A. 61/295, U.N. Doc. A/RES/61/295.

United Nations (2010). *G.A. Res. 64/292*, 5, 8, U.N. Doc. A/RES/64/292, ratified on July 28, 2010. [hereinafter UN G.A. Res. 64/292].

United Nations (2014). Human right to water and sanitation/International decade for action 'water for life' 2005–2015. http://www.un.org/waterforlifedecade/human_right_to_water.shtml.

United Nations (2015). *The Millennium Development Goals Report* 2015. http://www.un.org/millenniumgoals/2015_MDG_Report/pdf/MDG%202015%20rev%20(July%201).pdf.

United Nations (2018). *Goal 6: Sustainable Development Knowledge Platform*. https://sustainabledevelopment.un.org/sdg6.

United Nations Economic and Social Council, Subcommittee on the Promotion and Protection of Human Rights (2003). *Substantive Issues Arising in the Implementation of the International Covenant on Economic, Social, and Cultural Rights: General Comment No. 15 (2002)*. UN Doc. E/C.12/2002/11 [hereinafter UN General Comment 15].

Vörösmarty, C.J., McIntyre, P.B., Gessner, M.O. et al. (2010). Global threats to human water security and river biodiversity. *Nature* 467: 555–561.

Weissbrodt, D. and de la Vega, C. (2007). *International Human Rights Law: An Introduction*. Philadelphia: University of Pennsylvania Press.

Wilder, M. (2012). Exploring the textured insecurity and the human. *Environment Magazine* 54: 5–17.

World Health Organization and UNICEF (2015). *Progress on Sanitation and Drinking Water: 2015 Update and MDG Assessment*. Geneva: WHO/UNICEF.

World Health Organization and UNICEF (2017). *Progress on Drinking Water, Sanitation and Hygiene: 2017 Update and SDG Baselines*. Geneva: WHO/UNICEF.

Zetland, D. (2010). Water Rights and Human Rights. *Forbes*, March 25, 2010.

Ziganshina, D. (2008). Rethinking the concept of the human right to water. *Santa Clara Journal of International Law* 6 (1): 127.

11

Policy Processes in Flood Risk Management

Edmund C. Penning-Rowsell[1,2], Joanna Pardoe[3], Jim W. Hall[4], and Julie Self[5]

[1] *School of Geography and the Environment, University of Oxford, UK*
[2] *Flood Hazard Research Centre, Middlesex University, London, UK*
[3] *London School of Economics, UK*
[4] *Environmental Change Institute, University of Oxford, UK*
[5] *Government of Alberta, Canada*

11.1   Introduction

Understanding policy processes is no less important in flood risk management (FRM) than in other water policy domains. We would hope, through this understanding, to enhance the progress towards more effective policy change and more sustainable FRM strategies in the future than in the past, those being characterized by holistic processes geared towards risk reduction (Sayers et al. 2014). We can hope to avoid policy deficiencies and maximize the chances of policy effectiveness, assisted by the very substantial literature on the policy process, within the political science field, that is oriented in these directions.

In this chapter we summarize three important theories about the policy process, being a set of procedures whereby policy is written (usually with government decisions), debated, approved, implemented and evaluated (although policy evaluation is not covered here in detail).

We also summarize four case studies where the policy process has resulted either in policy change or in relative stasis. Each one illuminates differences in how the policy process is pursued, those involved, the influence of contextual factors, and what it tells us about policy process theories. The cases have been chosen because colleagues have researched them in detail, as published elsewhere (e.g. Sultana et al. 2008; Solik and Penning-Rowsell 2017), meaning that more detailed analyses can be found via those papers of the intricacies of the policy processes involved and its supporting evidence base. The fourth case study –Tanzania – is newly researched here, chosen to focus on a developing country with a particular attitude towards its flooding problems.

We would stress that the policy process is not straightforward. Policy change is not inevitable. Political factors can interfere with what otherwise might be straightforward and rational evaluations of current policies and their possible improvement. Economic and financial considerations may dictate an absence of moves towards what is judged as

Water Science, Policy, and Management: A Global Challenge, First Edition. Edited by Simon J. Dadson, Dustin E. Garrick, Edmund C. Penning-Rowsell, Jim W. Hall, Rob Hope, and Jocelyne Hughes.
© 2020 John Wiley & Sons Ltd. Published 2020 by John Wiley & Sons Ltd.

optimal. Social interests may mean that policy change is either avoided owing to adverse distributional consequences, or pursued in spite of them. Environmental conditions and forecasts of the impacts of policy change will obviously be important in this field. Although the theories of the policy process described here have endured over many years, they should not be seen as templates or 'recipe books' for future policy endeavour: they are largely descriptions or explanations of the policy process, and possible influencing factors, rather than prescriptions or expectations for it. In many circumstances they may have little predictive power.

11.2 Flood Risk: Global and Local Scales

In most countries and regions of the world, FRM policy is closely related to the flood risk likely to be found there. Global and national flood risk assessments are carried out to determine high-risk areas where further investment and support with policy and planning would be beneficial (Winsemius et al. 2013; Ward et al. 2015). This has been driven in Europe by the EU Floods Directive and elsewhere by the Hyogo and Sendai frameworks (Fuchs and Thaler 2018). Riverine and coastal FRM for many countries is often guided by their national flood risk assessments, usually available as flood maps to raise awareness of flood risk and to guide the choice of risk-reducing options (e.g. spatial planning in areas at risk). Global assessments allow for the ranking of high-risk countries; the top 10–15 are often Asian countries where large populations are exposed to severe flooding (e.g. Luo et al. 2015). This is particularly problematic in low-lying coastal areas where flood risks arise from coastal and riverine sources and as well as heavy rainfall associated with cyclones (Oppenheimer et al. 2014).

Global flood risks are expected to be exacerbated by climate change conditions. The IPCC reports anticipate more frequent and intense precipitation events over land (Kirtman et al. 2013). In addition, sea-level rise associated with glacier loss and changes in sea ice levels would also increase the severity of erosion and flood risk for low-lying coastal communities, which will become increasingly difficult to manage and protect (Kirtman et al. 2013). The IPCC highlights particular challenges for developing and rapidly urbanizing areas not previously occupied with such density (Oppenheimer et al. 2014).

Despite agreement on projections that imply greater flood risk under climate change overall, considerable uncertainty remains with regards to the precise degree and spatial distribution of this enhanced risk, particularly for riverine and surface water flooding driven by heavy precipitation (Kirtman et al. 2013). Policy-makers and investors often have to rely on climate projections that are unable to account fully for the complexities in various local and regional climate and riverine systems (Ward et al. 2015). Estimates of flood risk have relied upon assumptions that the future will resemble the past, and this assumption of 'stationarity' is increasingly being challenged both by observations and from evidence from climate models. Projections for some parts of the global south, in particular, still provide a mixed picture, with some models projecting considerably wetter conditions whilst others project drier climates (Rowell 2011; Lazenby and Todd 2018). Thus, in seeking to reduce annual average damages, account for exacerbated risk and ensure an efficient use of resources, the necessary planning can often lack the necessary scientific support. This uncertainty presents a considerable challenge to sensible national or regional policy and investment (Penning-Rowsell and Korndewal 2018).

But FRM philosophies have shifted over the last two decades. Since the 1990s there has been a move in many countries away from notions of 'flood protection' towards 'flood risk management', now for example the remit of England's Environment Agency. A repeated recommendation has been to deal comprehensively with flooding, including with portfolios of measures to reduce both exposure and vulnerability, rather than focusing entirely upon defences that can never provide completely reliable protection from all floods. Alongside the resulting policy reforms, there has been notable innovation in flood risk analysis, moving us away from local studies of flood hazard to analysing all aspects of risk at increasingly broad scales and using innovative approaches, as in the Dutch *Room for the River* approach (van Herk et al. 2015). Such innovation has been made possible by the widespread availability of a range of better information (e.g. digital terrain data) and new techniques (e.g. 2D mathematical modelling). It has been made necessary to gain public involvement, co-production of FRM plans, and the implementation of often controversial risk reduction measures.

11.3   Three Theories of the Policy Process

Understanding FRM policy and its evolution or otherwise can be assisted using the lenses of three well-known theories of the policy process (see below). These have been reviewed and refined many times (e.g. True et al. 2009; Birkland 2016; Jones et al. 2016), but their fundamental characteristics have not significantly changed over the last 30 years. This endurance helps to secure our attention (as amplified by Penning-Rowsell et al. 2017), but each has some shortcomings as applied to our field, as our case studies will reveal.

11.3.1   Punctuated Equilibrium

Baumgartner and Jones's (1993) idea of 'punctuated equilibrium', imported from evolutionary biology (Prindle 2012), attempts to understand how groups of actors influence policy priorities; the idea postulates episodic periods of rapid policy evolution.

This model explicitly articulates both policy change and policy stability, emphasizing that in terms of time, the latter dominates. As with the Advocacy Coalition Framework (ACF), discussed below, time is an important factor but the 'punctuated equilibrium' model sees ongoing periods of relative stability occasionally 'punctuated' by periods of public interest, media scrutiny and action (John 1998), thereby emphasizing 'the forces that create both incrementalism in many circumstances and rapid changes in others' (Baumgartner and Jones 1993, p. 4; True et al. 2009). Tackling here an inherently episodic phenomenon such as floods, this set of ideas has immediate appeal.

By focusing on shifts in the rate of policy change, Baumgartner and Jones explored how certain ideas become institutionalized in decision-making. Key individuals in institutions seek to ensure that certain issues dominate the agenda-setting process by enlisting the support of elite beliefs in the political process, ensuring favourable media coverage and public opinion. The outcome is generally incremental policy change or stasis. Large-scale policy punctuations spring from either a change in preferences or a change in attentiveness within individual or collective decision-making (True et al. 2009, p. 163). The result can be periods of crisis exploitation by actors within the policy subsystem to significantly alter levels of political support for new or refreshed public policies (Boin et al. 2009). The same forces that create stability also create instability. The difference is the extent to which new participants are attracted to, and affected by, an issue and how these new actors are mobilized by policy entrepreneurs.

This is useful here for a number of reasons. Firstly, it explicitly recognizes that whilst institutions are important, they are not dominant, and that the interests of individuals and groups of individuals in networks and coalitions are important in defining issues and setting agendas. These, when 'punctuated' by new ideas, allow new interest groups to alter the agenda-setting process (or when old ideas that just happen to catch political attention and trigger appropriate solutions – sometimes to old problems – at the right time). Secondly, the significance of policy entrepreneurs is important in problem definition and negotiation. Thirdly, the central notion is that ideas are critically important in the definition and negotiation of issues in the agenda-setting process. Finally, the 'punctuated equilibrium' model alone focuses explicitly on both policy stability and policy change, recognizing the importance of the rate of change and that the policy process can often remain stalled. This 'two speed' character is a useful component for analysing flood policy transition (see Lane et al. 2013).

11.3.2 Multiple Streams

John Kingdon's multiple streams approach (MSA) (Kingdon 2003) is concerned with understanding how issues materialize, how they get the attention of policy-makers, how they are framed as ideas in policy agendas, and why ideas 'have their time' (Parsons 1995, p. 192). Kingdon is specifically interested in the agenda-setting process, recognizing the role of individual actors, institutions and external events on the relationships between solutions, problems, issues and ideas. Policies are formed or not formed, and agendas are set or not set, as a result of three separate and distinct 'streams': problems, policies and politics (summarized and critiqued by Jones et al. 2016).

The *problem stream* is where the public and policy-makers become focused on something requiring attention: here, floods and flood risk reduction. The policy solutions to the problem emerge in the form of ideas in the *policy stream*. These ideas float in a policy 'primeval soup', dropping on and off the policy agenda, following Cohen et al.'s (1972) garbage can model of organizational choice. Policy communities, and policy entrepreneurs within them (see Meijerink and Huitema 2010), seek to ensure that certain ideas progress to, and remain on, the agendas of governments and their agencies. The third *political stream* determines how the emerging problems are defined and advanced. Here, therefore, public opinion, political activism, the media and government personnel are all important in influencing the definition of the problem and assessing potential solutions, as is a fundamental political will, or the will of decision-makers, to embrace change.

For policy change to occur, an idea needs to 'catch on' and dominate the policy agenda. This occurs when there is a 'policy window', brought about or responded to when the three streams align (i.e. they have congruency). A problem is recognized, and policy communities and entrepreneurs – often dominant in subnational issues (see Cairney and Zahariadis 2016) and perhaps working in collectives (Meijerink and Huitema 2010) – can press for their ideas to form solutions to the problem at a time when there is political receptivity. Such a 'window of opportunity' does not occur often or stay open for long, due, for example, to the issue attention cycle curtailing its longevity (Downs 1972; John 1998). Moreover, not all such 'windows' result in policy change: 'serendipity plays an important role in failures and successes' (Jones et al. 2016, p. 2).

11.3.3 Advocacy Coalitions

The Advocacy Coalition Framework (ACF) literature offers a viable alternative by offering no beginning and end to what is seen as a continuous and quintessentially competitive policy process. Sabatier and colleagues have stressed the importance of alliances and bargaining in coalitions, rather than a mere network of actors (Sabatier and Weible 2014; Weible and Sabatier 2017). Sabatier (1993) also differs from Kingdon (1984, 2003) in seeing the role of ideas, information and learning as a fundamental part of the political stream and a major force for change.

Sabatier regards the policy process as a function of policy subsystems involving a number of advocacy coalitions, each holding distinct beliefs. Each subsystem comprises two or more coalitions of actors, involving participants from a range of organizations and institutions, inside and outside government, who share common and normative beliefs. These fundamental norms and beliefs in turn lead to policy strategies reflecting core values (Sabatier 1988). The core beliefs are highly resistant to change from anything other than significant perturbations; these are either external (Sabatier 1993) or internal (Weible and Sabatier 2009) to the subsystem. In addition, there are secondary belief systems, such as about the policy instruments necessary to attain fundamental policy positions and/or implementation results. These secondary aspects are much more susceptible to change as a result of learning and negotiation within and between advocacy coalitions (Parsons 1995; Weible and Sabatier 2009, p. 196).

Through a negotiative process, each advocacy coalition adopts strategies for furthering their core beliefs, which emerge as ideas, and are negotiated in conflict with other advocacy coalitions, with mediation by 'policy brokers'. The subsystems operate in a wider institutional context involving relatively stable parameters such as cultural and social values on the one hand and more dynamic external events on the other, such as changes in the economy, public opinion and societal demographics (or in our case major flood events).

11.4 Four Contrasting Case Studies of the Policy Process

11.4.1 South Africa: 1994–2002 and Beyond

To explore the FRM policy process and its character, Johnson and her colleagues developed a framework drawing on a range of policy evolution ideas, specifically analysing FRM policy evolution in Britain (Johnson et al. 2005; Penning-Rowsell et al. 2006). That framework contrasted the influence of 'catalytic' change with intervening 'incremental' change – an analysis of the latter subsequently deepened by Lane et al. (2013) – and sought to demonstrate the importance of taking advantage of Kingdon's 'windows of opportunity' to put in place interventions towards better FRM. Our research in South Africa initially followed the same themes, but used a combination of punctuated equilibrium concepts and multiple streams perspectives, the latter encouraging us to look for congruence between Kingdon's three problem, policies and politics 'streams' (Solik and Penning-Rowsell 2017).[1]

In South Africa, the end of apartheid in 1994 made it necessary to reform legislation that was then in place, given that the majority of the population had been largely ignored or marginalized in and by many pre-existing statutes. Thus the end of apartheid, along with the major floods that occurred in the northeast of the country in 2000, created a significant policy window within which the country's flood policy was changed along with numerous other laws. Alongside these catalytic changes were incremental changes due to regional flooding and contextual changes as South Africa emerged from isolation to become a fully-fledged member of the international community.

The June 1994 floods in Cape Town (a regional-scale flood event) served as a spark that drove reform in disaster risk management. Subsequently the Inter-ministerial Committee developed its Green and White Papers on Disaster Management. These floods attracted significant public comment and constant review (the 'problem domain' within Kingdon's (2003) multiple streams framework) and, although necessary, this can stall the legislative process; there was evidence of this happening in South Africa. However, the major flood in 2000 punctuated the stalemate and provoked the necessary reaction to push for legislative closure. Due to the flood, a conference was called by the National Council of Provinces in May 2000 to debate the final details of the Bill. The Bill gained momentum, was tabled and approved by Cabinet in August 2001 and promulgated in January 2002.

Up until the promulgation of the law, Johnson's framework can therefore successfully explain the dynamics of policy change, as represented by the Disaster Management Act of 2002 clearly following 'catalytic' events and promoting the convergence of the policy and political situations. However, the implementation stage of the policy process is of particular relevance here. Implementation comprises putting the legislation and its policies into practice, and using those policies to make decisions that affect activities and, ultimately, the issue being addressed by that policy (here, it is flood risk): policy success should be viewed as the successful implementation of that policy, not its initiation. In order for policy implementation to take place, various complex and interconnected steps need to be taken; for example, funding must be allocated, personnel assigned and rules of procedure developed.

The Disaster Management Act of 2002 (hereinafter DMA 2002) laid the current foundations for disaster management in South Africa and includes guidance on floods, droughts and fires. The DMA (2002) describes disaster management as 'an integrated and co-ordinated process (for) preventing, reducing and mitigating the risk and severity of disasters, ensuring emergency preparedness, rapid and effective response to disasters and post-disaster recovery' (Disaster Management Act 2002 [South Africa]). In addition, the post-apartheid national Constitution (1996) brought in a new political paradigm where there was shift from the national, centralized control of state-run activities to a focus on more localized management. When the new DMA 2002 was promulgated,

---

1 Citations of this paper and those describing the policy processes in Bangladesh and the UK in this chapter are not repeated here except where quotations are given.

it accorded new powers to each of the national, provincial and local levels, but the 'most important sphere for effective implementation of disaster management is local government where most operational activities relating to disaster management will occur' (Botha et al. 2011, p. 24).

But implementation at the local level after 2002 was slow if not non-existent. We judge that there are some overarching and longer-term generic issues that led to such difficulties.

Firstly, local governments are often faced with numerous issues on a day-to-day basis which require their immediate attention. Therefore they face a daunting task of trying to balance these pressing and immediate socioeconomic challenges, and introducing longer-term policies into their planning – such as for disaster management – that will only pay off well into the future. Secondly, there is often a lack of human and political capital available locally to implement central government directives, representing significant obstacles to change. Finally, there is often a lack of integration between the various levels of local and central government.

The DMA 2002 required a multidisciplinary approach where all levels of government needed to work together to ensure adequate risk management, but there is evidence to suggest that this did not take place in South Africa (Solik and Penning-Rowsell 2017, pp. 57–60). More generally, Templehoff et al. (2009, p. 102) have concluded that the 'lack of clear disaster risk management guidelines and procedures from National Government has in many instances caused an implementation bottleneck at local government level'.

The inability of various agents within government to perform the tasks required by legislation contributes significantly to any implementation deficit, and policy entrepreneurs must ensure that sufficient capacity exists to administer the changes they propose. During the policy formulation, process steps should be taken to improve capacity further where responsibilities are newly dictated. In South Africa, more attention should therefore have been given to the implementation aspects of the DMA 2002, not least because the 1999 White Paper on Disaster Management had already highlighted key elements and requirements (Botha et al. 2011, p. 22):

1) An effective and comprehensive disaster management strategy;
2) Coordination and clear lines of responsibility for those involved in disaster management;
3) Government capacity, particularly of local government in rural areas, to implement disaster management; and
4) Integration of civil society into effective disaster management activities, particularly those concerned with risk reduction.

The moves following 2002 in South Africa necessitated an implementation stage requiring significant institution building, as a clear part of the policy process. Progress in this respect has been made since then, but the speed of change was not as anticipated in 2002 but much more protracted. This example illustrates the need for a full understanding of policy processes and their impacts ahead of proposed changes, so that the wherewithal for policy implementation is properly considered. It also shows that windows of opportunity opened by Kingdon's congruence of streams may only be the first steps towards policy change; such congruence is necessary but not sufficient for policy change to be implemented and thus realised.

11.4.2 Advocacy Coalitions in Bangladesh and the Role of Donor Agencies

Bangladesh is one of the most exposed countries in the world to flood risk. In a densely populated country, large numbers of vulnerable people live in a deltaic environment that experiences regular flooding during the monsoon season. But there have also been a number of particularly severe events over the past 50 years. These events are marked by significantly larger numbers of fatalities, more areas affected and larger crop production impacts than the regular flood events to which the population is well-adjusted. The research here, reported by Sultana et al. (2008), used Sabatier's ACF to investigate a significant policy change away from an engineering-only approach as a result of an intense struggle between fundamentally different core beliefs within the two broad coalitions involved.

In 1954, 1955 and 1956, successive major floods affected an estimated one third of present-day Bangladesh. These events effectively opened a 'window of opportunity' for FRM policy change (Kingdon 2003; Johnson et al. 2005). The influx of donors and other external agents accelerated efforts to install large-scale embankments. This approach shifted slightly after major flooding in 1974 which resulted in extensive inundation and an estimated 2000 deaths. The damage that the flood caused to the main rice crop compounded the death toll through widespread famine. This famine is estimated to have caused up to 1.5 million deaths both through loss of crops and the consequent inflation of food prices.

The floods in 1974 mobilized donor support to address food security risks. In this, support for flood control measures was also advocated, although donors were shifting support towards smaller-scale embankment projects rather than large-scale structures. This represented another acceleration of a general shift from large-scale to smaller-scale development projects.

Following decades of investment in structural (engineered) measures, 1987 and 1988 saw the worst floods on record, with the 1988 flood covering 62% of the country, affecting 45 million people. These floods catalysed debate that resulted in questions being raised about the effectiveness of embankment investments and thus opened another window for debate on alternative approaches.

Responding to the major floods in 1987 and 1988, a new coalition of actors rose to challenge the dominant perspective of engineered solutions advocated by the long-standing 'engineering coalition'. In this respect, the flood management policy landscape evolved in the 1990s to become characterized by a battle between two policy advocacy coalitions, the 'engineering coalition' versus the 'environmental coalition', both of which comprised alliances of different actors with shared beliefs on FRM approaches. The floods of 1987 and 1988 provided a perturbation that allowed the balance of power that the engineering coalition held to be challenged by a group that believed instead in less structural and more participatory management approaches.

The engineering coalition initially comprised politicians, consultants and donor funding agencies, emphasizing embankments to mitigate flood risk. In contrast, the 'environmental coalition' comprised journalists, NGOs and civil society organizations, which advocated instead a FRM approach that saw humans and nature as symbiotic rather than needing to be separated. This coalition recognized a need for functional floodplains that support livelihoods but which are managed to reduce risks to urban areas where people are encouraged to adapt to and live with the natural flood conditions.

Emboldened by the rise in democratic momentum on the wider political landscape, the environmental coalition challenged the Flood Action Plan that had been initiated following the 1987 and 1988 floods. That Plan detailed 26 studies and pilot projects that were backed by a consortium of donors. Concerned that structural measures had failed to fully prevent flooding and recognizing the limitations and impacts of these works, particularly the impacts on poorer people with livelihoods dependent on a functional floodplain, the environmental coalition challenged the belief in structural approaches that were underpinning the Flood Action Plan. With momentum from the national democratic movement, the legitimacy of the engineered approach was also challenged and community perspectives became an important element of the coalitions' concerns.

The Flood Action Plan concluded in 1995, at which point the donor consortia that was crucial to funding structural engineering measures decided that they could not support the planned embankments. Shifting from the engineering coalition to the environmental coalition, the donors tipped the balance towards the latter. Local community participation became an important element of planning processes, enshrined in the Bangladesh Water Development Board Act of 2000. In addition, the impacts of the embankments on fishing were incorporated into the National Environment Action Plan and the National Water Policy, which also both emphasized local community participation in identifying solutions and managing the natural environment accordingly.

The shift in approach to flood management in Bangladesh can be seen as a product of a shift in the balance of power between two policy advocacy coalitions that was catalysed by a range of environmental, contextual and behavioural factors. The major flood events created a perturbation and opened a window to discuss the limitations of the dominant engineered approach. Combined with shifts in the wider political landscape where democracy increased the recognition of local communities, the context of decision-making and values and beliefs provided an enabling environment for the donor community to switch allegiance and ultimately tip the balance of power in favour of the environmental coalition. One important difference here from the situation described by Sabatier and his ACF is that the changes were not negotiated between the two coalitions, but decided by that shift in the balance of power created by the defection of the donor communities from supporting the engineering approach and its core beliefs.

11.4.3 Flood Risk Management in Tanzania: The President as Policy Entrepreneur

Despite flood events and the scale of damage and displacement that these events cause, combined with increasing concerns that climate change and rapid urbanization are increasing the frequency and severity of flood events (Hallegatte et al. 2013), Tanzania lacks a dedicated policy to tackle flooding. Instead, Tanzania addresses FRM through wider disaster management policy, but progress in this respect has at times virtually stalled. This stasis leads us to consider the concepts underlying the punctuated equilibrium model of policy process, which alone, as described above, sees equilibrium rather than change as the dominant state of affairs.

Flooding was first addressed in policy in Tanzania as part of the 1990 National Disaster Relief Coordination Act. This Act established the institutional framework for disaster management, within which flooding is included as a key disaster risk. The Act established the Disaster Management Department within the Prime Minister's Office and initiated the Tanzania Disaster Relief Committee to coordinate disaster response efforts (Daly et al. 2015). However, the provisions in the 1990 Act addressed only disaster response and recovery. There were no provisions for anticipatory actions to reduce risks or prepare for hazard events.

The 2004 Disaster Management Policy addressed some of the gaps in the National Disaster Relief Coordination Act, introducing provisions for anticipatory actions to prepare for and mitigate risk. However, this policy was reviewed in 2010 through efforts promoted by the international donor community to better align the policy with international disaster frameworks, particularly the Hyogo Framework for Action, and address climate change concerns (Fisher and Mwase 2011; Daly et al. 2015). Whilst a new policy has yet to be agreed, a new Disaster Management Act was enacted in 2015.

Tanzania's Disaster Management Act 2015 provides an updated approach to disaster risk management that addresses preparation, mitigation and risk reduction, in addition to provisions for responding to emergency events, including provisions for early warning and evacuation orders. One of the key innovations of this Act is that it enables the establishment of a Disaster Management Agency which replaces the Disaster Management Department as the 'national focal point for the coordination of disaster risk reduction and management' (Disaster Management Act 2015 [Tanzania], p. 9).

Across the government ministries, and particularly in the case of flood or disaster risk management, a dependence on external support through international donor communities and agencies, such as the UN, has been driven by a lack of domestic resources and attention allocated to issues of flooding. As such, new policies on disaster management have been catalysed, not by flood events or other disasters, but rather by the interests of the international community.

The review of the 2004 Disaster Management Policy and subsequent Disaster Management Act of 2015 were both the result of a perceived need to align domestic policy with international frameworks, which are promoted and monitored by the international aid agencies (Fisher and Mwase 2011; Daly et al. 2015). As such, the policy landscape for flood and disaster risk management in Tanzania is characterized by an equilibrium that is punctuated by international policy processes and donor demand. Donors and international aid groups work together to deliver progress towards their shared values (Fisher and Mwase 2011; United Nations Tanzania 2015). In this manner, these external interests form an advocacy coalition which has yielded considerable power. These groups not only influence policy formulation but also carry out progress monitoring (Kalugendo 2015) and provide funding and training for staff (United Nations Tanzania 2015).

But political factors cannot be ignored. In 2015, the new President, John Magufuli, was elected on a strong anti-corruption agenda with an emphasis on economic development and on creating an environment to support the development of the domestic economy (Anyimadu 2016; Ng'wanakilala 2016; Paget 2017). Magufuli is focused on domestic interests and has declined invitations to attend international gatherings, limiting his international travel only to neighbouring countries[2] (Mugarula and Lugongo 2016). The President is not known for environmental concerns and he has yet to make reference to climate change. As such, flooding is portrayed as a problem of informal housing settlements whereby communities have settled in contravention of urban planning regulations. As

---

2 Since his election and to 2018, Magufuli only travelled to Rwanda, Uganda, and Kenya.

Magufuli emphasizes strict adherence to laws and such policies, the Government's response to criticisms of a lack of emergency aid and relief during recent flooding in 2017 was to comment that such support would be counter-productive as it would encourage at-risk communities to continue living in such risky informal settlements. Instead, the government highlighted the need for such communities to relocate and for better mapping and risk assessment to help identify – on a longer timescale and looking ahead – where land might be unsuitable for habitation due to flood risk (Domasa 2018).

So this is now the official government policy, and the President has a highly domestic focus and a strong voice. President Magufuli is now the main driver of policy development and implementation in a highly centralized system (Anyimadu 2016), and the influence of the international advocacy coalition has waned. The low position that environmental issues, particularly climate change and its associated flood risk, has on the President's current agenda explains that whilst policies are in place, implementation has stalled. Specifically, two and a half years since its conception, the Disaster Management Agency had not yet been established, leaving the Disaster Management Department in place. In addition, only 15 out of 182 Districts had prepared their emergency preparedness and response plans, as required under the Act (Domasa 2018).

Tanzania's resources are concentrated in areas that are seen as a greater priority than disaster risk management. In this respect, it is apparent that as elements of the contextual and behavioural factors are shifting, as a result of the new political landscape, the development of further progress on a new Disaster Management Policy and implementation is likely to be restricted until the President sees it as a clear priority.

Where flood events are often catalysts for policy shifts, recent flood events in Tanzania have highlighted that the contextual and behavioural factors, particularly the norms, values and beliefs that shape how flood victims are viewed, have changed. With a new power balance, events that would have previously been used by the international advocacy coalition to encourage better resourcing and policy revisions are now used as windows of opportunity to endorse the President's anti-corruption agenda by emphasizing the illegality of the informal settlements. Deflecting criticism of a lack of effective policy that would normally spark a policy revision and greater emphasis on emergency response resourcing, the Government instead uses flooding to highlight the need to abide by planning regulations.

This Tanzanian example therefore highlights the important role that a range of contextual, environmental and behavioural factors play not only in determining policy shifts and developments, but also in creating elements of policy inertia. As the punctuated equilibrium model suggests, the balance between these factors can change suddenly, and these shifts are not only catalysed by hazard events but also by wider political change and the shifting power relations that this can create. In this case the policy entrepreneur is somewhat different in terms of their power from those envisaged by Baumgartner and Jones (1993), but the Tanzanian president's key role here is not dissimilar to those they envisaged as being important in agenda setting.

11.4.4 Flood Insurance in the UK: Six Decades of Relative Policy Stability

Institutions, policies and approaches towards FRM in the UK have undergone a substantive transition over the past 60+ years, affecting the dominant philosophy, the type of risk reduction measures, and implementation strategies. The move from 'land drainage' and 'flood defence' to FRM reflects a fundamental shift in the dominant beliefs, values and attitudes of society towards the flood problem (Tunstall et al. 2004). Flood crises have also created 'windows of opportunity' for promoting and fostering policy change, especially regarding spatial planning (Johnson et al. 2005; Penning-Rowsell et al. 2006).

But despite all these changes, our research has shown that there have been relatively few changes over many years in flood insurance institutional structures, governance arrangements and policy stance (Penning-Rowsell et al. 2014). This slow process contrasts with the many profound UK societal changes that have occurred since the 1960s, affecting most areas of government intervention and insurance markets. From the early days of the 'Gentleman's Agreement', through a series of 'Statement of Principles', to the modern arrangement termed 'Flood-Re', flood insurance has been provided to those most at risk (indeed generally to all-comers), with subsidies generated by widespread insurance of those at less or even no risk. Such arrangements have been agreed between the insurance industry and government, and have endured for over 60 years (Table 11.1).

Many windows of opportunity created by major flooding or financial crises have not significantly affected the pace or direction of policy change here. This is not to suggest there have been no changes within the insurance market over the period of the Statement(s) of Principles (Table 11.1). Each has represented a progressive tightening of the competitive environment and increasing pressure on the government from the insurance industry to maintain the status quo, whereby domestic flood insurance is provided profitably by private insurance companies and the government provides capital schemes to reduce risk and hence the exposure of insurers to major losses. So whilst the market may have changed in this respect, we would argue that the overall policy direction remains unaltered, as agreed between the government and the insurance industry.

The important agents of recent change bringing about Flood-Re have been, firstly, threats to existing household insurers from new entrants unencumbered by agreements to insure all-comers. Secondly, the potential for climate change to increase flood losses considerably and unpredictably provides a threat to insurance company profitability, if not its solvency. Thirdly, technological changes have made exposure more explicit and pricing risk both easier and less expensive. Insurers are now better able to assess flood risk, and understand their exposure. Their customers (and their government) can see much more clearly now the cross-subsidies involved and also the threat of risk-reflective pricing greatly increasing premiums to those at most risk (and insurance penetration thereby declining). The Flood-Re scheme therefore seeks to continue the pattern so well established in the past: private insurance provision, cross-subsidies to retain company solvency, and government commitment to risk reduction via capital expenditure on largely engineering projects. The measures involved to seek to manage risk may change, but the overall policy direction remains unaltered, at least until the 25-year Flood-Re scheme is supposed to end in 2039.

If this policy stability appears abnormal in FRM (as reflected in several of the examples cited in this chapter), this is counterintuitive to most theorists of policy process and change, who argue that under 'normal conditions' policies are indeed relatively stable. That is the 'punctuated equilibrium' situation: stasis is normal; rapid change is unusual. This relative stasis occurs where it is in the interest of powerful elites, advocacy groups, actor networks or policy communities; a set of shared core beliefs and values results in

Table 11.1 Changes in policy agreements between the UK Government and insurance industry: 1961 to the present.

Period/date	Policy agreement	Characteristics of the policy agreement
Pre-1961	Market-driven – no specific policy agreement	Flood insurance was a part of composite policies from 1922, but total loss insurance not available until 1929 (Arnell et al. 1984) and was optional. Penetration was initially low.
1961–2001	Gentleman's agreement between Government and insurance industry	Insurers made an informal, yet strong commitment with Government to continue to provide cover to all permanently inhabited UK domestic properties. Coverage was enhanced when insurance on buildings was made compulsory for mortgage holders in the early 1970s.
20 February 2001	ABI Memorandum to Government	The first challenge of the 1961 agreement was to provide universal cover, but maintains a two-year commitment to continuing availability, requiring a government commitment to increased funding for flood defences and improving spatial planning in areas at risk.
January 2003	*Statement of Principles* policy agreement between UK Government and the ABI	This replaced the temporary memorandum agreement issued in 2001 with the ABI and Government re-emphasizing the commitments on both parts, but indicating insurers are more likely to consider the refusal of insurance in high-risk areas, as a last resort.
January 2006	Updated *Statement of Principles* policy agreement	This reinforced the minimum standard of protection of 1 in 75 years established in 2002, but strengthened the requirements for government to commit to reducing risk through flood defence investments.
January 2009	Updated *Statement of Principles* policy agreement	The final Statement excluded all newly built properties and strongly reinforced the message that policy terms and premiums should be based on risk. Announced this was the final time the Statement would be renewed.
Post-July 2013	Ending of the *Statement of Principles*	An 'understanding' between UK Government and insurers about a 'Flood-Re' solution which would mean flood insurance remains available but also affordable – but more clearly links premiums to risk. Retains the existing industry-maintained cross-subsidy favouring those at high risk.

Source: Penning-Rowsell et al. (2014). Reproduced with permission of Taylor and Francis.
ABI, Association of British Insurers.

a common understanding of the policy domain, the main policy problems, and the desirability and feasibility of different policy options and their continued deployment (Huitema and Meijerink 2009).

These groups can resist change, creating a policy equilibrium (Baumgartner and Jones 1993), serving thereby to enhance further their individual and collective interests and power in the strategic management of the policy process. Change is promoted, but that process of change is carefully controlled. The overriding importance of the London location for – and the profitability of – the insurance industry, both to government and to the insurers, explains the extraordinary 60-year stability of the overall policy direction.

11.5   Conclusions

Our four examples have highlighted how different processes can lead to change (or no change). Taken together, the three policy process theories have also been useful in helping to disentangle some of the complex social and political elements that influence policy change, and pointed the relevant research to factors explaining the policy process that might otherwise have been missed or given insufficient emphasis. Through these lenses, we are better able to see which forces exert power and how they are able to capitalize on environmental and other contextual conditions to achieve what they see as policy improvements.

But we can also see that no one theory is uniquely useful alone. The case of flood insurance in the UK presented elements of both the multiple streams and the advocacy coalition theories, and the emphasis of relative stasis reflects the punctuated equilibrium concept. In Bangladesh we see that coalitions can have core beliefs that are not unchanging because partners can shift within those coalitions. Also, the relative power of advocacy coalitions to influence policy change also changes, as demonstrated by a change in government in Tanzania and the lesser importance then of international donors. We also find that the Tanzanian policy landscape for flood and disaster risk management has been radically changed by a change in the political situation. In the case of UK flood insurance, the many windows of opportunity for change were relatively unimportant in comparison with the stability created by the mutual support provided by government and the insurance industry in maintaining a relatively unchanged situation.

But we also need to see some limitations to these ideas. The negotiation element of the ACF approach was not characteristic of the situation in Bangladesh. In South Africa the timescale within the window of opportunity concept was different, where change was announced in one key window, but not implemented for several decades; the window of opportunity was open for a great deal longer than the multiple streams concept envisages. These theories seem, perhaps, to be best suited to country contexts where policy processes are based on established democratic arrangements and therefore its relatively slow-moving political scene. In our two developing country contexts we see a more complicated political landscape embedded in often rapidly changing conditions, and this is further complicated by the role of external actors such as NGOs and donors who bring their own agendas. The policy process theories are not unhelpful here but perhaps need further development.

We can also learn about policy issues themselves. The policy process challenges as outlined in our four cases offer some practical lessons in supporting the design of better processes and better FRM policy and policy outcomes in the future. And, whilst relating just to FRM here, these lessons can extend across a range of policy areas. For example, the support of government is important for policy success (an obvious conclusion from the UK example, as also in very different circumstances in Tanzania). Where the influence of policy coalitions goes beyond policy content to, for example, resourcing the implementation of that policy, it is important to ensure buy-in to the policy and its implementation from the responsible government and any other implementing agencies (a lesson from South Africa).

Policy teams must also ensure their policy is realistic and timelines are appropriate in the context of competing priorities and the capacity of implementing agencies to ensure the policy is meaningful and delivered 'on the ground' (another lesson from South Africa). This necessitates that policy processes are inclusive of those directly and indirectly affected (as the Bangladesh example shows). Finally, it is self-evident but often ignored that policy design needs to clearly consider the practicality of the decisions and actions proposed during the policy development and its debate, not after the policy change has been formally approved.

Acknowledgements

Joanna Pardoe's work on this chapter was supported by the Future Climate for Africa UMFULA project, which is funded by the Natural Environment Research Council (NERC) and Department for International Development (DFID) under grant number NE/M020398/1.

References

Anyimadu, A. (2016). *Politics and Development in Tanzania: Shifting the Status Quo, Research Paper*. London: Chatham House.

Arnell, N.W., Clark, M.J., and Gurnell, A.M. (1984). Flood insurance and extreme events: the role of crisis in prompting changes in British institutional response to flood hazard. *Applied Geography* 4 (2): 167–181.

Baumgartner, F.R. and Jones, B.D. (1993). *Agendas and Instability in American Politics*. Chicago: University of Chicago Press.

Birkland, T.A. (2016). Policy process theory and natural hazards. In: *Oxford Research Encyclopaedia of Natural Hazard Science* (ed. S.L. Cutter). Oxford: Oxford University Press https://oxfordre.com/naturalhazardscience/view/10.1093/acrefore/ 9780199389407.001.0001/acrefore-9780199389407-e-75?print=pdf.

Boin, A., 't Hart, P., and McConnell, A. (2009). Crisis exploitation: political and policy impacts of framing contests. *Journal of European Public Policy* 16 (1): 81–106.

Botha, D., van Niekerk, D., Wentink, G. et al. (2011). *Disaster Risk Management Status Assessment of Municipalities in South Africa*, 1e. Pretoria: South African Local Government Association (SALGA).

Cairney, P. and Zahariadis, N. (2016). Multiple streams analysis: a flexible metaphor presents an opportunity to operationalize agenda setting processes. In: *Handbook of Public Policy Agenda-Setting* (ed. N. Zahariadis), 87–105. Cheltenham: Edward Elgar.

Cohen, M.D., Marsh, J.G., and Olsen, J.P. (1972). A garbage can model of organisational choice. *Administrative Science Quarterly* 17 (1): 1–25.

Daly, M.E., Yanda, P.Z., and West, J.J. (2015). *Climate Change Policy Inventory and Analysis for Tanzania*. CICERO Report 2015:05. Oslo: CICERO.

Domasa, S. (2018). Tanzania: Floods Cost Tanzania US $2 billion Annually. *Tanzania Daily News* (28 February). http://allafrica.com/stories/201802280677.html.

Downs, A. (1972). Up and down with ecology – the "issue-attention cycle". *The Public Interest* 28: 38–50.

Fisher, M. and Mwase, N. (2011). *Strengthening National Disaster Preparedness and Response Capacity – End of Programme Evaluation*. Dar es Salaam: UNICEF Country Office https://www.unicef.org/evaldatabase/index_66795.html.

Fuchs, S. and Thaler, T. (2018). *Vulnerability and Resilience to Natural Hazards*. Cambridge: Cambridge University Press.

Hallegatte, S., Green, C., Nicholls, R.J., and Corfee-Morlot, J. (2013). Future flood losses in major coastal cities. *Nature Climate Change* 3: 802–806.

Huitema, D. and Meijerink, S. (eds.) (2009). *Water Policy Entrepreneurs. A Research Companion to Water Transitions around the Globe*. Cheltenham: Edward Elgar.

John, P. (1998). *Analysing Public Policy*. London and New York: Continuum.

Johnson, C.L., Tunstall, S.M., and Penning-Rowsell, E.C. (2005). Floods as catalysts for policy change: historical lessons from England and Wales. *International Journal of Water Resources Development* 21 (4): 561–575.

Jones, M.D., Peterson, H.L., Pierce, J.J. et al. (2016). A river runs through it: a multiple streams meta-review. *Policy Studies Journal* 44 (1): 13–35.

Kalugendo, F. (2015). *United Republic of Tanzania: National progress report on the implementation of the Hyogo Framework for Action (2013–2015)*. http://www.preventionweb.net/english/hyogo/progress/reports.

Kingdon, J.W. (1984). *Agendas, Alternatives and Public Policies*. Boston: Little, Brown.

Kingdon, J. (2003). *Agendas, Alternatives and Public Policies*, 2e. New York: London.

Kirtman, B., Power, S.B., Adedoyin, J.A. et al. (2013). Near-term climate change: projections and predictability. In: *Climate Change 2013: The Physical Science Basis. Contribution of Working Group I to the Fifth Assessment Report of the Intergovernmental Panel on Climate Change* (eds. T.F. Stocker, D. Qin, G. Plattner, et al.), 953–1028. Cambridge: Cambridge University Press.

Lane, S., November, V., Landstrom, C., and Whatmore, S. (2013). Explaining rapid transitions in the practice of FRM. *Annals of the Association of American Geographers* 103: 330–342.

Lazenby, M.J. and Todd, M.C. (2018). Future precipitation projections over Central and Southern Africa and the adjacent Indian Ocean: what causes the changes and the uncertainty? *Journal of Climate* 31: 4807–4826. https://doi.org/10.1175/JCLI-D-17-0311.1.

Luo, T., Maddocks, A., Iceland, C. et al. (2015). *World's 15 Countries with the Most People Exposed to River Floods*. World Resources Institute http://www.wri.org/blog/2015/03/world%E2%80%99s-15-countries-most-people-exposed-river-floods.

Meijerink, S. and Huitema, D. (2010). Policy entrepreneurs and change strategies: lessons from 16 case studies of water transitions around the globe. *Ecology and Society* 15 (2): 1–21.

Mugarula, F. and Lugongo, B. (2016). The untold cost of JPM's stance on foreign travels. *The Citizen* (April 13). http://www.thecitizen.co.tz/magazine/politicalreforms/The-untold-cost-of-JPM-s-stance-on-foreign-travels/1843776-3157154-1284pm9/index.html.

Ng'wanakilala, F. (2016). Tanzania's Magufuli takes anti-corruption drive to ruling party. *Reuters World News* (December 14). https://www.reuters.com/article/us-tanzania-corruption/tanzanias-magufuli-takes-anti-corruption-drive-to-ruling-party-idUSKBN1431VG.

Oppenheimer, M., Campos, M., Warren, R. et al. (2014). Emergent risks and key vulnerabilities. In: *Climate Change 2014: Impacts, Adaptation, and Vulnerability. Part A: Global and Sectoral Aspects. Contribution of Working Group II to the Fifth Assessment Report of the Intergovernmental Panel on Climate Change* (eds. C.B. Field, V.R. Barros, D.J. Dokken, et al.), 1039–1099. Cambridge: Cambridge University Press.

Paget, D. (2017). Magufuli has been president for two years: how he's changing Tanzania. *The Conversation* (2 November). http://theconversation.com/magufuli-has-been-president-for-two-years-how-hes-changing-tanzania-86777.

Parsons, W. (1995). *Public Policy: An Introduction to the Theory and Practice of Policy Analysis.* Cheltenham UK and Massachusetts USA: Edward Elgar.

Penning-Rowsell, E.C., Johnson, C., and Tunstall, S.M. (2006). 'Signals' from pre-crisis discourse: Lessons from UK flooding for global environmental policy change? *Global Environmental Change* 16: 323–339.

Penning-Rowsell, E.C., Johnson, C., and Tunstall, S.M. (2017). Understanding policy change in FRM. *Water Security* 2: 11–18.

Penning-Rowsell, E.C. and Korndewal, M. (2018). The realities of managing uncertainties surrounding pluvial urban flood risk: an *ex post* analysis in three European cities. *Journal of Flood Risk Management* 1–12. https://doi.org/10.1111/jfr3.12467.

Penning-Rowsell, E.C., Priest, S., and Johnson, C. (2014). The evolution of UK flood insurance: incremental change over six decades. *International Journal of Water Resources Development* 30 (4): 694–713. https://doi.org/10.1080/07900627.2014.903166.

Prindle, D.F. (2012). Incorporating concepts from biology into political science: the case of punctuated equilibrium. *Policy Studies Journal* 40: 21–43.

Rowell, D.P. (2011). Sources of uncertainty in future changes in local precipitation. *Climate Dynamics* 39 (7–8): 1929–1950.

Sabatier, P.A. (1988). An advocacy coalition framework of policy change and the role of policy-oriented learning therein. *Policy Sciences* 21: 129–168.

Sabatier, P.A. (1993). Policy change over a decade or more. In: *Policy Change and Learning: An Advocacy Coalition Approach* (eds. P.A. Sabatier and H. Jenkins-Smith), 13–39. Boulder, CO: Westview Press.

Sabatier, P.A. and Weible, C.M. (eds.) (2014). *Theories of the Policy Process*, 3e. Boulder, CO: Westview Press.

Sayers, P., Galloway, G., Penning-Rowsell, E.C. et al. (2014). Strategic flood management: 10 'golden rules' to guide a sound approach. *International Journal of River Basin Management* 13 (2): 137–151.

Solik, B. and Penning-Rowsell, E.C. (2017). Adding an implementation phase to the framework for flood policy evolution: insights from South Africa. *International Journal of Water Resources Development* 33 (1): 51–68.

Sultana, P., Johnson, C., and Thompson, P. (2008). The impact of major floods on flood risk policy evolution: insights from Bangladesh. *International Journal of River Basin Management* 6 (4): 339–348.

Templehoff, J., van Niekerk, D., van Eeden, E. et al. (2009). The December 2004–January 2005 floods in the garden route region of the Southern Cape, South Africa. *Journal of Disaster Risk Studies* 2: 93–112.

True, J.L., Jones, B.D., and Baumgartner, F.R. (2009). Punctuated-equilibrium theory: explaining the stability and change in public policy making. In: *Theories of the Policy Process* (ed. P.A. Sabatier), 155–187. Boulder, CO: Westview Press.

Tunstall, S.M., Johnson, C.L., and Penning-Rowsell, E.C. (2004). Flood hazard management in England and Wales: from land drainage to FRM. *World Congress on Natural Disaster Mitigation Proceedings Volume 2,* New Delhi, India (19–22 February 2004), 447–454. Institute of Engineers.

United Nations Tanzania (2015). Emergencies, Disaster Preparedness and Response. http://tz.one.un.org/what-we-do/overview/8-emergencies-disaster-preparedness-and-response?showall=1 (accessed 15 July 2018).

van Herk, S., Rijke, J., Zevenbergen, C. et al. (2015). Adaptive co-management and network learning in the Room for the River programme. *Journal of Environmental Planning and Management* 58 (3): 554–575.

Ward, P.J., Jogman, B., Salamon, P. et al. (2015). Usefulness and limitations of global flood risk models. *Nature Climate Change* 5: 712–715. https://doi.org/10.1038/nclimate2742.

Weible, C.M. and Sabatier, P.A. (2009). Coalitions, science, and belief change: comparing adversarial and collaborative policy subsystems. *Policy Studies Journal* 37 (2): 195–212. https://doi.org/10.1111/j.1541-0072.2009.00310.x.

Weible, C.M. and Sabatier, P.A. (2017). *Theories of the Policy Process*, 4e. Boulder, CO: Westview Press.

Winsemius, H.C., Van Beek, L.P.H., Jongman, B. et al. (2013). A framework for global river flood risk assessments. *Hydrology and East System Sciences* 17: 1871–1892. https://doi.org/10.5194/hess-17-1871-2013.

12

The Political Economy of Wastewater in Europe

Heather M. Smith[1] and Gareth Walker[2]

[1] *Cranfield Water Science Institute, Cranfield University, UK*
[2] *Insight Data Science, San Francisco, USA*

12.1 Introduction

On the global development stage, wastewater and sanitation have long lagged behind drinking water in terms of the relative attention and investment they receive (Murray et al. 2011). Whilst a 'human right to water' was first debated in the 1990s (McCaffrey 1992), the UN did not affirm 'human right to water and sanitation' until 2010, and a UN resolution that included distinct provisions for the latter was adopted only in 2015 (UNGA 2010, 2015). In Europe, as we argue in this chapter, the picture is somewhat different. For many EU member states, wastewater has been a stronger driver of investment than drinking water in recent decades, influenced in large part by the need for compliance with European legislation. However, despite this economic significance, wastewater is rarely subjected to the same political limelight as drinking water, and it has therefore assumed something of a hidden role within the wider governance of the European water sector.

This socio-political schism between drinking water and wastewater is particularly evident in recent debates around public vs. private models of service delivery. Academically, the 'public–private debate' has largely run out steam – successive studies of the relative performance of public and private models have yielded only mixed results (Porcher and Saussier 2017), and it can be argued that the success or failure of particular systems is more to do with context-specific governance arrangements than the choice of public or private models *per se* (Barraqué 2012). However, socially and politically, the debate is still very much alive.

European nations have seen the public–private debate reignite in the past decade, in part due to the fallout from the 2007 economic crash. On the one hand, the European financial establishment has favoured the liberalization of public services (Barraqué 2012) and pushed for a number of austerity-driven measures in several countries that have included the transfer of utility services to private operators, on the grounds of fiscal responsibility and the need to attract investment (particularly for wastewater). On the other hand, other EU member states have been caught up in the recent global 'wave'

Water Science, Policy, and Management: A Global Challenge, First Edition. Edited by Simon J. Dadson, Dustin E. Garrick, Edmund C. Penning-Rowsell, Jim W. Hall, Rob Hope, and Jocelyne Hughes.
© 2020 John Wiley & Sons Ltd. Published 2020 by John Wiley & Sons Ltd.

of remunicipalizations, in which a large number of cities have opted to transfer public services (particularly water and energy services) back to the public sector. These trends have ignited emotive rhetoric on all sides of the issue. There have been large-scale protests and political actions in countries such as Ireland and Italy – the latter held a referendum in 2011 in which voters roundly rejected privatization of water services (Kennedy 2011). The debate has even re-entered the public discourse in the UK, where much of the sector is fully privatized, as a result of the Labour Party's 2017 manifesto pledge to nationalize water, rail and energy services (James 2017).

This chapter argues two essential points: (i) that the politics of the European sector (and particularly these more recent debates around governance and ownership models) are still predominantly focused on drinking water, and would benefit from a closer engagement with the distinct technical and economic features of wastewater services; and (ii) that, conversely, the wastewater sector could benefit from enhanced social and political awareness and engagement. To illustrate these two points, the chapter first presents an overview of different governance models for water and wastewater service delivery, and examines the debate over their relative merits. It then presents a discussion of how wastewater has become a key economic driver for sector investment, particularly in the EU. Following this, the chapter outlines two case studies – Paris and Ireland. These cases were selected because they are two of the most highly politicized recent examples of dramatic shifts in governance arrangements in the European water sector. Additionally, both cases neatly illustrate the aforementioned 'schism' between water and wastewater, and the need for an enhanced understanding of wastewater within socio-political governance debates. Finally, the chapter concludes by discussing some key emerging trends in wastewater, particularly the shift towards circular economy approaches, and highlights how the potential value of the resources held within wastewater could help spur greater socio-political engagement and foster more creative thinking around models of management and ownership (in keeping with concurrent debates around drinking water).

12.2 Models of Service Delivery

Globally, there is a vast array of models for the ownership and operation of water and wastewater utility services, encompassing different roles and responsibilities for public and private sector actors. Whilst there is no widely agreed glossary, and the terminology is often contested, *fully public* models can be characterized as those where there is little or no private sector involvement, and where all aspects of utility services are fully owned and operated by governments (whether municipal, regional or national). In contrast, *fully private* models are those where the utility services (including all the assets) are fully owned and operated by private corporations. These models are relatively rare, with England providing one of the few examples in the world of a fully privatized sector (different models exist in Scotland, Wales and Northern Ireland).

Between these two extremes of the public–private spectrum lie a myriad of options. One broad family of arrangements can be described as *contractual* models, sometimes referred to as 'public–private partnerships' (PPPs). Again, there is no set definition of a PPP, and analysis of the concept has generated a vast literature (Hodge and Greve 2007). In this context it generally refers to situations where governments and private firms work together to produce and deliver water and wastewater services, and where the relationship is governed under some form of contractual arrangement. Those contractual arrangements can vary massively in terms of their scope: which aspects of the services that private firms have a hand in (e.g. design, design-build, operation); their longevity; the degree of decision-making authority that private firms have; the distribution of risk and reward between parties; their mechanisms for oversight; and many other factors. In France alone there are several types of contractual arrangements for private operators in water and wastewater services, including 'lease' contracts, 'concession' contracts, and 'gerance' contracts (Barraqué 2012; Porcher 2018). In all cases, however, ownership of the assets remains in (or, in the case of many design-build contracts, is transferred to) public control.

Another broad family can be described as *corporatized* models. Again, this term is often contested and there is no widely accepted definition. But where contractual models are characterized by the involvement of fully private firms in the development or delivery of services, corporatized models generally involve the creation of utility organizations that are, in essence, a form of public–private hybrid, or 'public enterprise'. Typically, the utility organization is owned by a government (whether municipal, regional or national), but it has separate legal and financial status, and a resulting degree of autonomy from government, and its organizational structure tends to more closely mirror that of a private corporation (as opposed to a government department). As with contractual models, the specifics of each corporatized arrangement can vary significantly, particularly in terms of the extent to which they 'behave' like private firms, and the degree of autonomy from government that they actually possess – indeed, in cases where autonomy is limited, the line between a corporatized model and a fully public one can be murky. Corporatized utilities may also incorporate more innovative forms of ownership arrangements, such as the Welsh model wherein the utility (Dwr Cymru/Welsh Water) is owned by a single-purpose, not-for-profit company (Glas Cymru), which has no shareholders and is run exclusively for the benefit of customers.

The question of which model(s) of water utility ownership and management produces optimal social, economic, and environmental outcomes remains a contested and emotive topic in contemporary politics and civil society. The essential role of water and sanitation in public health, economic productivity, and environmental integrity have made the so-called public–private debate a lightning rod for wider political debates. The involvement of the private sector in water and wastewater services is often premised on arguments of full cost recovery, access to capital, and the operational efficiencies anticipated by appropriate regulatory and contractual incentives. Conversely, proponents of public services point to market failures stemming from water's geography as a flow resource integral to both social and environmental processes, and the failure of commoditized models of water services to address the social imperatives of affordability and universal access. As previously mentioned, whilst the issue has generated substantial analysis in the academic literature (Bakker 2003; Swyngedouw 2005; Lorrain and Poupeau 2016), this has resulted in few firm conclusions. Many have acknowledged that the political debate is therefore primarily an ideological one, which can overshadow the real practical challenges of providing effective and affordable services (Wolff and Palaniappan 2004).

The European context is noteworthy for two widely identified trends tending to either side of this debate. On the one hand, highly visible and contested privatizations have been driven by policy arising from the European debt crisis of 2009. Following several mostly southern European states being unable to refinance or service national debt, a triumvirate of the European Commission (EC), the European Central Bank (ECB), and the International Monetary Fund (IMF) (often collectively referred to as the 'Troika') imposed strict public austerity conditions on financial aid to affected states. The transfer of utilities to private ownership and management was often included in these measures.

Concurrently, as mentioned previously, several northern European states have witnessed a trend towards remunicipalization of utility services – particularly the water sector in France, and the energy sector in Germany (Hall et al. 2013) – in which the perceived failures of private contractors, and the opportunities granted by the expiration of service contracts, have allowed governments to take services back into public ownership and control. Kishimoto et al. (2015) identified over 200 cases of water remunicipalization, in 37 (mainly high-income) countries, that occurred between 2000 and 2015. The number that occurred post-2010 was more than double the number that occurred in the previous decade. Nearly half of the identified cases occurred in France. Such regime shifts are thought to be primarily driven by municipalities' expectations over prices and leakage rates, although socio-political factors – notably political ideologies, and mimicry of the decisions of neighbouring municipalities – are known to play a strong role (Pérard 2009; Porcher and Saussier 2017).

What is particularly notable about the remunicipalization trend is that it often implies a shift from a contractual model, not to a fully public model, but to a *corporatized* model (McDonald 2015). In France, local authorities wishing to 'reclaim' their water services from private delegations are required to create corporatized, semi-public bodies – national regulation actually prevents them from returning to a fully public 'direct labour' model (Barraqué 2012). This can present a bit of an ideological paradox for some – whilst remunicipalizations are widely celebrated as 'taking back' public control, corporatization has often been dismissed as an extension of neoliberal philosophy and 'the next best thing' to privatization (Furlong et al. 2018). Indeed, support for corporatization has often been premised on the same efficiency and fiscal responsibility arguments as support for private sector involvement. However, more recent examples of corporatized water utilities have challenged the assumption that they should necessarily behave just like private firms, and demonstrated their capacity for proactively working towards social objectives (McDonald 2015). The Paris case, discussed below, is an illustration of this.

Conspicuous in its absence from most of the public discourse surrounding these trends has been any acknowledgment of wastewater and sanitation services. Indeed, the debate has been almost entirely characterized as a struggle for hegemony over drinking water services. This is partly the result of the structure of the sector – in many countries, water and wastewater are managed by separate utilities. In the 235 cases of remunicipalization noted by Kishimoto et al. (2015), 13 were identified as pertaining to wastewater/sanitation, whilst a further 11 appeared to address both drinking water and sanitation. It is clear that the remunicipalization trend is predominantly related to drinking water. Whilst on average wastewater collection and treatment represent the larger proportion of asset values and domestic charges for utility services as a whole, in general they generate little public or political interest. We explore this further in the sections below.

12.3 Wastewater as a Driver of Investment and Cost Recovery

From the late nineteenth century onwards, the water sector in Europe has witnessed a significant and widespread shift in its drivers of investment, from water resource security, to wastewater treatment and environmental integrity. Whilst modern European water utilities generally find their origins in the industrial revolution of the eighteenth and early nineteenth centuries and its associated demand for water and sanitation services, the late twentieth century witnessed widespread de-industrialization and a growing premium placed on environmental quality. European Directives, most significantly the Urban Waste Water Treatment Directive (UWWTD) of 1991 and the Water Framework Directive (WFD) of 2000, have mandated minimum standards for the discharge of wastewater to the environment and established catchment-wide targets for environmental quality.

The material consequences for present-day EU utilities are significant. For example, following a comparative analysis of 29 cities across EU member and candidate states, Juuti and Katko (2005) observed a general decline in specific (per capita) and net water demand in urban centres from the 1970s onwards (Figure 12.1). They concluded that

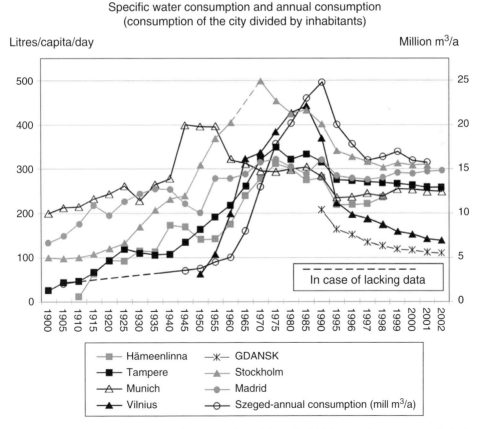

Figure 12.1 Per capita consumption (with exception of Szeged which is annual net consumption) of water in selected European cities over time. *Source:* Courtesy of Juuti and Katko (2005).

this decline was driven by socioeconomic change, as well as the introduction of volumetric charging policies for water services. Furthermore, they noted that very few utilities had any form of chemical or biological treatment of wastewater in place until the 1960s onwards, after which investment in wastewater treatment was primarily driven by environmental policy and legislation. Indeed, the EC estimated that the cost of compliance for the UWWTD alone reached approximately €45 billion by 2006. To place this in context, a recent analysis of EU water utility infrastructure investment estimated €45 billion to be the current annual investment rate for all EU states (EurEau 2017). Perhaps more importantly, a separate analysis demonstrated the dominance of wastewater investments in EU utility capital expenditure, and projected the share of wastewater investments to grow through to 2025 (Figure 12.2).

The most direct impact on European citizens has been the growing role of wastewater in utility revenue collection. An analysis in 2017 of revenue collection for EU water and wastewater services placed wastewater collection and treatment as the dominant driver in many EU member states, and as the marginally dominant driver of costs overall (EurEau 2017). This trend was confirmed by independent research carried out by the International Water Association, who further noted that the political view of 'water as a human right' (but not wastewater) has led to a global trend of divergence between tariff structures for water and wastewater, with higher Value Added Tax (VAT) often applied to wastewater (IWA 2018).

Beyond simply pushing up service charges, investment in wastewater also introduces a highly socialized set of costs and benefits. Whereas drinking water services directly benefit their consumers, the majority of wastewater treatment costs are driven by downstream environmental quality requirements. Article 16 of the WFD calls for member states to take into account the so-called 'polluter pays' principle, namely that charges

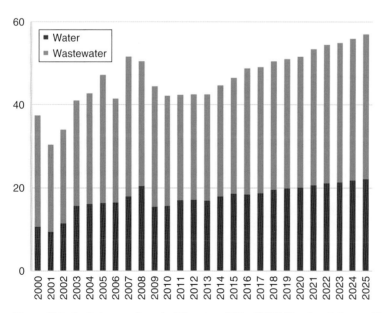

Figure 12.2 Capital expenditure by EU water utilities ($US billion/year). *Source:* Bluefield (2016). Reproduced with permission of Bluefield Research.

should be levied on polluters which are proportionate to the economic damage caused. The majority of wastewater charges are therefore intended to disincentivize or mitigate the environmental and social externalities of treatment and disposal, rather than provide a direct benefit to the consumer. As a result, a wide array of levies and taxes have been implemented across member states, varying from those integrated into taxation systems, to charges based on volume of water consumed, to rates applied to specific industries at the point of abstraction (Valero et al. 2018).

12.4 Case Studies – Paris and Ireland

As previously mentioned, these two cases were selected because they are two of the most highly politicized recent cases of governance reform in the European water sector. Paris is often referred to as a flagship of the remunicipalization trend, whilst the scale and fervour of the protests that followed the creation of Irish Water made headlines around the world. The cases (described below) illustrate two key points. First, they underscore the ideological nature of the 'public–private debate' by demonstrating that the success or failure of a particular utility model is more to do with contextual factors than the choice of model *per se*. The central irony of these two cases is that they involve very similar utility models (publicly owned, arm's length corporatized utilities), but due to widely different social and political contexts, one was applauded as the restoration of public control over services, while the other was derided as the loss of it. Secondly, and more importantly, the two cases illustrate that wastewater is often overlooked in these socio-political debates, despite its economic significance. As we will discuss, this oversight may present a significant risk to delivering effective wastewater policy in future.

12.4.1 Paris

In June 2017, representatives from Eau de Paris received a UN Public Service Award for 'Guaranteeing Access to Water for All through a Public Management Water System' (UNDESA 2017). The award came less than a decade after the creation of Eau de Paris, in what has become one of the most famous cases of remunicipalization in the global water sector. What was remarkable about the transformation in Paris' water governance was that it occurred in the home city of the world's largest two water and wastewater corporations, Veolia and Suez, which together had dominated the city's services for decades. Although the two companies were keen to describe the shift as a political manoeuvre that had nothing to do with their performance, it was nonetheless widely regarded as a highly symbolic defeat in the 'heartland' of privatized water.

In France, water and sanitation services are the responsibility of local authorities, and they have the option of managing those services directly 'in house', or delegating part or all of those services to private contractors (under various models). There are some 31 000 individual water and wastewater utilities in France. Figures from 2011 show that approximately 71% of the French population was served by private drinking water operators, whilst 56% was served by private wastewater firms (Porcher 2018). Recently, there has been a notable shift in favour of remunicipalization, although privatization has hardly vanished. Between 1998 and 2014, there were 290 cases of remunicipalization in the country, and 236 cases of privatization – although remunicipalizations affected a much larger portion of the population (Porcher and Saussier 2017). The Paris transition has played a significant role in this trend due to the high population served and the fact that numerous cities have followed their example (De Clercq 2014).

As in many old cities, the history of Paris' water and wastewater services is complex and fragmented, and private firms have long had a hand in it. Throughout most of the nineteenth century, water and wastewater services were managed together by the regional Seine county (Paris plus 80 suburban communes). In the late 1860s, the suburban communes opted to delegate their water services to Veolia (under its previous guise as *Compagnie Générale des Eaux*). In Paris itself, water supply and distribution became municipally managed, whilst the function of billing customers for services was similarly contracted out to Veolia. Wastewater services, meanwhile, remained under the auspices of Seine county (Grafton et al. 2015).

In 1984, then-mayor Jaques Chirac signed a 25-year contract with Veolia (awarded the right bank of the Seine River) and Suez (awarded the left bank) to manage the city's water supply, distribution and billing. Under those contractual arrangements, tariffs increased dramatically, by more than 265% from 1985 to 2009 (more than three times the inflation rate over the same period). The companies justified this in part by pointing to a significant reduction in leakage rates as evidence of their investment in the network. Network losses fell from 22% in 1985 to just 3.5% in 2009 (with the most significant drop occurring after a contract renegotiation in 2003). Still, the reported profitability of the companies' Paris operations caused many city officials to question their justification for the price increases (Barraqué 2012; Pigeon et al. 2012).

One of the less visible but ultimately more influential changes that occurred during the same period was the near-complete loss of transparency. The city became almost completely dependent on the two companies for any technical and financial information about the system. Claims about the network's technical performance became difficult to verify independently (Pigeon et al. 2012). Procurement trails for any commissioned works were opaque at best, and there were allegations that the companies were subcontracting works to their own subsidiaries at inflated prices. Even the companies' profits from the Paris services were difficult to establish reliably, as there were discrepancies between what was declared to the city and what was reported publicly to shareholders (Pigeon et al. 2012; Valdovinos 2012).

The idea of remunicipalization first arose in 2001, when Socialist mayor Bertrand Delanoë was elected. Green Party representative Anne Le Strat quickly became the champion of the cause, as she was appointed as CEO of SAGEP (the body ostensibly in charge of overseeing the contracts with the two water companies). It quickly became apparent that the city would have to wait until the contract period expired to complete the process, so the intervening years were largely spent renegotiating the contract terms, taking steps to improve transparency, and laying the logistical and regulatory groundwork for the eventual transition. Remunicipalization finally occurred in 2010, when the newly created Eau de Paris was awarded control of the city's entire drinking water system (from raw water to end user). Eau de Paris is a *régie à personnalité morale et autonomie financière* – a semi-autonomous public company, with separate legal status and its own budget and structure, owned and largely controlled by the city.

One of the particularly noteworthy things about this transition was that, although it was undoubtedly political (Barraqué 2012), there was very little public drive for it (Pigeon et al. 2012). The politicians (particularly Le Strat) were pursuing positions that were largely consistent with their respective parties' stances on public services, and Delanoë promised to bring water services back under public control as part of his re-election campaign in 2007 (which he won comfortably). So the elections themselves can be seen as tacit public support for remunicipalization. However, there was no widespread public or civil society interest in the subject. Delanoë's promise came as a surprise to some, and even after his re-election, many assumed it was a negotiating tactic and that the companies would still be able to renew their contracts after the initial period expired. The relative lack of public interest is partly due to billing – very few Parisians pay directly for their services (which are often wrapped up in rent and building charges).

What sets Eau de Paris apart from many previous corporatized models of water utilities is their proactive pursuit of social objectives. This was due in large part to Anne Le Strat herself, who served as the utility's first president until 2014, and freely describes herself as something of an activist in support of public management of services (Kishimoto et al. 2015). The utility invested in social tariff programmes, which occurred alongside a change in French law in 2013 that made it illegal to cut off household water supplies for non-payment of bills. Eau de Paris also championed the idea of more 'open governance' through the inclusion of stakeholder representation in their own structures. The utility's board includes representation (with voting rights) from its own staff, independent water management 'experts', an environmental NGO, a consumers' association, and a member of the Paris Water Observatory. The Observatory was created at the same time as Eau de Paris, as a means of ensuring the *public* transparency and accountability of the utility's operations. All official documentation related to water management must be submitted to the Observatory before being considered by the municipal council, to allow the chance for public comment. Although engagement with the Observatory is still limited, it is credited with reinvigorating the interest of civil society in the governance of water management (Kishimoto et al. 2015).

The Paris case highlights the contrasts between the relative political attention attributed to wastewater vs. drinking water. For instance, the wastewater sector was largely exempt from these pendulum swings of politics and governance. As previously mentioned, Paris' sanitation services remained governed at the regional level. When Seine county was broken up in the 1960s, Paris joined together with its three new neighbouring counties to form the Greater Paris Sanitation Authority (SIAAP) which continues to manage wastewater treatment facilities and storm water at the regional level. Control over sewerage systems (wastewater collection) fell to the municipalities. Due to these institutional arrangements, wastewater services in Paris were never privatized to the same extent as drinking water, but private firms (the same ones ousted from the drinking water system) continue to have a strong hand in the system, including recent upgrade, operation and maintenance contracts. This is likely due in large part to the fact that SIAAP, as an arm's length public body at county level (and governed by a board with representatives from four different counties), is not strongly affected by municipal electoral party politics. However, it is also indicative of a relative political and public disinterest in wastewater.

Barraqué and Le Bris (2007) noted that, across France, national expenditure on water and wastewater services in France was over 50% driven by wastewater collection and treatment investments. In Paris, it has been argued that the increases in domestic tariffs that occurred between 1985 and 2010 were driven in large part by increases in levies and taxes designed to cover the costs of wastewater management, specifically for implementing the UWWTD (Barraqué 2012). This went largely unmentioned in the political debates surrounding the tariffs and the perceived unfair pricing practices of the private operators. One of the early success storeys for Eau de Paris was its ability to drop water tariffs quickly by 8%, despite the costs of the governance transition, and whilst maintaining network performance – a symbolic victory against the claims of the private operators (Pigeon et al. 2012). When the decrease in water tariffs was announced, SIAAP (the regional wastewater authority) quietly used the opportunity to increase wastewater tariffs by a similar amount, again to support compliance with the UWWTD (Barraqué 2012) – a development that went largely unnoticed. Eau de Paris also vowed to keep prices stable despite any loss of revenues that might occur through its promotion of water efficiency measures and resulting decreases in water consumption rates. SIAAP opposed those water efficiency measures, on the grounds that decreasing consumption undermined the effectiveness of its treatment plants and cut into its own revenues, as wastewater tariffs are calculated on the basis of water consumption (Grafton et al. 2015). Again, these protests garnered little attention, despite the fact that the private water operators had been criticized for adopting similar stances against efficiency.

Examples of this political disconnect between water and wastewater can also be seen elsewhere in France. For instance, Bordeaux, a city that notably followed Paris' example in remunicipalizing its water services, has just done the opposite (fairly quietly) with its wastewater services in awarding a large concession contract to Veolia (Veolia Group 2018). The contract is for a notably short timeframe (seven years) and involves a 'partnership governance' proposal with the city – something the company has likely implemented in response to criticisms of its transparency that arose during the Paris remunicipalization debates.

12.4.2 Ireland

In late October 2014, around 120 000 people took to the streets of the Irish Republic (O'Doherty 2014). They withstood foul weather to express their anger over the creation of a national water and wastewater utility (Irish Water) and the introduction of tariffs for water and wastewater services. The protests puzzled many around the world – although the domestic tariffs were new, they were amongst the cheapest in Europe, and many saw them as entirely sensible for such a vital utility sector (Worstall 2014). But for the Irish, the introduction came on the heels of a period of severe austerity, leaving the country ripe for widespread unrest, and the tariff introduction was labelled by many as (pun intended) the drop that burst the dam (O'Toole 2014).

Less well reported was the fact that the Irish water and wastewater sector was in severe difficulty, with failing assets and infrastructure in many regions. Prior to the creation of Irish Water, water and wastewater services were the responsibility of local authorities, and although PPPs had a strong role in the sector (particularly design–build–operate contracts – see Reeves 2011), domestic services were financed entirely by the central government (including any capital investment), recouped largely through general taxation. That left a legacy of chronic under-resourcing in the sector, which was becoming manifest in the system's deterioration well before the economic collapse of 2007/2008. Leakage rates were estimated at 40% in 2011. In 2013, around 23 000 people were on Boil Water Notices because their water was microbiologically unfit for consumption. The Environmental Protection Agency found that around a third of the country's drinking water treatment plants were at risk of failure in terms of national and European quality parameters. The wastewater sector was even worse, and due to Ireland's failure to meet the requirements of the Urban Wastewater Treatment Directive in 71 areas, the EC undertook an Infringement Case against the country in 2011 (Bresnihan 2016).

The creation of Irish Water is, in the immediate sense, an offshoot of the financial crisis and resulting bailout and austerity measures – the EU and the IMF had pressured the country into creating an arms-length, self-financing utility that removed the burden of financing the sector from the national government's balance sheets. But it is also a response to the more long-standing issue of under-investment. An Irish Minister later stated publicly that the Irish Water model was specifically intended as a means to attract new forms of financing for the sector, as the utility would be able to borrow money directly from international creditors. The introduction of tariffs was a key part of that process, and essential for making the utility appear attractive to investors (Bresnihan 2016).

However, these issues received relatively little attention in the Irish media. Irish Water is a publicly owned utility, and its creation likely lessened the role of the private firms in the country's water and wastewater services (given the previous reliance on PPPs). Nonetheless, Irish Water has been, in many ways, portrayed as synonymous with (or a step towards) privatization, and subjected to the same criticisms that had been reserved for private utilities in other countries, such as profiteering motives, and a lack of transparency and oversight. A poll in 2014 revealed that a third of respondents intended not to pay the tariffs when they were introduced the following spring. The utility and its tariffs became key subjects of the 2016 national election, and after hotly contested negotiations, water tariffs were eventually suspended under the new government. The idea of universal domestic charges was eventually abandoned entirely in 2017, to be replaced with a new system of 'excess water' charges in January 2019 – so that only those households using water above a certain threshold will be required to pay a tariff (Irish Water 2018). This has brought the financial viability of the utility into question, as it is now much more heavily dependent on business and commercial tariffs.

In contrast with the open approach of Eau de Paris, Irish Water was frequently accused of having little transparency, and for having a 'ham-fisted' approach to outreach and engagement (particularly in relation to its explanation of the rationale behind the introduction of tariffs). From its inception, it was plagued with data protection issues, errors in communication, and criticisms of the knowledge of its staff. There were also criticisms of what were seen as excessive and unjustified remunerations for top executives. The utility was quickly labelled as a 'toxic brand' (Bagnall 2014). Despite these challenges in its approach, there is little doubt that Irish Water was caught up in, and became a focal point for, wider anti-austerity sentiment. The tariffs themselves were simply labelled as 'the latest austerity measure' rather than cost recovery mechanisms to support needed investment in ailing infrastructure. The timing of the national election also played a role, as more than 100 candidates vying for seats were keen to align themselves with the protest movement (*Independent* 2016). All in all there is reason to believe that, even if the utility had employed a progressive and fully transparent approach to its governance and public engagement, it may still have become a focal

Table 12.1 Irish Water's proposed network capital costs, 2019 (€m).

Network capital costs	IW request 2019
Water national programmes	129
Wastewater national programmes	62
Other infrastructure national programmes	17
National programmes total	208
Water capital maintenance	94
Wastewater capital maintenance	32
Capital maintenance total	127
Water projects	196
Wastewater projects	273
Projects total	469
Total network capital costs	*803*

Source: CRU (2018), p. 24. Reproduced with permission of CRU.

point for wider anti-austerity fervour. The strength of that wider sentiment is exemplified by the fact that the level of Irish Water's tariffs was minute compared with the scale of previous tax increases that Irish taxpayers had already absorbed.

As with the Paris case, discussion of wastewater is conspicuously absent from these political and social debates about the utility and its tariffs, despite its economic significance. For Irish Water, addressing an ageing drinking water distribution network was still the key driver of projected capital expenditure from 2017 to 2021 (Table 12.1). However, wastewater is the largest driver of planned *new* infrastructure investments (as evidenced by the relatively high projected investment in wastewater 'projects'). Indeed, these proposed investments were a direct response to the EC's decision to pursue court action against Ireland in relation to their failure to comply with treatment requirements under the UWWTD. Were it not for the exceptional investment required to revive Irish Water's drinking water distribution network, this emphasis on wastewater would mirror that of neighbouring UK Water and Sewerage Companies (WaSCs), who are projected to spend approximately £16.5 billion on wholesale water, versus £20.9 billion on wholesale wastewater, from 2015 to 2020. A further consequence of projected investments in new wastewater treatment plants is that Irish Water expects its sludge production to increase by 80% by 2040.

12.5 Discussion and Conclusion

12.5.1 The Hidden Role of Wastewater

Whilst we accept that current incarnations of the 'public–private debate' are something of a straw man, we argue that it is still worthwhile understanding the more nuanced landscape of models for service delivery, and critically examining the motivations (both economic and political) for shifting between different models.

Fundamentally speaking, debates around these models are primarily concerned with the distribution of costs and benefits for services. If the cost burden on service users is seen as high, whilst the benefits are seen as inflating the profits of private firms and transformed into rewards for executives and shareholders, then there is a greater likelihood of outcry. What recent remunicipalization trends perhaps illustrate is a reconsideration of the role of utility services in society. Rather than background functions that are largely forgotten, they can be harnessed to work proactively towards social objectives, such as greater equality and the active engagement of citizens in the structures of governance. This is evident in the emergence of more progressive corporatized utilities (McDonald 2015), including examples such as Eau de Paris. It is also evident to a degree in some of the adaptations that companies such as Veolia and Suez have undertaken to respond to the remunicipalization trend, including the adoption of more open governance measures, lowering their profits, and the more active involvement of users.

What remains clear, however, is that such trends and debates still revolve primarily around drinking water. Wastewater services have generally not experienced the same amplitude of 'pendulum swings' between public and private models (Hall et al. 2013) that have been seen in drinking water systems. In many countries, including India and China, private firms have struggled to get a foothold in the wastewater sector due a lack of viable business models and limited cost recovery options (Murray et al. 2011). Likewise, as stated previously, recent remunicipalization trends have largely impacted on drinking water systems rather than wastewater. Whilst there have certainly been public outcries over high-profile failures of private companies at managing wastewater – such as Thames Water's recent debacle of sewage spills in the UK (Stern 2017) – these do not tend to draw the same level of fervour as calls to 'take back' public control of drinking water. This is notable because, as this chapter has argued, wastewater is a significant driver of the need for investment and cost recovery across the sector. This hidden significance of wastewater can become problematic if the emotive politics that engulf drinking water eventually undermine the potential to generate investment in wastewater infrastructure (which may occur in Ireland due to the turnaround in tariff arrangements). And as the wastewater landscape evolves, both technologically and financially (as discussed in the next section), there will be an even greater need for proper recognition of the role of wastewater services and the potential models for their delivery.

12.5.2 Emerging Needs and Opportunities in Wastewater

In Europe, as previously discussed, there was considerable capital investment in the wastewater sector through the 1990s and early 2000s, firstly to meet the requirements of the UWWTD, and later the WFD. However, whilst these investments have achieved marked environmental improvements in many member states, improved understanding of the impacts of many contaminants (resulting in even stricter discharge requirements) means that the need for further capital investment has hardly waned. In the UK, due to the implementation of the WFD, many more wastewater treatment facilities are facing strict discharge consents for phosphorus, which are difficult to achieve through the most commonly used end-of-pipe technologies. Whilst innovative regulatory systems (such as catchment-level consents) are being trialled as a means of delivering more cost-effective environmental management, it is difficult for WaSCs to avoid the need for investment in treatment – Severn Trent alone spent an estimated £120 million on phosphorus removal in the most recent spending period (Vallely 2016). On top of this, there are the looming spectres of stricter discharge requirements for trace metals, pesticides, pharmaceutical products, and other emerging contaminants, which could have major implications for investment needs.

Concurrently, there has been a decided global shift in emphasis towards 'circular economy' approaches in wastewater management. The 2017 World Water Development Report, the UN's flagship report on global water, was entitled 'Wastewater: the untapped resource' (UNESCO 2017). A recent World Bank blog proclaimed that wastewater is a 'critical component of a circular economy' (Rodriguez 2018). These are indicative of a growing global recognition of the scale and value of the resources inherent in wastewater – from recycled water, to energy, to nutrients (particularly phosphorus), to a host of other material products (such as cellulose and plastics). In Europe, a whole package of circular economy measures have been introduced by the EC, including new quality standards for recycled water, and plans to ease the pathway for wastewater-derived fertilizer products to be sold on the single market (EC 2017). There are numerous indications that European policy-makers are keen to find ways to facilitate and encourage circular economy approaches in the water and wastewater sector, which could significantly shift the financial landscape around wastewater collection and treatment. In the UK, there has long been interest in the potential for energy recovery from wastewater (through biogas), which is seen as a key potential source of renewable energy (Gooding and Booth 2017).

To date, the impacts have been modest. For many of the potential resources in question, the recovery process is either expensive (as with recycled water) or still in relatively early technological stages. The market value of some recovered products (such as struvite) is not well established, so large-scale investment in their recovery remains limited. There are also numerous governance challenges to be navigated, including a regulatory grey area around sludge management (Smith and Fantinel 2017). The entire regulatory framework for water and wastewater is largely built around a linear model that presumes the eventual treatment and disposal of wastewater. 'Closing the loop' and transforming the waste into marketable resources, whilst protecting public health and the environment, requires a rethink of that landscape, as it implies marrying previously disparate regulatory frameworks.

Despite these hurdles, there is considerable interest in the potential value to be captured from wastewater, and in understanding how the related costs and benefits can/ should be distributed. This involves re-examining the delivery models for water and wastewater services as a whole, including the role of public and private actors. In particular, fragmentation in the management of services (especially between water and wastewater systems, which are often managed by separate utilities) can undermine the adoption of more circular approaches, which are better served by a more holistic approach to service planning. Additionally, as previously mentioned (in the Irish case in particular), private firms and corporatized utilities are often seen as better placed for attracting capital investment as they can seek financing directly from the open market. Whilst investment in wastewater collection and treatment can be seen as unattractive due to the limited options for cost recovery and investment returns, the ability to recover marketable resources can alter that picture significantly. Conversely, if service users come to understand the potential value of resources in their wastewater, they may begin to question whether the revenues from those resources (as with drinking water tariffs) should accrue primarily in private hands. It is therefore worth wondering whether the shift towards circular economy approaches might finally awaken public sentiment around a 'human right to wastewater' in Europe.

References

Bagnall, R. (2014). 'Irish Water bosses now quizzed on 'toxic brand'. *Evening Echo* (30 October). https://web.archive.org/web/20141104120333/http://www.eveningecho. ie/2014/10/30/irish-water-bosses-now-quizzed-toxic-brand (accessed 31 August 2018).

Bakker, K. (2003). A political ecology of water privatization. *Studies in Political Economy: A Socialist Review* 70 (1): 35–58. https://doi.org/10.1080/07078552.2003.11827129.

Barraqué, B. (2012). Return of drinking water supply in Paris to public control. *Water Policy* 14 (6): 903–914.

Barraqué, B. and Le Bris, C. (2007). *Water Sector Regulation in France*. CESifo DICE Report 5(2): 4–1.

Bluefield (2016). $526B Municipal Water CAPEX Increase Pushes Europe's Utilities Forward. Bluefield Research (21 September). https://www.bluefieldresearch.com/ ns/2016-09-21 (accessed 10 October 2018).

Bresnihan, P. (2016). The bio-financialization of Irish water: new advances in the neoliberalization of vital services. *Utilities Policy* 40: 115–124. https://doi.org/10.1016/j. jup.2015.11.006.

CRU: Commission for Regulation of Utilities (2018). *Irish Water Revenue Control 2019*. Dublin: CRU. https://www.cru.ie/wp-content/uploads/2016/12/CRU18097-IW-Revenue-Control-2019-Consultation-Paper.pdf (accessed 10 Oct 2018).

De Clercq, G. (2014). Paris's return to public water supplies makes waves beyond France. *Reuters*. https://www.reuters.com/article/water-utilities-paris/pariss-return-to-public-water-supplies-makes-waves-beyond-france-idUSL6N0PE57220140708 (accessed 31 August 2018).

EC (2017). Report from the Commission to the European Parliament, the Council, the European Economic and Social Committee and the Committee of the Regions on the Implementation of the Circular Economy Action Plan. Brussels: European Commission. http://ec.europa.eu/environment/circular-economy/implementation_report.pdf (accessed 31 August 2018).

EurEau (2017). Europe's Water in Figures: An Overview of the European Drinking Water and Waste Water Sectors. Brussels: The European Federation of National Water Services. https://www.danva.dk/media/3645/eureau_water_in_figures.pdf (accessed 10 October 2018).

Furlong, K., Acevedo, T., Arias, J., and Patiño, C. (2018). Rethinking water corporatisation: a 'negotiation space' for public and private interests, Colombia (1910–2000). *Water Alternatives* 11 (1): 187–208.

Gooding, A. and Booth, C. (2017). Insights and issues in the uptake and development of advanced anaerobic digestion within the UK water industry. In: Sustainable Development and Planning IX (eds. C. Brebbia, J. Longhurst, E. Marco and C. Booth), 783–792. Southampton: WIT Press. http://eprints.uwe.ac.uk/33405 (accessed 31 August 2018).

Grafton, Q., Daniell, K.A., Nauges, C. et al. (eds.) (2015). Understanding and Managing Urban Water in Transition. Volume 15 of Global Issues in Water Policy. Dordrecht: Springer Netherlands.

Hall, D., Lobina, E., and Terhorst, P. (2013). Re-municipalisation in the early twenty-first century: water in France and energy in Germany. *International Review of Applied Economics* 27 (2): 193–214. https://doi.org/10.1080/02692171.2012.754844.

Hodge, G.A. and Greve, C. (2007). Public–private partnerships: an international performance review. *Public Administration Review* 67 (3): 545–558. https://doi.org/10.1111/j.1540-6210.2007.00736.x.

Independent (2016). Election candidates join thousands in Dublin protest against water charges (20 February). https://www.independent.ie/breaking-news/irish-news/election-candidates-join-thousands-in-dublin-protest-against-water-charges-34470717.html (accessed 31 August 2018).

Irish Water (2018). *Irish Water Charges Plan,* Reg_PP_IW-WCP-001. https://www.water.ie/docs/Water-Charges-Plan-7-Feb-2018.pdf (accessed 31 August 2018).

IWA (2018). International Statistics for Water Services 2016. International Water Association. http://www.iwa-network.org/wp-content/uploads/2016/10/Water_Statistics_SCREEN.pdf (accessed 18 October 2018).

James, W. (2017). UK's Labour pledges infrastructure nationalization, credit card cap. Reuters (25 September). https://www.reuters.com/article/us-britain-politics-labour-debt/uks-labour-pledges-infrastructure-nationalization-credit-card-cap-idUSKCN1C01X0 (accessed 10 October 2018).

Juuti, P. and Katko, T. (eds.) (2005). Water, Time, and European Cities. European Commission. http://www.watertime.net/docs/WP3/WTEC.pdf (accessed 10 October 2018).

Kennedy, D (2011). Italy nuclear: Berlusconi accepts referendum blow. BBC News (14 June). www.bbc.co.uk/news/world-europe-13741105 (accessed 31 August 2018).

Kishimoto, S., Lobina, E., and Petitjean, O. (eds.) (2015). Our Public Water Future: The Global Experience with Remunicipalisation. Amsterdam: Transnational Institute. https://www.tni.org/files/download/ourpublicwaterfuture-1.pdf (accessed 31 August 2018).

Lorrain, D. and Poupeau, F. (eds.) (2016). Water Regimes: Beyond the Public and Private Sector Debate. Abingdon and New York: Routledge.

McCaffrey, S.C. (1992). A human right to water: domestic and international implications. *Georgetown International Environmental Law Review* 1: 1–24.

McDonald, D. (2015). What is corporatization? The 'new' look of water and power utilities. Municipal Services Project. https://www.municipalservicesproject.org/blog/what-corporatization-%E2%80%98new%E2%80%99-look-water-and-power-utilities (accessed 31 August 2018).

Murray, A., Mekala, G.D., and Chen, X. (2011). Evolving policies and the roles of public and private stakeholders in wastewater and faecal-sludge management in India, China and Ghana. *Water International* 36 (4): 491–504. doi: 10.1080/02508060.2011.594868.

O'Doherty, C. (2014). 120,000 Irish protest as the good boys of Europe turn bad. *Irish Central* (2 November). https://www.irishcentral.com/opinion/cahirodoherty/120000-irish-protest-as-the-good-boys-of-europe-turn-bad (accessed 31 August 2018).

O'Toole, F. (2014). The Irish Rebellion Over Water. *New York Times* (19 December). https://www.nytimes.com/2014/12/20/opinion/fintan-otoole-the-irish-rebellion-over-water.html?_r=0 (accessed 31 August 2018).

Pérard, E. (2009). Water supply: public or private? *Policy & Society* 27 (3): 193–219. https://doi.org/10.1016/j.polsoc.2008.10.004.

Pigeon, M., McDonald, D., Hoedeman, O., and Kishimoto, S. (eds.) (2012). Remunicipalisation: Putting Water Back into Public Hands. Amsterdam: Transnational Institute. https://www.tni.org/files/download/remunicipalisation_book_final_for_web_0.pdf#page=24 (accessed 31 August 2018).

Porcher, S. (2018). In hot water? Issues at stake in the regulation of French water public services. *EUI Working Paper RSCAS 2018/24*. Florence: Robert Schuman Centre for Advanced Studies - European University Institute.

Porcher, S. and Saussier, S. (2017). 'Get what you pay for': the story underneath remunicipalizations in the water sector. *Academy of Management Proceedings* 2017 (1) doi: 10.5465/ambpp.2017.12843abstract.

Reeves, E. (2011). The only game in town: public private: partnerships in the Irish water services sector. *The Economic and Social Review* 42 (1): 95–11.

Rodriguez, D. (2018). Wastewater treatment: a critical component of a circular economy. *The Water Blog* (16 April). Washington, DC: World Bank. http://blogs.worldbank.org/water (accessed 31 August 2018).

Smith, H.M. and Fantinel, F. (2017). Realising the circular economy in wastewater infrastructure – the role of governance. *Proceedings of the International Symposium for Next Generation Infrastructure* (11–13 September 2017, London). http://isngi.org/conference-outputs (accessed 31 August 2018).

Stern, J. (2017). Thames Water fined £20m for sewage spill. BBC News (22 March). www.bbc.co.uk/news/uk-england-39352755 (accessed 31 August 2018).

Swyngedouw, E. (2005). Dispossessing H_2O: the contested terrain of water privatization. *Capitalism Nature Socialism* 16 (1): 81–98. https://doi.org/10.1080/1045575052000335384.

UNDESA (2017). *The Future is Now: Accelerating Public Service Innovation for Agenda 2030*. Report of the 2017 United Nations Public Service Forum. The Hague (22–23 June 2017). New York: UN Department of Economic and Social Affairs. http://workspace.unpan.org/sites/Internet/Documents/UNPAN97975.pdf (accessed 31 August 2018).

UNESCO (2017). *Wastewater: The Untapped Resource*. United Nations World Water Development Report 2017. Paris: UNESCO.

UNGA (2010). Resolution adopted by the General Assembly on 28 July 2010 A/RES/64/292. The human right to water and sanitation. http://www.un.org/ga/search/view_doc.asp?symbol=A/RES/64/292.

UNGA (2015). Resolution adopted by the General Assembly on 17 December 2015. 70/169. The human rights to safe drinking water and sanitation. A/RES/70/169. https://undocs.org/A/RES/70/169.

Valdovinos, J. (2012). The remunicipalization of Parisian water services: new challenges for local authorities and policy implications. *Water International* 37 (2): 107–120. doi: 10.1080/02508060.2012.662733.

Valero, L., Pajares, E.M., and Sánchez, I.M.R. (2018). The tax burden on wastewater and the protection of water ecosystems in EU countries. *Sustainability* 10 (1): 212. https://www.mdpi.com/2071-1050/10/1/212/pdf.

Vallely, L. (2016). Putting the focus on phosphorus. *Utility Week* (29 February). https://utilityweek.co.uk/putting-the-focus-on-phosphorus (accessed 31 August 2018).

Veolia Group (2018). The City Of Bordeaux Appoints Veolia To Handle Its Wastewater Treatment And Rainwater Management Services. *Water Online*. https://www. wateronline.com/doc/the-city-of-bordeaux-wastewater-treatment-and-rainwater-management-services-0001 (accessed 31 August 2018).

Wolff, G.H. and Palaniappan, M. (2004). Public or private water management? Cutting the Gordian knot. *Journal of Water Resources Planning and Management* 130 (1) https://doi.org/10.1061/(ASCE)0733-9496(2004)130:1(1).

Worstall, T. (2014). The New Irish Water Charges Look Excellent; So Why The Protests? *Forbes* (1 November). https://www.forbes.com/sites/timworstall/2014/11/01/the-new-irish-water-charges-look-excellent-so-why-the-protests/#68294891bbbb (accessed 31 August 2018).

13

Drought Policy and Management

Rachael McDonnell[1,2], Stephen Fragaszy[3,4], Troy Sternberg[5], and Swathi Veeravalli[6]

[1] Water, Climate Change and Resilience Strategic Program, International Water Management Institute, Colombo, Sri Lanka
[2] International Centre for Biosaline Agriculture, Dubai, UAE
[3] Water Directorate, New Zealand Ministry for the Environment,, Wellington, New Zealand
[4] National Drought Mitigation Center, School of Natural Resources, University of Nebraska, USA
[5] Environmental Change Institute, University of Oxford, UK
[6] US Army Corp of Engineers, USA

13.1   Introduction

News images of near-empty reservoirs and parched agricultural lands have become familiar symbols, effectively conveying that, globally, drought is the most common and costliest natural hazard in terms of human, monetary and environmental impacts (Guha-Sapir et al. 2016). If not properly prepared for, drought can amplify poverty and depress livelihoods, and can act as a threat multiplier. For instance, in northern China and Mongolia, drought exacerbates the impact of extreme cold, aridity and low plant productivity on communities. The significant complexity and risk for communities and countries is reflected in drought's spatial and temporal global distribution, extent and intensity. Exposure and vulnerability vary across societies, with serious impacts for agriculture, water supply and human wellbeing most pronounced, but still spatially variable. These factors present great challenges for drought management, monitoring and policy.

The literature on drought highlights the evolution and global experiences of the complexities and challenges of drought policy and management. It is clear from studies around the world that over-arching national drought policies encompassing broader risk management measures are still the exception rather than the norm (World Meteorological Organization (WMO) and Global Water Partnership (GWP) 2014). International best-practice was however, established through the Integrated Drought Management Programme in 2013 (www.droughtmanagement.info), which stresses the importance of policy, governance, monitoring and planning at regional, national and local levels. To appreciate the challenge that drought presents, contextual understanding is essential, including environmental dynamics, human wellbeing and economic drivers.

This chapter begins with an assessment of drought as a hazard, then examines social and physical implications and considers ways to improve resilience. Case studies

Water Science, Policy, and Management: A Global Challenge, First Edition. Edited by Simon J. Dadson, Dustin E. Garrick, Edmund C. Penning-Rowsell, Jim W. Hall, Rob Hope, and Jocelyne Hughes.
© 2020 John Wiley & Sons Ltd. Published 2020 by John Wiley & Sons Ltd.

highlight the links between drought and climate change, expanding water demand and resource conflict that reflect the complexity of drought today. These show examples of progress being made in the move from crisis-led responses to more proactive policy development and risk management, but that much more remains to be done.

13.2 Drought, Aridity, Water Scarcity, and Desertification

A major challenge for drought management is that droughts are notoriously difficult to define, characterize and track, unlike hazards such as floods or earthquakes. Yet declaring that an event is happening, so triggering mitigation actions, is a crucial part of the drought management process.

The basic definition of drought is straightforward: a period with a significant negative deviation from the normal hydrological conditions for an area (Mishra and Singh 2010). In practice, however, drought is challenging to define because its definition is location and context-specific: drought in northwest Europe is very different to events in Chile or Tunisia. Drought definitions need to relate to local environmental and social contexts, and a range of social, economic, agricultural and environmental factors controlling drought's spread through those systems and its impacts (Kallis 2008). The definition of drought in one context does not reflect that in others. The slow-onset, often creeping nature, and wide area of drought effects compound the challenge of drought definition. If at different times drought struck the same location with the same climatic intensity, duration and extent, the impacts and effects could be different simply because the social conditions there are different (Wilhite et al. 2014).

Drought assessments typically use four common typologies to frame definitions: meteorological, agricultural, hydrological and socioeconomic (Mishra and Singh 2010) (Table 13.1). The first three rely on biophysical indicators, whereas socioeconomic drought incorporates hydro-social feedbacks and reflects drought impacts and socioeconomic context and settings (Budds et al. 2014; Van Loon et al. 2016). Using these typologies facilitates the identification of relevant drought indicators and indices, points of drought onset, impacts, and potential mitigation efforts for diverse water users at different timescales. Their usage also highlights the fact that operational drought definitions are contextual and that stakeholders must develop definitions with reference to local environmental and social conditions, for specific purposes, and in relation to the impacts and timeframes they wish to assess.

Increasingly researchers and government officials reach beyond meteorological drought assessments and utilize wider convergence of evidence-based approaches embodied in the more complex drought typologies to understand how drought relates to key themes (Hayes et al. 2012). The Middle East – North Africa (MENA) case study in this chapter highlights the development of such combined index approaches that government agencies increasingly use in the operational context of national or subnational drought policy and management.

Not only is drought difficult to define, but at times the term is used interchangeably with water scarcity, aridity and desertification (Table 13.2). Whereas aridity is a permanent landscape feature, drought is a stochastic one. Whilst drought contributes to water scarcity and desertification, it is a temporary feature that affects the hydro-social framework – the way water and society make and re-make each other – and

Table 13.1 Drought types.

Drought type	Definition (Mishra and Singh 2010)	Time-frames for assessment of drought	Primary indicators (WMO and GWP 2016)	Commonly used indices (WMO and GWP 2016)
Meteorological	Lack of precipitation over a specific time period	Shortest time-frame, typically one to six months	Primarily precipitation	Standardized Precipitation Index (SPI) is the most widely used index (Hayes et al. 2011)
Agricultural	Reduced soil moisture and resultant crop impacts	Agricultural season (typically up to one year)	Precipitation, soil moisture, potential evapo-transpiration, temperature, wind, crop condition	Soil moisture anomaly (SMA), soil moisture deficit index (SMDI) normalized difference vegetation index (NDVI), standardized precipitation and evapotranspiration index (SPEI), Palmer Drought Severity Index (PDSI) (WMO and GWP 2016)
Hydrological	Inadequate surface and/or groundwater for established water uses	May be inter-annual depending on hydrological system	Streamflow, snowpack, precipitation, reservoir levels, groundwater levels	Surface water supply index, aggregate dryness index (ADI), low flow index, streamflow drought index, groundwater resource index (WMO and GWP 2016)
Socioeconomic	Failure of water systems to meet water demands	Sector-specific	Water demand, reservoir levels, groundwater abstraction, precipitation, and social, economic and environmental feedbacks	Assessment frameworks rather than specific indices since they include socioeconomic and potentially land use impacts, factors and responses (e.g. Mehren et al. 2015; Van Loon et al. 2016)

Table 13.2 Characterizing aridity, water scarcity and desertification.

	Aridity (Mishra and Singh 2010)	Water scarcity	Desertification (UNCCD 1994)
Description	Dryness is a permanent physical condition of the landscape	Water supply does not meet demand	Land degradation in arid, semi-arid, and dry sub-humid areas resulting from various factors, including climatic variations and human activities
Definition type	Physical	Relative to human needs	Ecological process
Causal factors	Climate	Climate/hydrology; environmental, social, economic, increase in demand and infrastructure factors	Climate/hydrology; land, water, and agricultural management practices
Common definitions	Typically precipitation <200 mm/year	Water supply to demand ratio; water availability <500m^3/year/capita (Falkenmark et al. 1989)	Vegetation cover and type; soil characteristics

human–environment interactions (Budds et al. 2014). Making connections but distinctions between these four terms is a precondition for establishing effective drought monitoring and management programmes, particularly in vulnerable semi-arid to sub-humid climate transition zones.

Conceptually we can distinguish between drought and water scarcity through consideration of time-frames and the physical basis for hydrological stress. Socioeconomic drought results from implications of meteorological, agricultural or hydrological drought in a socioeconomic system. In contrast, water scarcity reflects an underlying mismatch between available water resources and the water needs of people, the environment and the economy. Thus, water scarcity might occur as a result of infrastructure degradation or conflict and forced migration, which can transfer resource insecurity from one place to another, just as it may occur from long-term growth and resource usage.

Indices of national water scarcity such as the Water Stress Index typically incorporate data on water resource availability and quality features, supply infrastructure, social access to water, and environmental characteristics. Drought research increasingly aims to disentangle the climatic and anthropogenic components of drought's spread through socio-environmental systems to understand how anthropogenic activities affect drought events (Van Loon et al. 2016).

Societal responses to drought can lead to long-term water scarcity or can build resilience. Post-drought development pathways may legally and economically lock in long-term reliance on unsustainable resource use (Wilson 2013), for example systemic groundwater over-abstraction that ultimately produces long-term water scarcity (Hornbeck and Keskin 2014). However, responses can also 'build back better', given foresight around anticipated long-term constraints (UNISDR 2015).

13.3   Climate Change and Drought

Long-term, natural climate variation and anthropogenic climate change further compli-cate the drought context, with some areas moving to more arid conditions relatively abruptly, whilst in others drought dynamics have changed. Regional studies have con-cluded that in the period since 1950, some areas have been undergoing a slow but steady warming and drying period as evidenced by a shift in observed and modelled historical Standardised Precipitation Index (SPI) values (Göbel and De Pauw 2010). The IPCC (2014) concluded with medium confidence that in the immediate past (1950–2012), some areas have experienced more intense and longer droughts, including southern Europe and West Africa, whereas others have faced shorter and less intense drought, especially central North America and northwestern Australia.

Historical climate reconstruction studies have highlighted shifts in drought occurrence and severity in the past and provide useful analogues to anticipated scenarios of climate change in the future. For example, in the southwest United States, past droughts accom-panied by hemispheric temperature changes that favour particular atmospheric circula-tion patterns and amplify regional drought conditions are most suitable to compare with future scenarios (Woodhouse et al. 2010). Historical reconstructions have also been used to contextualize recent droughts. Cook et al. (2016) analysed more than 900 years (1100–2012) of Mediterranean drought variability using spatiotemporal tree ring reconstruc-tion of the June–July–August self-calibrating Palmer Drought Severity Index (PDSI). They were able to show that the recent 15-year drought in the Levant (1998–2012) is the driest in the record. The consequences for Jordan, Lebanon and Syria, at a time of increasing resource scarcity, have been highly challenging to water and food security.

Given the media's heightened tendency to attribute modern drought intensity to human-induced climate change, it is important to acknowledge how complicated this issue is. Observation and modelling studies, however, clearly suggest that human-induced climate change increases the probability of drought event occurrence, but in specific locales under specific conditions. Further, anthropogenic climate forcing is expected to continue to exacerbate these trends significantly (e.g. Bergaoui et al. 2014; Droogers et al. 2012).

Going forward, the IPCC (2014) concluded with medium confidence that in the twenty-first century, central and southern Europe, the Mediterranean region, central North America, Central America and Mexico, northeast Brazil and southern Africa will face more frequent and intense droughts. Elsewhere there is low confidence in projec-tions due to inconsistent model signals. Whilst model agreement on temperature changes are conclusive – the length, frequency and intensity of heat waves will increase over most land areas – confidence is far more localized in the direction of precipitation shifts. In relation to Köppen climate classifications that broadly reflect temperature and precipitation conditions, Rubel and Kottek (2010) show that projected temperature increases will lead to shifts from warm temperate to arid environments and major expansions of the hot arid environment, especially in mountainous areas such as Central Asia and the Mediterranean region. This means for some areas that the development of well-prepared drought management plans and resilience building is now crucial.

A changing climate presents challenges for drought policy and management. Droughts are defined relative to local average conditions. Issues of stationarity and anthropogenic climate change are therefore important in establishing the drought context for an area, as many drought indices use long-term averages for baseline conditions. Whilst some meteorological records go back more than 100 years, in many parts of the world they rarely stretch back beyond 30 years. The observed record is obstructively short in many areas where long-term climatic patterns lead to significant shifts in norms over decades or centuries.

Policy decisions based on poor understanding of historical meteorological and hydrological variability (and conversely poor foresight on future variability) can result in long-term unsustainable resource use and otherwise suboptimal outcomes that increase vulnerability to drought. The Colorado River allocation regime, which was based on 'average' flows from a particularly humid multi-year period, is a case in point of this potential policy problem from a water supply and allocation standpoint (Garrick 2015). This is echoed in the Australian experience with drought declaration and payouts, described below from a specific drought legislation and policy perspective (Botterill and Wilhite 2005).

Historical precedent shows that typically the political will to create or amend drought legislation, policy and management regimes at the national scale emerges only from drought crises (Stone 2014). Given the difficulty and length of time required to change policy settings, the rapidity and intensity of climate change will increase the importance of flexible policies and emphasis on proactive risk-management approaches. The Ethiopian example of recent policy reforms resulting in effective drought management highlights the opportunities inherent in improved drought policy outcomes (Singh et al. 2016).

13.4 Drought Policy and Management Development

We here employ two complementary perspectives: first, drought policy and associated legislation and governance set the frameworks for drought management; and secondly, we set out the 'three pillars' that are central to the actual practice of drought management (Lowi 2003; WMO and GWP 2014).

13.4.1 Drought Legislation

Drought legislation can be stand-alone law, but is most commonly specific provisions included in wider water, agriculture, or disaster response legislation. Generally, the aims of drought laws include emergency relief and establishing the legal basis and roles for drought monitoring, policy-making and management, rather than specifying the actions themselves (Wilhite et al. 2014). Some examples illustrate these points and the distinction between drought laws and policies.

In the United States, federal law establishes the national drought monitoring system. The 2008 and 2014 Farm Bills authorize emergency drought relief payments through the Livestock Forage Disaster Program run by the Department of Agriculture, and that in turn relies on the mandated monitoring information. Spain's drought law delegates drought policy-planning to river basin organizations, and Mexico established a National Programme Against Drought (termed PRONACOSE) that has a broad applied research and development remit to inform the national water commission's policy-making (Martins et al. 2015). The European Water Framework Directive contains several provisions dealing with water allocation issues connected to water scarcity, but no legally binding requirements focused specifically on solving drought issues.

The lack of explicit legislation and at times formal drought policy, including for water infrastructure operational regimes, can leave water managers at the state, basin or utility level to fend for themselves in complex systems. This was the case in the Colorado basin in 2005 when the US federal government threatened to intervene if states did not act. To avoid federal intervention, basin states in 2007 developed formal integrated reservoir guidelines and associated mechanisms to manage shortages, and are currently negotiating a drought contingency plan to prevent and address water shortages in the basin (Sullivan et al. 2019). These types of formal arrangements typically require juridical mediation and entail significant policy changes from the status quo, with attendant costs for resource users and government agencies.

13.4.2 Drought Policies

Drought policies should establish the principles, governance and operating guidelines for responses to drought. There are few over-arching national drought policies in place, with limited provisions sometimes included in more general water policies (WMO and GWP 2014). Drought policy objectives are often centred on facilitating water management and reallocation during the event, sustaining impacted populations, and supporting early recovery. Some policies may also encourage the development of measures to build greater drought resilience and self-reliance in vulnerable economic sectors and population groups.

Wilhite et al. (2014) identified three primary drought policy pathways and linked policy intents (Table 13.3). Post-impact interventions are the most commonly adopted and usually come into play after a formal declaration of drought by the necessary authority, usually a committee or task force. Normally the interventions involve

Table 13.3 Drought policy intervention types.

Policy type	Examples	Policy intent	Challenges
Post-impact interventions	Water and feed provision for livestock; debt forgiveness for farmers; rural job-creation programmes; water rationing and pricing regimes	Relief measures for those affected by drought, reduce long-term effects	Implementation without reducing incentives for risk reduction measures; timeliness of interventions
Pre-impact programmes for mitigation	Drought early warning systems; surface water storage; irrigation efficiency; water demand management; pricing regimes	Reduce vulnerability and impact	Can lead to path-dependency on unsustainable resource use (e.g. groundwater over-abstraction)
Development of preparedness plans and policies	Organizational frameworks; institutional arrangements; operational plans and triggering technical definitions	Facilitate and expedite coordination, collaboration, and action	Requires strong institutional capacity and coordination to implement effectively

Source: After Wilhite et al. (2014). Reproduced with permission of Elsevier.

emergency assistance aimed at providing money or other aid to ease the impacts of the event. In many countries, agriculture is one of the hardest hit sectors, so assistance is often in the form of livestock feed, animal vaccination programmes, credit rescheduling, or compensation through insurance or government payouts for losses resulting from water reallocation or just lack of resources.

Justifying pre-impact programmes for mitigation and preparedness policies and plans can be difficult as the impacts of drought are often difficult to calculate in economic terms, with effects widely dispersed and persisting long after the next rains come. The costs of action can be classified into three categories: preparedness costs, drought risk mitigation costs, and drought relief costs. Most funds today go to the last of these, with far less investment on preparedness or mitigation actions.

Several examples illustrate this. In 1999–2001, Moroccan drought relief programmes cost about USD $318 million, about 2.3% of total government expenditure in 1999–2000. In 2007–2008, Australian expenditure on Exceptional Circumstances drought subsidies reached approximately AUD$380 million. Between 2008 and 2016, the US Livestock Forage Disaster Program granted USD $6.77 billion (MacLachlan et al. 2018). Given these large sums, it is hardly surprising that the evidence base increasingly highlights the benefits, and avoided costs, of shifting from crisis management to risk reduction and management frameworks by focusing on the development of pre-impact mitigations and preparedness policies and plans (WMO and GWP 2017).

13.4.3 Drought Governance

Drought policies typically outline intersectoral collaboration mechanisms for drought management. American, Australian (Botterill and Wilhite 2005) and Brazilian (Martins et al. 2015) examples of drought policy development and coordination highlight the importance of coalitions internal to government and also amongst sector groups to push national governments into action and ensure longevity and effectiveness of the resultant policies.

The reality in most countries, however, is that the highly contentious and political nature of drought management, as well as its cross-sector and environmental reach, leads drought management to be a hugely complex undertaking. By its very nature, drought risk management coordination – whether long-term or crisis mitigation – requires wide stakeholder involvement in policy processes, yet in most of the world, the spirit of such participatory governance is in its infancy. However, even in areas where this is difficult, such as in the MENA region (see below), research has identified a strong appetite from public officials to engage more widely and deeply on water resource issues including drought management. Often, cooperation on drought monitoring and other less contentious issues in the first instance can help to build the necessary trust and relationships to permit useful discussions and collaboration on the politically harder themes such as drought policies that have wide-ranging implications for water resource users.

13.5 The 'Three Pillars' of Drought Management

Turning to the actual management of drought, an important starting reference has been the development by the World Meteorological Organization (WMO) and the Global Water Partnership (GWP) of the Integrated Drought Management Program (www.droughtmanagement.info). This programme resulted from the High-Level Meeting on National Drought Policies in Geneva in 2013 organized by the WMO and the Secretariat of the UN Convention to Combat Desertification (UNCCD), and involving drought experts from 93 countries. The programme gives a structured approach to drought management planning and is based on 10-step plans and the three 'pillars' approach to drought management (WMO and GWP 2014). Such drought management often has two main foci – defining a set of actions and responses during an event, and building resilience during non-drought years.

13.5.1 Pillar 1: Drought Monitoring and Early Warning Systems

Drought monitoring consists of the generation of information, often mapped, that illustrates the extent and intensity of drought conditions. It can be used as a trigger for subsequent management actions, as in the US and Mexico. As indicated above, numerous indices exist, and in recent years there have been efforts to develop Composite Drought Indicators (CDIs). The US Drought Monitor combines approximately 100 different datasets from many government agencies, as well as more qualitative inputs from drought observers across the country, to generate the weekly drought map (www.droughtmonitor.unl.edu). In Jordan, Morocco, Lebanon and Tunisia, developments are in place to produce regular drought maps based on a CDI approach using four inputs: SPI, Normalized Difference Vegetation Index (NDVI), and land surface temperature and soil moisture anomalies.

These maps, or similar sets of hydrological data, are important as they are unbiased science-based evidence of the occurrence of drought and are useful in triggering drought management actions in a country or area without undue political influence. For example, the US Livestock Forage Disaster Program provides emergency relief when the US Drought Monitor indicates that drought severity thresholds have been breached for specific time-frames.

Whether used for specific management mechanisms or not, it is important that government and private sector stakeholders receive the monitoring information in a timely manner and trust the outputs, as they are often the starting point for actions such as mandated water infrastructure management regimes or livestock programmes. Reservoir management is time-sensitive for drought monitoring, given the multipurpose nature of dam storage and the intersectoral conflicts that can emerge as a result of water shortages. In large basins with significant variability in the source and volume of stream inflows, such as the Nile, or those reliant on snowpack, such as the Indus, these issues are particularly relevant to managing the effects of drought as they spread and progress through water resource and social/economic systems (see Chapter 19; Mankin et al. 2015).

13.5.2 Pillar 2: Drought Impact and Vulnerability Assessments

The second pillar, the vulnerability and impact assessment, brings together quantitative and qualitative data on drought history, impacts, and underlying causes of vulnerability across geographical regions and economic sectors, communities and the environment. Drought impacts, and their management, occur at all levels of society, from the household to the international arena when one considers commodity and food value chains.

Understanding the impacts of past droughts helps to identify the areas and sectors that face the greatest pressure and require the greatest support. In many countries this is the agricultural sector and the water and sanitation supply networks. Pressure on these sectors can lead to over-exploitation of poorly controlled water resources such as groundwater systems, or lead to consumers relying on water not supplied from public networks, such as from private tankers. Drought can also lead to health issues associated with poor water quality and limits to food supplies as well as psychological and heat stress. In some countries, drought affects energy supplies, particularly those reliant on hydroelectric power systems (van Vliet et al. 2016).

Understanding the root cause of drought vulnerability can improve the effectiveness of drought policies and resilience-building programmes by enabling targeted responses. But many questions are raised: is drought vulnerability primarily environmental or socioeconomic? Is it due to poor access to credit, markets and infrastructure, or property rights issues?

O'Brien et al. (2004) provided a useful vulnerability assessment template to understand the social, physical and environmental dimensions of drought vulnerability. These can be continental in scale: Carrão and Barbosa (2015) defined quantitative and spatially discrete indicators across all of South America for drought hazard (the frequency of events of a given magnitude), exposure (the presence and prevalence of affected populations, crops, sectors, ecosystems, etc.), and adaptive capacity (the ability of populations, crops, sectors, ecosystems, etc. to weather a hazard). These indicators can also focus more specifically on socioeconomic components, as Simelton et al. (2009) examined in their work on the vulnerability of Chinese food systems to drought.

These types of assessments highlight how dynamic drought vulnerability is, varying as a result of changing biophysical and social processes (O'Brien et al. 2005). These assessments inform targeted policy development and preparedness planning. For example, in Mexico vulnerability assessments have informed development of basin-specific preventive measures and drought mitigations as well as the targeting of national resources during drought events.

13.5.3 Pillar 3: Drought Preparedness Planning

The third pillar focuses on identifying and implementing actions to mitigate drought impacts. Drought preparedness plans and activities centre on the following themes: prioritizing impacts to target; identifying actions and interventions that will reduce impacts; and identifying triggers to phase in and out mitigation actions when drought is beginning and ending. Identifying triggers relies heavily on the indices and indicators from the drought monitoring system. Once these triggers are identified, agencies, and organizations in the country or region concerned need to develop collaborative strategies and authorities to implement actions. This pillar therefore includes the development of arrangements for roles and responsibilities of different agencies and stakeholders, whether at the river basin scale as in Mexico and Morocco, or at national or state level.

One of the primary objectives of this pillar is to design actions that can support a quick return to normal water supply conditions for people, the economy and the environment. International collaborative efforts aim to catalyse development of disaster risk management frameworks with the adoption of the Sendai Framework (2015). The core tenets are understanding disaster risk, strengthening governance to manage disaster risk, investing in disaster risk reduction for resilience, and enhancing disaster preparedness for effective response and 'building back' better. These aim to guide national policies towards risk reduction frameworks and improve crisis management responses.

13.5.4   A Range of Policy Instruments Including Insurance and Water Allocation Regimes

Governments around the world promote the development of – and provide subsidies for – agricultural drought risk insurance. There are two broad types: indemnity-based insurance protects against predefined losses, whilst index-based insurance protects against predefined risk events (Barnett et al. 2008). Both are informed by drought monitoring regimes that capture drought extent and severity. Insurance helps smooth farmers' incomes and governmental budget expenditure. It can support investment decisions at the macro- and micro-levels, for instance related to the adoption of new technologies or sectoral specialization. Finally, insurance can also support greater institutional transparency and contribute to the adoption of adaptive behaviours.

Water allocation to key economic sectors and/or users, and subsequent reallocation in shortage conditions, is always complex, political and emotive. During droughts, supplies decrease and demands can increase, necessitating compromises. Markets and centralized allocation regimes are but a few of the possible mechanisms for addressing these challenges.

In some areas, formal markets are used for water trading, such as in the Murray-Darling Basin, where some users can be paid by others not to consume their allocated water. Garrick (2015) found that water markets in the United States and Australia are able to facilitate water allocation to higher-value uses and for environmental flows during drought periods. In Spain, there is evidence that legal reforms to reorganize water user groupings and allocation frameworks during drought years have marginally improved overall economic outcomes compared with antecedent arrangements.

In other areas, water management rules during drought depend on meteorological, agricultural or other determinants. For example, in Tunisia, there are a series of water allocation rules based on when the drought begins in the agricultural year and the length of the event. Certain crop production systems have priority for irrigation – olives and fruit trees receive water before cereals and annual crops in most locations. These centralized approaches to water allocation planning during drought emerge from overarching legal and institutional regimes that attempt to minimize drought's overall social and economic impacts.

Two case studies here, assessed through the three pillars framework, illustrate key drought policy and management components. The Mongolian example highlights how drought can be subsumed by or conflated with another natural disaster, in this case extreme winter conditions. The MENA example focuses on the coordination challenges inherent in drought monitoring and management across national boundaries.

13.6   Drought in Mongolia

In Mongolia, drought frames the viability of the national nomadic lifestyle and culture. Indeed, it is thought to have instigated Genghis Khan's drive across the steppes, originally in search of better vegetation. Today it remains a serious event for the rural population, and is especially severe for mobile pastoralists who comprise a third of the population.

Any lack of precipitation in the arid and semi-arid region affects vegetation cover and directly impacts herders' livelihoods (Sternberg 2018). A key issue is the relation between drought and the country's worst natural hazard – extreme winter conditions (i.e. temperatures down to −40°C) known as *dzud*. Drought events are a recurring phenomenon that affects plant growth and highlights the sensitivity of livestock and livelihoods to environmental factors. Its importance suggests drought should be at the centre of Mongolian weather forecasting, government policy and international development assistance, whilst ensuring an incorporation therein of herders' traditional knowledge.

Pastoralists traditionally coped with their highly variable landscape through migration in search of better pasture and by changing livestock numbers (reduced births, selling animals) and herd composition (e.g. shifting from goats and cattle to sheep and ideally-suited two-humped camels). Low precipitation (<120 mm) during droughts results in little pasture vegetation. Yet extreme events were considered positive as the weak animals died, leaving the best breeding stock. During the long Soviet period (colloquially regarded as 'good for animals, bad for people') the government restricted animal numbers, created designated reserve pastures in drought conditions and provided livestock transport and fodder reserves. In the chaotic post-Soviet transition to a market economy, the government role evaporated. This meant herders became responsible for all emergency strategies without external organization or funding. Thus drought quickly becomes problematic and may lead to herders migrating >300 km to find adequate vegetation. Whilst mobility is customary, the new costs of petrol, transport and fees to access previously communal land is a major burden. Aid programmes often imported from Africa have limited relevance to the steppe and have skewed government perspectives to attract funding. Alas, Mongolian culture, landscape and climate factors are quite different – international projects adapted poorly and difficult-to-measure drought gave way to easy-to-see extreme winter as the country's disaster focus.

13.6.1 Pillars 1 and 2: Drought Monitoring, Impacts, and Vulnerability

The challenges of drought identification, spatio-temporal scale, intensity and impact create real problems in Mongolia. The vast territory, with the lowest population density on the planet, experiences variable conditions, levels of vulnerability and differences in hazard awareness. The landscape is composed of mountain and steppe environments, with the Gobi Desert the predominant feature (2.9 million km^2). In the capital, located in the centre of the country, there are worries about drought implications for the settled urban population, now more than 40% of the residents. Pastoralists are concerned with what is considered climate warming and the increased severity of dry events. A vital issue is the changing policy and socioeconomic forces since the country's transition from communism to liberal democracy in 1991. As herding evolved from subsistence to an income-driven activity, customary mitigation strategies have shifted dramatically to reflect economics-based decision-making.

Drought is regarded as a threat multiplier, in this case exacerbating winter cold, ice and snow that leads to massive livestock mortality, most recently in 2009–2010. This feared association results in policy that focuses on *dzud* relief rather than drought mitigation. Recent research (Sternberg 2018) disassociates the two hazards; in fact *dzuds* have a very low probability (<9%) of occurring following a drought. The result is that policies and strategies are geared towards *dzud* relief and ignore drought impacts. This scenario finds that meteorological observation and limited government funds are poorly directed as the two hazards are separate events. Drought, the more common hazard, continues to affect the nation significantly, yet receives less investment, research and attention.

13.6.2 Pillar 3: Drought Preparedness, Mitigation, and Response Strategies

Whilst drought is often severe in Mongolia, government action is fragmented and poorly documented. A *dzud* is represented by mounds of snow-covered dead animals in the media; the impacts of drought are less dramatic but equally deadly. Herders' traditional knowledge is expected to compensate for moisture deficiencies and changes in vegetation, although external conditions have shifted rapidly. Funding concentrates on the capital where drilling more boreholes and trucking-in water lessens drought impacts. This illustrates the relative political weakness of rural communities in maintaining national attention and the 'hazard fatigue' of the government and international donor agencies. The country lacks drought-specific laws, and weak governance at national and local levels makes policy and engagement a reactive, ineffective process. Rural water sources, predominantly hand wells, receive little government support; disaster response is minimal. The elusiveness of drought leads to neglect and greater exposure of rural livelihoods to its damaging effects.

Mongolia's evolving traditional drought experiences are shared in other regions, particularly in drylands dependent on environmental factors (e.g. weather, vegetation, rangeland productivity) for farming and livestock-raising. Several regions in Africa, India, China and the Americas are affected by severe drought; Kelley et al. (2015) stressed the role of poor governance, policy neglect and gross drought mismanagement as key contributory factors to the Syrian civil war. In democratic Mongolia, drought expertise and attention is atomized between ministries, the National Agency for Meteorology, academics, the Academy of Sciences and, importantly, the international community. The limited focus and continued efforts to understand drought dynamics (Sternberg et al. 2011), other physical disasters (*dzud*; earthquakes; thawing permafrost), economic crises (IMF bailout) and political conflict with China (border closures; debt; interference) stretch government capacity and ability to engage with recurring disasters.

New approaches to drought engagement are emerging, often drawn from global drylands, and those in Mongolia include the Meteorology Office's weather app for herders with mobile phones, international projects working on a livestock early warning system, and livestock risk insurance trials. Yet the *de facto* strategy remains the age-old herding practices and wisdom derived from monitoring animal behaviour, bird flight patterns, vegetation composition and solar observations. For greater drought mitigation this combination needs greater clarity, awareness and support to make a difference for both city-dwellers and environmentally dependent rural livelihoods alike.

13.7 The Example of the Middle East and North Africa Region

The underlying water scarcity conditions of the Middle East North Africa region, allied with complex geopolitics, ensures that drought affects people through numerous aspects of everyday life. With agriculture, predominantly rainfed, being important for food security and for import/export trade balances, the effects of these dry events are particularly pronounced for many of the countries. The current conflicts between transhumance communities over grazing lands during a third consecutive year of drought in Morocco has unfortunately led to fighting and deaths. For urbanites, it interrupts daily routines and increases the cost of living by initiating and increasing the frequency and length of water cuts from utility networks, often necessitating purchase of water from tanker trucks.

Water quality degradation, poor infrastructure, and lack of alternative water sources can lead to human health problems even where water tankers can meet the demand shortfall. More widely, drought can disrupt social cohesion. Rural communities in particular report far more conflicts between neighbouring farmers and types of water users and far fewer weddings during drought years because of prohibitive costs. Similarly, drought can drive smallholders to leave the land and move to cities; it contributes to the rural exodus driving rapid urbanization challenges in the region.

Historically, pastoralists used nomadism and mixed agricultural systems as drought risk management mechanisms. When drought hit, smallholders could allow livestock to fend from reduced cereals output once the next season's seed was assured and also move further afield for fodder. The extension of cultivation into former rangelands and broader shift to largely single or dual-income systems has reduced the viability of these coping mechanisms. Likewise, farmers face the twin challenges of low yields and low quality produce for most crops. As a result of drought, farmers can lose access to lucrative export markets and simultaneously face stiffer competition locally for reduced outputs.

Across the MENA region, individuals managed drought risks and impacts on their own farms, herds and businesses, and the state attempted to minimize political and social fallout from the crisis. In the post-World War II era, the state role has increasingly shifted to reducing impacts on individuals through costly crisis management interventions. This case study illustrates recent work in the MENA region that has a short-term goal to develop, validate and deploy drought monitoring tools (Bijaber et al. 2018), and a long-term aim to improve overall drought policy and management to facilitate drought risk management at the political, sectoral and personal level. It highlights the policy and governance challenges inherent in drought monitoring and management efforts to support this objective (Bijaber et al. 2018).

13.7.1 Pillar 1: Technical and Institutional Drought Monitoring Challenges

Stakeholder needs assessments in Morocco, Tunisia, Lebanon, and Jordan have highlighted the wide range of drought-related data that various agencies collect regularly. They also highlighted the difficulty that the primary drought monitoring agencies face in obtaining, synthesizing and analysing these data to support policy development, implementation, and wider drought management efforts, all themes discussed more widely in Chapters 1 and 5 of this book.

Several common barriers exist and contribute to this challenge in each project country:

- Data often must be purchased even when government agencies are the producers and users
- The timeliness and consistency of data collection, collation, formatting, and transfer is often not satisfactory
- The lack of data platforms and IT systems to facilitate data-sharing, and rigidity of protocols surrounding information sharing

- A culture of data ownership – information retained is power over those who need it
- The political sensitivity of drought data and the information produced.

These data- and information-sharing challenges reflect and contribute to difficulties in institutional coordination and governance. Indeed, in most cases they reflect institutional characteristics and arrangements more than any lack of technical capability and capacity or inexperience.

13.7.2 Pillars 2 and 3: Drought Management Institutional Coordination Challenges

Similar to the Mongolian example, in the MENA region, drought monitoring and management remits and expertise are spread over numerous agencies and include regional and international institutions. Numerous governmental ministries, agencies and authorities, and sometimes non-governmental organizations, conduct official drought monitoring that is used to support political decision-making and subsequent interventions. They collect information on a range of meteorological, hydrological, agricultural, and, to a lesser extent, socioeconomic and environmental indicators.

National meteorological offices, which are often part of the ministries responsible for transportation, typically collect climate, meteorological and precipitation data. Ministries or agencies in charge of water resources management collect, or oversee collection of, hydrological data, in some cases including data on precipitation. River basin authorities and in some cases dam management agencies typically have significant autonomy from the ministry in which they are housed or report to, and in the case of Morocco they are local government entities. Where they are prominent, these authorities undertake much of the hydrological data collection and policy development. Various types of entities, from central government agencies to private sector firms, are responsible for municipal water supply, and they collect and house relevant water supply and socioeconomic data.

Ministries of agriculture oversee the collection of a wide range of agricultural data by local offices, including in some cases precipitation and soil moisture data. Often they collaborate with ministries of defence or national research organizations, which typically house the agencies or units with the most experience and capacity in remote-sensing activities. Ministries with environment, agricultural commerce, health and finance roles collect data that are used to monitor drought impacts after they have begun, and in limited cases to inform and characterize descriptions of drought onset.

The result is a dizzying array of organizations involved in drought monitoring data collection across a range of governance scales. It is no small wonder that, overall, stakeholders describe institutional coordination and collaboration challenges as a critical obstacle to improving drought policy development and management here. In addition to the governance challenge of corralling the many entities involved in drought monitoring and management to collaborate effectively (a comparatively wide range of organizations oversee and undertake drought management interventions), stakeholders described several common barriers that increase the difficulty of effective institutional interactions:

- A lack of legal and policy clarity over agency roles in drought monitoring and management activities leads to significant gaps and overlaps

- Formal requirements for collaboration: in many cases, agencies cannot coordinate, collaborate, or share data without formal conventions or collaboration mechanisms
- Personal rather than institutional relationships often frame inter-institutional interactions
- Institutional coordination typically begins in earnest after drought impacts are already causing a crisis.

The MENA countries tend to have highly centralized policy decision-making structures in relation to drought monitoring and management. However, critical monitoring data are produced at the local level and then transferred to central agencies. Non-governmental and local government stakeholders widely reported in this research a dissonance in central agencies' expectations of timely information delivery to them and their tendency to provide little information and guidance in return. Since in many cases local government agencies cannot act independently and must take directives from the centre, this issue leads to delays in coordination at the local level and therefore delays in interventions down the drought management line of command.

13.7.3 Building Resilience – The Moroccan Drought Insurance Example

In some cases MENA initiatives have overcome these drought policy, governance and management challenges, such as through the development of index-based drought insurance in Morocco (Troy 2013). In the 1990s, the Moroccan government established a centralized calamity fund intended to cover drought losses in rainfed cereals areas. The programme never reached half its intended coverage of 300 000 ha and was subsequently replaced by a public–private partnership to develop multi-risk insurance coverage (including for drought) of cereals, legumes and oil-seed crops. Nevertheless, in less than 10 years the area insured has reached over 1 million ha, about a third of the country's cultivated cereals area, and in the drought year of 2016, indemnification reached nearly US$100 million.

13.8 Discussion

13.8.1 Case Studies Synthesis

Despite having different environmental and socioeconomic contexts, the MENA and Mongolia examples show striking similarities in how institutions deal with drought, climate data, planning and mitigation efforts in sometimes suboptimal ways. Given the importance of timely interventions in determining the cost and effectiveness of drought management efforts, improvements in this area are badly needed.

The case studies also provide insight into how drought crisis response too often takes centre-stage compared with proactive drought policy, legal and management planning. Whilst drought interventions are a major policy focus in the MENA region, much like *dzud* response in Mongolia, policy-makers often overlook connections with related governance and socioeconomic issues. In Mongolia, this plays out as drought monitoring and management being under-resourced, and in the MENA region, it connects more generally to the institutional and nested governance collaboration challenges highlighted here.

However, in both regions there are positive innovations coming from national government initiatives and often in partnership with international networks of expertise. In Mongolia and in the MENA region, the development of new drought monitoring and early warning systems is positive, as is the exploration, and in some cases successful implementation, of drought insurance.

13.8.2 Future Directions for Research

This chapter highlights some future directions for drought management research. There is a need to assess the benefits of the development of alternative mitigation strategies. It also shows the increasing need to evaluate drought management through integrated bio-physical and anthropogenic lenses.

In the absence of a prolonged crisis and without robust evidence, policy-makers are unlikely now to undertake a major legislative or policy initiative (WMO and GWP 2017). As such, research is needed to provide rigorous evidence of the economic and social benefits of shifting to risk management frameworks. Whilst a growing body of evidence shows improved outcomes, information quantifying the costs and benefits would be invaluable to support and frame new policy developments.

Drought monitoring and management challenges will increase in the future due to climate change, increasing water scarcity, and linked economic and demographic growth. Increasingly, research investigates drought as a natural phenomenon affected by human responses through water and land management (Van Loon et al. 2016). The three pillars approach (WMO and GMP 2014) highlights the necessity of interdisciplinary and multisectoral engagement in devising drought monitoring and management regimes, and increasingly the academic community is recognizing the importance of understanding the cumulative effects of myriad human–environment interactions in the context of drought monitoring and management.

13.9 Conclusions

As a drought overview, this chapter introduces several themes for consideration. It identifies drought as a natural climate and environmental phenomenon that, as a deviation from normal hydrological conditions, is a local/regional event that can occur in any landscape. Likewise, it situates drought severity and impacts within relevant socioeconomic and infrastructural contexts.

Drought's spatial and temporal distribution, typologies, intensity, and potential to initiate human misery and socioeconomic damage, frame societal drought risk attitudes. Drought dynamics are further exacerbated by climate change, growing populations, limited mitigation capacity, and a focus on drought crisis response rather than strengthened monitoring and risk mitigation. As highlighted here, current technical tools, such as satellite monitoring, provide much relevant information. Whilst physical factors can be measured and much data are available, the challenge is to use all sources of information in establishing effective drought legislation and policy as well as management and monitoring programmes.

Information sharing, and institutional coordination challenges, are surmountable barriers. Indeed, current efforts to produce new drought monitoring tools are themselves helping to break the otherwise continuing silos. Stakeholders clearly recognize the synergies possible from improved technical and institutional governance arrangements and emphasize how they can facilitate and expedite political decision-making by reducing the political difficulty of official drought declaration and related interventions. Ongoing development of seasonal forecasting in a number of different regions will further engender collaboration as this information is required by water managers and users, whether a drought is predicted or not.

The main challenge is converting potential into practice for drought policy, monitoring and management. The growing capabilities in several countries offer chances for peer-learning and legal and policy comparison, with a critical mass of countries moving towards drought risk management. Awareness, involvement and engagement of water researchers, stakeholders and practitioners are key to reducing risks from the ever-present drought hazard.

Acknowledgements

This book chapter was made partly possible through the support of the Office of Science and Technology, Bureau for the Middle East, US Agency for International Development, under the terms of Award No. AID-ME-IO-15-003. The opinions expressed in this publication are those of the authors and do not necessarily reflect the views of the US Agency for International Development.

References

Barnett, B., Barrett, C., and Skees, J. (2008). Poverty traps and index-based risk transfer products. *World Development* 36 (10): 1766–1785.

Bergaoui, K., Mitchell, D., Zaaboul, R. et al. (2014). The contribution of human-induced climate change to the drought of 2014 in the southern Levant region. *Bulletin of the American Meteorological Society* (Special Supplement, Explaining Extreme Events of 2014) 96 (12): S66–S70.

Bijaber, N., El Hadani, D., Saidi, M. et al. (2018). Developing a remotely sensed drought monitoring indicator for Morocco. *Geosciences* 8: 55. https://doi.org/10.3390/geosciences8020055.

Botterill, D. and Wilhite, D. (eds.) (2005). *From Disaster Response to Risk Management: Australia's National Drought Policy*. Dordrecht: Springer Netherlands.

Budds, J., Linton, J., and McDonnell, R. (2014). The hydrosocial cycle. *Geoforum* 57: 167–169.

Carrão, H. and Barbosa, P. (2015). *Models of Drought Hazard, Exposure, Vulnerability and Risk for Latin America*. Technical Report. EUROCLIMA II Desertification, Land Degradation and Drought (DLDD), and bio-physical modeling for crop yield estimation in Latin America under a changing climate. Brussels: European Commission Joint Research Centre. http://www.droughtmanagement.info/literature/EC-JRC-IES_Drought-Hazard-Exposure-Vulnerability-and-Risk-Latin%20America_2015.pdf#page=10andzoom=100,0,360.

Cook, B., Anchukaitis, K., Touchan, R. et al. (2016). Spatiotemporal drought variability in the Mediterranean over the last 900 years. *Journal of Geophysical Research Atmospheres* 121 (5): 2060–2074. https://doi.org/10.1002/2015JD023929.

Droogers, P., Immerzeel, W., Terink, J. et al. (2012). Water resources and trends in the Middle East and North Africa towards 2050. *Hydrology and Earth System Sciences* 16: 3101–3114. https://doi.org/10.5194/hess-16-3101-2012.

Falkenmark, M., Lundquist, J., and Widstrand, C. (1989). Macro-scale water scarcity requires micro-scale approaches: aspects of vulnerability in semi-arid development. *Natural Resources Forum* 13 (4): 258–267.

Garrick, D. (2015). *Water Allocation in Rivers Under Pressure: Water Trading, Transaction Costs and Transboundary Governance in the Western US and Australia.* London: Edward Elgar Publishing.

Göbel, W. and De Pauw, E. (2010). *Climate and Drought Atlas for parts of the Near East: A baseline dataset for planning adaptation strategies to climate change.* Final report. GIS Unit, International Center for Agricultural Research in the Dry Areas (ICARDA).

Guha-Sapir, D., Hoyois, P., Wallemacq, P., and Below, R. (2016). *Annual Disaster Statistical Review 2016: The Numbers and Trends.* Brussels: CRED.

Hayes, M., Svoboda, M., Wall, N., and Wildman, M. (2011). The Lincoln declaration on drought indices: universal meteorological drought index recommended. *Bulletin of the American Meteorological Society* 92: 485–488.

Hayes, M., Svoboda, M., Wardlow, B. et al. (2012). Drought monitoring: historical and current perspectives. National drought mitigation center faculty publications. In: *Remote Sensing of Drought: Innovative Monitoring Approaches* (eds. B.D. Wardlow, M.C. Anderson and J.P. Verdin), 1–19. Boca Raton, FL: CRC Press/Taylor & Francis.

Hornbeck, R. and Keskin, P. (2014). The historically evolving impact of the Ogallala Aquifer: agricultural adaptation to groundwater and drought. *American Economic Journal: Applied Economics* 6 (1): 190–219. https://doi.org/10.1257/app.6.1.190.

IPCC (2014). *Climate Change 2014: Synthesis Report. Contribution of Working Groups I, II and III to the Fifth Assessment Report of the Intergovernmental Panel on Climate Change* [Core Writing Team, R.K. Pachauri and L.A. Meyer (eds.)]. Geneva: IPCC.

Kallis, G. (2008). Droughts. *Annual Review of Environment and Resources* 33: 85–118.

Kelley, C.P., Mohtadi, S., Cane, M.A. et al. (2015). Climate change in the Fertile Crescent and implications of the recent Syrian drought. *Proceedings of the National Academy of Sciences* 112 (11): 3241–3246.

Lowi, T. (2003). Law vs. public policy: a critical exploration. *Cornell Journal of Law and Public Policy* 12 (3): 294–301.

MacLachlan, M., Ramos, S., Hungerford, A., and Edwards, S. (2018). *Federal Natural Disaster Assistance Programs for Livestock Producers, 2008-16.* Economic Research Service – Economic Information Bulletin Number 187.

Mankin, J., Vivirolo, D., Singh, D. et al. (2015). The potential for snow to supply human water demand in the present and future. *Environmental Research Letters* 10 (11): 114016. https://doi.org/10.1088/1748-9326/10/11/114016.

Martins, J., Engle, N., and De Nys, E. (2015). Evaluating national drought policies: a comparative analysis of Australia, Brazil, Mexico, Spain and the United States. *Parcerias Estratégicas* 20 (41): 57–88.

Mehren, A., Mazdiyasni, O., and AghaKouchak, A. (2015). A hybrid framework for assessing socioeconomic drought: linking climate variability, local resilience, and demand. *Journal of Geophysical Research: Atmospheres* 120: 7520–7533.

Mishra, A. and Singh, V. (2010). A review of drought concepts. *Journal of Hydrology* 391 (1–2): 202–216. https://doi.org/10.1016/j.jhydrol.2010.07.012.

O'Brien, K., Leichenko, R., Kelkar, U. et al. (2004). Mapping vulnerability to multiple stressors: climate change and globalization in India. *Global Environmental Change* 14: 303–313.

O'Brien, K.L., Eriksen, S., Schjolden, A., and Lygaard, L. (2005). *What's in a word? Interpretations of vulnerability in climate change research*. CICERO Working Paper 2004:04. Oslo: CICERO.

Rubel, F. and Kottek, M. (2010). Observed and projected climate shifts 1901–2100 depicted by world maps of the Koppen-Geiger climate classification. *Meteorological Zeitschrifft* 19: 135–141.

Simelton, E., Fraser, E., Termansen, M. et al. (2009). Typologies of crop-drought vulnerability: an empirical analysis of the socio-economic factors that influence the sensitivity and resilience to drought of three major food crops in China (1961–2001). *Environmental Science and Policy* 12 (4): 438–452.

Singh, R., Worku, M., Bogale, S. et al. (2016). *Reality of Resilience: Perspectives of the 2015–16 Drought in Ethiopia*. Resilience Intel 6. London: Building Resilience and Adaptation to Climate Extremes and Disasters (BRACED).

Sternberg, T. (2018). Investigating the presumed causal links between drought and dzud in Mongolia. *Natural Hazards* 92 (1): 27–43.

Sternberg, T., Thomas, D., and Middleton, N. (2011). Drought dynamics on the Mongolian steppe, 1970–2006. *International Journal of Climatology* 31 (12): 1823–1830.

Stone, R. (2014). Constructing a framework for national drought policy: the way forward – the way Australia developed and implemented the national drought policy. *Weather and Climate Extremes* 3 (6): 117–125.

Sullivan, A., White, D., and Hanemann, M. (2019). Designing collaborative governance: insights from the drought contingency planning process for the lower Colorado River basin. *Environmental Science and Policy* 91: 39–49.

Troy, B. (2013). *Assurance et développement agricole: nouvelles dynamiques en Algérie, au Maroc et en Tunisie. Fondation pour l'agriculture et la ruralite dans le monde*. Document de travail no. 5. http://www.fondation-farm.org/zoe/doc/farm_201312_doctrav5_assuagrimaghreb.pdf.

UNCCD (United Nations Convention to Combat Desertification) (1994). *United Nations Convention to Combat Desertification in Those Countries Experiencing Serious Drought and/or Desertification Particularly in Africa: Text with Annexes*. https://treaties.un.org/pages/ViewDetails.aspx?src=TREATYandmtdsg_no=XXVII-10andchapter=27andclang=_en.

UNISDR (United Nations International Strategy for Disaster Reduction) (2015). *Sendai Framework for Disaster Risk Reduction 2015–2030*. http://www.wcdrr.org/uploads/Sendai_Framework_for_Disaster_Risk_Reduction_2015-2030.pdf.

Van Loon, A., Gleeson, T., Clark, J. et al. (2016). Drought in the Anthropocene. *Nature Geoscience* 9: 89–91.

van Vliet, M., Sheffield, J., Wiberg, D., and Wood, E. (2016). Impacts of recent drought and warm years on water resources and electricity supply worldwide. *Environmental Research Letters* 11: 124021.

Wilhite, D., Mannav, N., Sivakumar, V., and Pulwarty, R. (2014). Managing drought risk in a changing climate: the role of national drought policy. *Weather and Climate Extremes* 3: 4–13. https://doi.org/10.1016/j.wace.2014.01.002.

Wilson, G. (2013). Community resilience: path dependency, lock-in effects and transitional ruptures. *Journal of Environmental Planning and Management* 57 (1): 1–26. https://doi.org/10.1080/09640568.2012.741519.

Woodhouse, C., Meko, D., MacDonald, M. et al. (2010). A 1,200-year perspective of 21st century drought in southwestern North America. *Proceedings of the National Academy of Sciences* 107 (50): 21283–21288. https://doi.org/10.1073/pnas.0911197107.

World Meteorological Organization (WMO) and Global Water Partnership (GWP) (2014). *National Drought Management Policy Guidelines: A Template for Action, Integrated Drought Management Programme (IDMP) Tools and Guidelines Series 1* (ed. D.A. Wilhite). Geneva: WMO and Stockholm: GWP.

World Meteorological Organization (WMO) and Global Water Partnership (GWP) (2016). *Handbook of Drought Indicators and Indices, Integrated Drought Management Programme (IDMP) Tools and Guidelines Series 2* (eds. M. Svoboda and B.A. Fuchs). Geneva: WMO and Stockholm: GWP.

World Meteorological Organization (WMO) and Global Water Partnership (GWP) (2017). *Benefits of Action and Costs of Inaction: Drought Mitigation and Preparedness – a Literature Review* (N. Gerber and A. Mirzabaev). Integrated Drought Management Programme (IDMP) Working Paper 1. Geneva: WMO and Stockholm: GWP.

Part III

Water Management

We see water management here as interacting with the development of science-informed water policy. It is more concerned with year-by-year issues rather than overall strategy, but our concern here is not with the day-to-day management of, say, a water treatment plant. Water management concerns decision-making, reflecting priorities, so as to achieve objectives set within policies and the goals dictating those objectives.

It is not just the water that is managed, as in a drought or flood situation, but it is also the management of the available financial resources, the skills within the organizations involved, the interface with other relevant agencies, and the management of any risks, uncertainties or adverse impacts. Many such activities are routine, but many others demand innovative approaches and site-specific solutions. Around the world, different organizational structures have grown up to seek to achieve effective water management, and these structures are fundamental to guiding the processes of water management towards sustainable solutions.

Many activities of governments, agencies and institutions require management, be it defence, health, education or transport. But water management has certain key characteristics that make its management particularly challenging.

Firstly, water management usually requires substantial capital investment, to build structures to store water, deliver it to communities and provide them with safety from drought or flooding. Pollution abatement facilities also require significant capital investment.

Secondly, most water management plans and their projects have long lives, stretching to decades after they are initiated or constructed; they may also take many years to design and implement. Time dependency, therefore is built into water management: what can be done today is constrained by what has happened in the past. Relevant time dimensions also include the future, and significant climate change tomorrow must influence the management plans in place today.

Thirdly, water projects tend to have significant impacts on other sectors of government and private sector activities – for example on food supply, environmental protection, or health services – and these cross-sectoral linkages involve multidimensional impacts needing careful management. They are often one of the compelling reasons for state management of water resources rather than private sector domination, or at least a hybrid public–private arrangement.

Water Science, Policy, and Management: A Global Challenge, First Edition. Edited by Simon J. Dadson, Dustin E. Garrick, Edmund C. Penning-Rowsell, Jim W. Hall, Rob Hope, and Jocelyne Hughes.
© 2020 John Wiley & Sons Ltd. Published 2020 by John Wiley & Sons Ltd.

Finally, many water management plans and projects across the world have significant cross-boundary dimensions, crossing regional boundaries within states and crossing the boundaries of states themselves. These situations require cooperation in the management of the relevant resources and may involve some relaxation of sovereignty, bringing significant international political difficulties.

Many of these challenges are illustrated in the chapters that follow. Water management inevitably involves managing risk, whether it be risks to the supply of clean water or the risks of flooding and the damage caused. For that management we need principles and tools linking risk assessment via modelling to risk management via risk-reducing interventions (Chapter 14) to promote water security (Chapter 18). Given the need for cost recovery, water management needs guidance from the economics involved and then exploiting the financial mechanisms that can be developed (Chapter 15). Environmental impacts obviously also need to be considered and managed, and trade-offs fine-tuned between performance and investment using science-based innovations, for example in managing wastewater treatment (Chapter 16). Maximizing the delivery of clean water to large urban populations is another challenge, using formal internationally agreed goals as management tools within dedicated governance arrangements (Chapter 17). Finally, particular challenges exist in enhancing water security in international and transboundary catchments (Chapter 19), because insecurity here can adversely affect the livelihoods of millions of people but be the subject of political decisions not directly connected with water management itself.

14

Water Resource System Modelling and Decision Analysis

Jim W. Hall[1], Edoardo Borgomeo[2], Mohammad Mortazavi-Naeini[3], and Kevin Wheeler[4]

[1] *Environmental Change Institute, University of Oxford, UK*
[2] *International Water Management Institute, Colombo, Sri Lanka*
[3] *Land and Water Division, NSW Department of Primary Industry, New South Wales, Australia*
[4] *School of Geography and the Environment, University of Oxford, UK*

14.1 The Challenge of Sustainable Water Supply

Water resources globally face intensifying challenges (Hall et al. 2014). The demands on water resources are increasing, whilst changes to the hydrological cycle brought about by climate change mean that the amount of water available in the environment is increasingly unpredictable. Agriculture accounts for about 70% of freshwater withdrawals globally. Demand for increasing supplies of food, fibre and biofuels, driven by a growing and more wealthy global population, is increasing pressure on water resources. There is great potential to use water more efficiently in agriculture if financial, behavioural and institutional obstacles can be overcome. Meanwhile, rapidly expanding urban areas are competing for water supplies. Many of the fastest growing cities are in arid locations, on coasts where over-pumping of groundwater is leading to ground subsidence and saltwater intrusion, or on the lower reaches of highly polluted rivers. Water is also used in the energy sector, notably for cooling thermoelectric power plants and in hydroelectric reservoirs. Whilst the latter source of energy is often considered 'renewable', large volumes of water can be lost through evaporation from the surface of reservoirs, and the local environmental and social impacts can be profound. Whilst in some regions (e.g. the European Alps) potential hydropower sites have almost all been exploited, there is still great potential for hydropower development in other parts of the world, particularly notable in regions where there are large unmet energy demands such as sub-Saharan Africa.

Overall we see growing demands for water, including in locations where water resources are already unsustainably overexploited (Vörösmarty et al. 2000). For example, groundwater levels in many parts of the world (e.g. in northern China) have lowered dramatically (see Chapter 3), whilst surface water withdrawals often deprive aquatic and riparian habitats from the volume and timing of flows that need to be sustained

Water Science, Policy, and Management: A Global Challenge, First Edition. Edited by Simon J. Dadson, Dustin E. Garrick, Edmund C. Penning-Rowsell, Jim W. Hall, Rob Hope, and Jocelyne Hughes.
© 2020 John Wiley & Sons Ltd. Published 2020 by John Wiley & Sons Ltd.

(Vörösmarty et al. 2010). Excessive and unsustainable water withdrawal means that the aquatic environment will not be able to continue to provide many ecosystem services (e.g. fisheries, healthy recreation opportunities and cultural services), nor will these water bodies be resilient enough to provide extra water at times of extreme need (see Chapter 5).

Plans to exploit water resources have always had to cope with the inherent variability in the availability of surface water and groundwater. Water availability varies on time-scales from hours to years, often with strong seasonal variability and in many locations significant variability from year to year (intra-annual variability). Human civilizations have long dealt with temporal variability through the construction of storage reservoirs, and with spatial variability through distribution infrastructure such as pipelines, aqueducts and canals. Technological innovation, notably in desalination and waste water re-use, is providing further capability to provide water in locations that are naturally arid, albeit at a considerable cost and energy use.

Although the implications of climate change are uncertain, particularly at local scales, climate science predicts that a warmer atmosphere will be able to hold more water, whilst higher temperatures will also increase evapotranspiration from the land and reduce snow cover. Unpredictable changes in weather patterns at a range of spatial scales will modify these general phenomena. Unfortunately, climate models are notoriously bad at reproducing large-scale hydrological phenomena at local scales. The precipitation predicted by most climate models is significantly biased when compared with observations (Eden et al. 2012; Chapter 2, this volume). At large scales, there are some disturbing discrepancies, for example between the observed drying trend in the Horn of Africa compared with a prediction of increasing precipitation from coupled ocean–atmosphere GCMs (Yang et al. 2014).

Faced with increasing demands and uncertainties, decision-makers may despair at the prospect of having to make long-term choices about the allocation of water resources and investment in water supply infrastructure. These are indeed difficult decisions, which impact upon future generations and potentially lock-in patterns of development. However, there are grounds to be optimistic that sustainable choices can be made, in particular (i) through deliberately building flexibility into decisions regarding institutional and infrastructure design and (ii) by adopting methodologies that help planners and decision-makers to navigate trade-offs and explore the robustness to future uncertainties.

The notion of 'flexibility' might seem surprising given that we are talking, amongst other things, about concrete and steel infrastructure that can be very costly to adapt. But there is a growing recognition of the extent to which modification of the operation of infrastructure systems can yield very different outcomes (Wheeler et al. 2018a). Infrastructure operators may tend to rely on a pre-established set of rules, but there should be nothing to stop them from adaptively managing systems based on changing environmental conditions. Similarly, institutional arrangements like water rights need to be flexible enough to adapt to changing circumstances, for example using 'water shares' arrangements that allocate different amounts of water based on observed flows, and also allow trading which can reallocate scarce water to the most productive uses (Dinar et al. 1997).

Recognition of the deep uncertainties associated with climate change has stimulated a flourishing of methodologies for decision-making under uncertainty (Lempert et al. 2006). Indeed, water resources decision-making has become one of the best-known testbeds for these methodologies, which have now reached maturity in research terms and are increasingly being applied in practice (Simpson et al. 2016). As we explain in the following section, all of these methods are underpinned by simulation models of hydrology and the operation of water resources systems (withdrawals, storage, allocation to users and return flows). Simulation models of this type have become the work-horses of water resources planning and management. In that sense, the vision of the Harvard Water Programme has taken hold in practice. Nonetheless, major challenges remain such as navigating the trade-offs between multiple objectives, dealing with multiple sources of uncertainty and providing meaningful and influential input into contested political processes. In this chapter we explore these challenges. First, we briefly set out our understanding of the water resource system before exploring methodologies for dealing with multiple objectives, hydraulic variability and uncertainty. We examine how system modelling can be embedded in decision-making, before touching on some of the frontiers of water resources systems analysis techniques.

14.2   The Water Resource System Problem

In simple terms, the water resources planning and management problem involves deciding what operational, institutional, or infrastructure interventions to make in a river basin in order to modify the flow of water[1] through that basin to meet a set of defined objectives. This conceptualization might be seen as supposing that some pristine 'natural' state pre-exists human intervention, but in practice most river basins have been subject to some form of human intervention, sometimes for hundreds, if not thousands, of years. Thus, depending on the location and objective of the analysis, a water resource systems problem can be framed as an intervention in a system that has already been coupled with human effects.

Figure 14.1 illustrates the flow of water through the river basin, with rainfall in the headwaters running off into streams and rivers. Reservoirs and other structures may be constructed to regulate flows, for example by providing storage that can be released for downstream users during times of low flow, storage to catch extreme floods before they cause destructive damage downstream, or storage and head for hydropower production. Structures downstream are used to divert water from the river, possibly also incorporating storage reservoirs, for municipal (in which case the water will be treated to the required standard) or agricultural uses. Agricultural uses will consume a large proportion of the water (which is evaporated or transpired by plants), but there may be some residual flows that return to the river system. Although water is also lost from municipal systems (e.g. through leakage from distribution pipes and sewers), large quantities of water flow through the system and are used to transport sewage waste for treatment at wastewater treatment works, before returning to the river. In large basins these cycles of water withdrawal, use, and return flow may happen several times. In addition, precipitation in downstream catchments may contribute to flows.

---

1 The approaches described in this chapter focus on 'blue' water resources (liquid water) in aquifers, rivers, lakes and dams. 'Green' water resources (soil moisture in soils) are not considered, although their importance for food production is recognized.

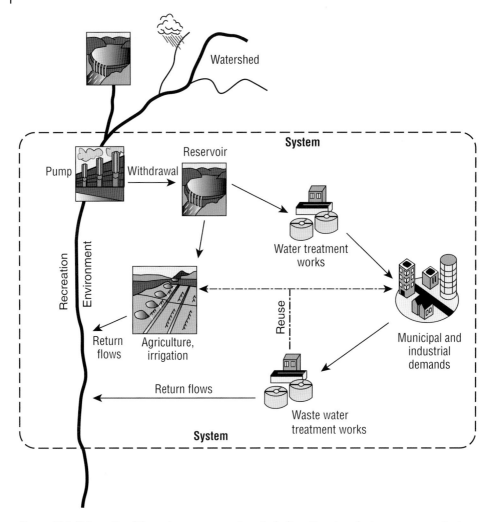

Figure 14.1 Schematic of the water resource system, including diversions for human uses and return flows. *Source:* Adapted from http://iora.nus.edu.sg/research/enviornmental.html. (*See color representation of this figure in color plate section*).

Figure 14.1 illustrates a surface-water dominated system, and does not depict the role of groundwater. Groundwater systems (see Chapter 3), which interact with (sometimes ephemeral) surface water, are inherently more complex. They depend upon particular geological characteristics which can only be partially observed (through borehole surveys and geophysical investigations). Groundwater levels respond depending on the permeability of the aquifer, which can vary over many orders of magnitude. The approaches to water resource systems modelling and decision analysis that are described in this chapter are much more developed for surface water systems, so more work is required to apply them to groundwater systems, particularly large and complex systems.

We identify two classes of decision-making within water resources systems:

1) *Operational decisions*, which deal, for example, with how infrastructure is operated and how water is allocated on a timescale of hours to months.

2) *Long-term planning decisions*, which deal with changes to the overall configuration of infrastructure and institutional arrangements, which typically take place over timescales of years.

Operational decisions depend heavily on observed data (e.g. from monitoring and sensors), adapting the system in response to these observations. Long-term planning decisions are more susceptible to the effects of change within the system (e.g. due to changing demands for water and catchment or climate changes) – thus these decisions require particular attention to be paid to the role of long-term uncertainties and scenarios.

These two classes of decisions interact, since long-term planning relies upon assumptions about how the system will be operated in future. Indeed, recent developments in water resource system modelling have demonstrated how the potential for adapting operational rules should be incorporated in the appraisal of longer-term options (Zeff et al. 2016).

Both classes of decision-making rely upon scientific and engineering understanding of the processes being enacted in the water resources system. The most fundamental of these scientific principles is the notion of water balance: there are 'flows' of water entering and leaving the system boundaries, so the 'stock' of water in the system is a function of the difference between these flows. Whilst the concept of water balance does not require a computer model to quantify, computer models have become central to quantifying the stocks and flows in water resources systems because of their capacity to simulate the multiple interacting entities in a complex system and to simulate the dynamic processes (e.g. routing water down river systems) that determine the timing of availability of water in the system. Simulation modelling has become so central to the epistemology of water resources planning and management that it would now be considered irresponsible for any major planning decision in a complex water resources system to be made without recourse to a computer model. These models are attractive in that they can deal with the interacting complexity of multiple entities (e.g. hydrology, storage and transfer infrastructure, multiple water uses, return flows) and can be used to explore a multitude of future scenarios and system configurations. Simulation models have facilitated the shift from a deterministic approach to prescribing management plans, to approaches that are more exploratory, attempting to expose the implications of uncertainties and trade-offs in order to identify robust plans, and understand and mitigate possible undesirable impacts.

14.3   Dealing with Multiple Objectives

Water is pervasive in the environment, society and the economy. Multiple values are therefore associated with water (Garrick et al. 2017). One approach to categorizing these values derives from the framework of ecosystems services (TEEB 2010), which are categorized as: provisioning services; regulating services; and cultural services. Water is associated with each of these categories of services (see Chapter 5), for example through provision of freshwater fish, regulation of flooding and assimilation of waste, and the aesthetic, recreational and spiritual values associated with rivers, lakes and wetlands.

We also recognize that there are beneficial and harmful outcomes associated with water resource systems (Sadoff et al. 2015). Water provides beneficial outcomes for people and the economy, e.g. by enabling agriculture, hydropower production and transport on inland waterways. On the other hand, water has hazardous impacts that materialize in floods, droughts, and times of harmful water quality.

A further inevitable complexity of decision-making regarding water resource systems is the existence of multiple actors with an interest in the system. Not only are there multiple values associated with water, but different actors will balance and trade-off those values in different ways.

A methodology for informing decisions regarding water resource systems therefore needs to be capable of properly reflecting multiple perspectives and values. The tools of multi-objective optimization are particularly powerful for exploring trade-offs between different attributes of water resource systems (Nicklow et al. 2010; Reed et al. 2013). These enable system states (e.g. in different possible future scenarios) to be presented in terms of their performance with respect to multiple objectives (see for example Figure 14.2). To compute system performance against relevant objectives requires a valuation process that translates state variables within the system depicted in Figure 14.1 into values of interest to decision-makers (e.g. monetary values and/or impacts on people and species).

Water resource systems are often of critical political significance, in particular where rivers cross national boundaries. In that context, there may be objectives of national

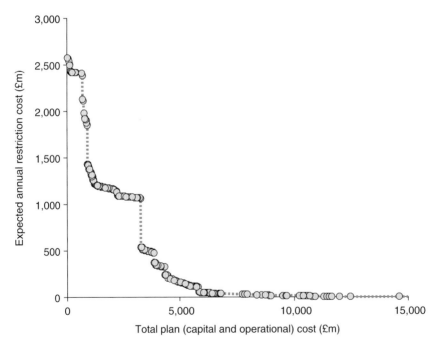

Figure 14.2 Example of presenting the trade-off between different objectives for a water resource system. This example shows the trade-off between the economic risk of water shortages (*y*-axis) and the cost of providing additional water supplies (*x*-axis). The economic risk of shortage is a quantification of the aggregate economic impact on water users due to restrictions on supply. Each dot on the graph shows a possible water resource management plan for this system. There are many other possible plans to the right of the curve (i.e. more costly plans that also incur higher risk), but the ones along the curve illustrate the inevitable trade-off between these two objectives. *Source:* Adapted from Borgomeo et al. (2018b). Reproduced with permission of John Wiley & Sons. (*See color representation of this figure in color plate section*).

security associated with water resource systems. National security concerns material-ize when there is the possibility that actions by a co-riparian state may interfere in the river basin in a way that threatens domestic water functions, such as agricultural pro-duction, water for people and livelihoods, hydropower production, flood control, or navigation. These impacts need not materialize for them to be publicly salient – the *risk* of them materializing may be enough to motivate political concern and even conflict.

The approach to water resource system modelling and decision analysis that we pro-pose involves carefully exposing the various perspectives and interests in water resource systems. These perspectives should be valued in ways that are meaningful for decision-makers, preferably through quantification, but we recognize that there will be uncer-tainty associated with these valuations, particularly for very different attributes of a water resource system. Our approach does not imply condensing all values and per-spectives into one objective function – instead we seek to explore trade-offs and the implications of uncertainties, although we recognize that proper scrutiny of trade-offs becomes increasingly challenging as the number of dimensions of objectives and uncer-tainties increases.

14.4   Variability and Risk

A fundamental and practically irreducible source of uncertainty in water resource sys-tems is their natural variability on a wide range of different timescales. Hydrological variability derives from variations in precipitation (rainfall and snow), evaporation and the movement of water through the landscape. These are complex processes, which before the advent of computers water resources planners had to condense into simple statistical descriptions. We now have access to computer simulation models that can explicitly represent hydrological variability.

There is a range of approaches to representing variability on the hydrological boundary conditions to a water resource system. In one set of methods, weather (which may be simulated from climate models or numerical weather prediction) is taken as the boundary condition, and hydrological models (rainfall-runoff models, groundwater models, or more general land surface models) are then used to simulate the passage of water though the landscape. This approach has the attraction of implic-itly dealing with multiple dependencies between variables, in space and over time. It also has the intellectual attraction of being a physically realistic representation of the passage of water through the landscape. However, this physically based approach, to represent the relevant physics and parameterize models, is an extremely challenging task, so there are major deficiencies in model outputs which are manifest as biases and errors.

The alternative is to endeavour to characterize the variables that are of most direct relevance to the water resource system, for example, river flows and groundwater levels at points where water is withdrawn from the system (see Figure 14.1). There is a long tradition of statistical analysis of streamflow variables and construction of statistical models that can simulate realistic streamflow with statistics well matched to observa-tions (Fiering and Jackson 1971; Matalas 1967). The disadvantage of these approaches is their limited capacity to represent the effects of upstream change (e.g. due to changed climatic conditions or land-use change). This is a major limitation for exploration of the effects of projected future changes, but also calls into doubt the validity of approaches that are calibrated against observations with (unknown) change signals. Given the pros and cons of these alternative approaches, a variety of methods for representing hydrological variability may be adopted and compared, to assess system sensitivity to assumptions and uncertainties (Borgomeo et al. 2015).

Variability in water resources systems means that beyond a few days or weeks (with the possible exception of the largest river systems and slowest responding groundwater systems) it is impossible to predict relevant variables (e.g. flow) in deterministic terms. The best that can be achieved is a good characterization of the statistics of the variable (ideally the full probability density function and dependence structure). Inevitably, therefore, the outcomes and impacts that may materialize in future can at best be described in probabilistic terms as risks. This leads to the notion that water resources planning should be based upon the analysis of risks (Gallagher et al. 2016; Hall and Borgomeo 2013), that is, should be based upon weighing up the probability and consequences of possible future states of the water resource system, under a range of possible scenarios and management options. As we have already noted, impacts of interest within the water resource system may touch upon several different dimensions of value, so risk will be a multidimensional construct, reflecting the multiple values associated with the range of possible system states.

14.5 Uncertainty and Decisions

Acknowledging hydrological variability naturally leads to decision-making based on analysis of risk. However, estimating risk is bound to be fraught with uncertainties, even in the present. Looking into the future will bring even more sources of uncertainty.

For risk estimates in the present day, the main sources of uncertainty relate to:

- Scarcity of data for empirically estimating the likelihood and consequences of system states.
- The possibility of non-stationarity in observed datasets (e.g. due to climatic or catchment changes), which mean that statistical estimates may be biased.
- Incomplete knowledge of the operation of water resource systems, which may be governed by well-specified rules, but is still subject to the vagaries of human choices.
- Imperfect models of system behaviour, such as hydrological models.

Looking into the future we can expect significant changes that will impact upon water resource systems, but the nature of those changes is uncertain, often severely so, due to reasons including:

- climate change;
- catchment change, e.g. urbanization, deforestation/afforestation, increase/decrease in extent of surface water ecosystems;
- changing human and economic demands for water, e.g. due to urbanization, changing agricultural demands and industry (including the power sector);
- changing policies and legal changes, e.g. changing regulations of water withdrawals due to changing environmental priorities/policies.

All of these uncertainties need to be incorporated into decision-making. For some uncertainties it may be possible to adopt the same probabilistic methods that are used for dealing with natural variability: for example the UKCP09 climate projections (Murphy et al. 2007, 2009) adopted a probabilistic approach to characterizing the uncertainty across a multimodel ensemble of climate projections (conditional on a given greenhouse gas emissions scenario). However, future uncertainties are mostly very difficult to quantify in probabilistic terms. In that sense, they can be thought of as 'severe', 'deep', or Knightian uncertainties, the latter of these terms referring to the economist Frank Knight who distinguished between risk-based decision-making (in which choices should be based upon the principle of maximizing expected utility) and decisions where the probabilities of future states of nature cannot be estimated – in which case there is no normative approach to making a choice (Knight 1921). A range of methods have emerged for decision-making under deep uncertainty (Simpson et al. 2016), which are all to some extent based on the notion that in the presence of uncertainty, it is desirable to seek choices that perform reasonably well across a range of possible future conditions, and so are *robust* to uncertainty (Hallegatte et al. 2012; Simpson et al. 2016). This emphasis upon robustness is in principle quite different to optimizing methods (e.g. maximizing expected utility), as the aim is to identify options that perform acceptably well (i.e. they satisfy a set of criteria) over the widest possible space of possible futures. This approach to decision-making is therefore also known as 'satisficing' (a term first coined by Herbert Simon) (Simon 1956).

In practice, the robustness of a decision can be enhanced by: (i) delaying decisions to acquire further information; and (ii) adopting flexible options that can be adapted to a range of possible circumstances. Neither of these approaches are cost-free. Delaying decisions means that water users will be subject to greater risks in the meantime before a decision is made. There is often a long lead-time associated with major water resources planning decisions (e.g. regulatory reforms or major infrastructure), so additional delay will push the eventual action a long time into the future, possibly beyond the point where risks become intolerable. Flexibility can be achieved, for example by focusing upon actions that can be scaled up and down incrementally (e.g. efforts to reduce water demand) or by purposefully designing infrastructure so that it can be modified in future. But flexibility is usually not cost-free, so adopting flexible solutions will typically involve foregoing other more efficient (but possibly less robust) solutions.

Despite the big difference in principle between optimizing and satisficing, in practice the tools of optimization are still helpful in robust decision-making – for searching option spaces and seeking options that maximize robustness.

Another methodological connection is with methods for sensitivity analysis. Sensitivity analysis investigates how the variation in the output of a numerical model can be attributed to variations of its input factors (Pianosi et al. 2016). In a sense, a robustness analysis is a form of sensitivity analysis, because it investigates how the performance of a decision (output) varies depending on different uncertainties and boundary conditions such as climate change and water demands (inputs), to identify a solution that is relatively insensitive to uncertainties about the future. Methods for robustness analysis therefore have a lot in common with methods for sensitivity analysis, for example, using structured Monte Carlo sampling to explore a range of possible future conditions.

Our methodology for water resources planning is therefore structured in the following layers:

Level 1:   At the core of the decision analysis will be a deterministic simulator of the system. This will take boundary conditions (e.g. climate simulations, water demands) and compute system state variables (e.g. the amount of water available for different users), and should also translate those system variables into quantities of relevance to decision-makers (e.g. the economic value of water restrictions).

Level 2:   The core deterministic simulator will be run multiple times with multiple time series of hydrological conditions (e.g. climate series or flow series) in order to test system response to hydrological variability. For each simulation, the outcomes of interest (e.g. the value of water shortages) will be calculated, so that over many simulations a probability-weighted aggregate estimate of risk can be computed.

Level 3:   Relevant sources of uncertainty, for example in future changes (see above), will be explored by sampling a range of possible future conditions (scenarios) and estimating risk (Level 2) in each of those scenarios. Thus, at Level 3 we construct a set of risk and cost estimates, the range of which will reflect the range of deep uncertainties.

Level 4:   A set of possible decision variables is explored (typically through optimization) to explore their effectiveness in reducing risk (Level 2) and their response to uncertainty (Level 3), and thus to identify robust decisions. As well as having an impact on the water resource system, each possible decision will have an associated cost, which will form part of the decision calculus (e.g. via cost–benefit or cost-effectiveness analysis).

14.6   Embedding Simulation Modelling in Practical Decision-making

In the preceding sections we have set out some of the reasons in principle why water resources planning and management decisions are difficult, and some of the responses to those challenges, in particular through the structured use of simulation models. We have set out a modelling framework (in four levels) for addressing those challenges, with a particular focus upon the role of uncertainty. Although there are variants upon this framework, it is becoming widely recognized as a template for water resources decision-making. Aspects of the approach have been taken up in practice, for example, a version of info-gap analysis is used in the planning of water resource systems in the southeast of England. However, despite their maturity, penetration of these methods into practice has not been as extensive, for example, the uptake of risk-based methods in the planning and design of flood protection systems. Here we reflect upon some of the reasons why this may be the case, and how these barriers might be overcome.

1) **The intuitive attraction of deterministic methods**: We have argued elsewhere (Hall et al. 2012) that traditional concepts of water resources planning, such as 'supply–demand balance' are not observable (i.e. the supply–demand balance in a system

cannot be directly measured) and are not particularly meaningful in dynamic and nonlinear systems. However, they have a persistent appeal because they can be read- ily understood in terms of a mental model of water flowing into a basin and then being withdrawn for a variety of (consumptive and non-consumptive) uses. Whilst risk-based concepts have taken hold in other sectors (notably flood risk manage- ment), they are still some way from being ubiquitous in the context of water resource system planning and management. The deterministic notions of 'supply–demand balance' are proving to be hard to displace. Nonetheless, a growing call for 'out- comes-based' management of water resources, for reporting of climate-related risks (TCFD 2017) and the legal duty in England to 'secure resilience' of water supplies, is now providing a powerful motivator for risk-based methods.

2) **Complexity**: One of the main practical complaints about simulation methods that involve a lot of repeated sampling is that they are time-consuming and computation- ally expensive. Computational costs have come down dramatically. Cloud comput- ing means that more or less unlimited computing power can be obtained on-demand without having to invest in in-house hardware. Nonetheless, there are still computa- tional obstacles, more associated with inadequate technical capacity within organi- zations rather than the existence of the necessary computer power. Notwithstanding the existence of scalable computer resources, some water resource system models are still too costly to use in a framework that requires tens of thousands of simula- tions, in which case reduced complexity models can be used for most sensitivity analysis and optimization purposes, possibly combined with a few runs with full complexity models to verify the final results (Haasnoot et al. 2014). Even if tractable computational methods can be found, there remains the challenge of communicat- ing complexity, e.g. trade-offs between multiple objectives and representations of uncertainty. There have been advances in visualization methods (e.g. for multidi- mensional Pareto surfaces; using parallel coordinate plots; decision pathway dia- grams), but these are still hard to comprehend for most non-specialists.

3) **Accurately representing the operation of real systems:** We mentioned in point (2) that operational models of water resource systems can be very complex, because they contain large numbers of entities (e.g. pumps, control structures) with specific control rules. This is a challenge purely from a modelling perspective, but perhaps more significantly, the organizations responsible for these systems may not be will- ing to reveal these control rules for reasons of commercial sensitivity or national security. This may require those who attempt to develop a system model, but lack direct access to the actual operational logic used, to infer rules from available obser- vations (e.g. reservoir and river elevations), which is usually an ill-posed problem.

4) **Integrating with real decision situations:** Although the methods described are aimed at informing practical decisions, there seem to be barriers to their uptake in practice, beyond the ones mentioned in (1) and (2) above. In particular, water resource systems models are configured to answer particular categories of questions, but in real decision-making contexts, the agenda and perceptions of the problem itself change rapidly based on political realities, in which case models may need to be reconfigured, if not entirely rebuilt to reflect this dynamic nature of a problem. The turnaround time for developing a new scenario or problem formulation can be much slower than the pace of decision-making, for example in multi-stakeholder water negotiations. This is in part a technological issue, but often requires the right

combination of professional skills. In particular, technical analyses to address politically charged issues require professionals who understand water resource systems
modelling, the need for transparency of methodologies for stakeholder acceptance,
and a willingness to engage with the political and institutional contexts of decision-
making (Wheeler et al. 2018b).

5) **Supporting negotiated decision-making amongst multiple parties:** We have
identified that water resources planning and management problems almost always
involve trade-offs between multiple actors. This raises questions of transparency
and trust from all of the parties involved. It can be difficult to identify models and
modellers who are considered to be impartial. Trust has to be built over many years
of in-depth interaction with all of the parties. Under these circumstances, obtaining
data, for example on operation rules, can be particularly problematic. Nonetheless,
models can provide a shared platform for negotiation where, ideally, the known facts
about the system can be agreed, thus bringing into focus areas of disagreement
where negotiation and trade-offs are required.

14.7   The Expanding Boundaries of Water Resource Systems

Although the water resource systems field that we have defined is broad, encompassing
climate, hydrology (surface and groundwater), infrastructure and human uses of water,
the innovations that we observe in water resource systems are pushing these boundaries
still further. These developments are consistent with the approach that we have set out,
in that they seek to better quantify the implications of water-related decisions. Many
can be addressed within the framework for modelling and decision analysis that we
have proposed.

14.7.1   New Data Sources

The proliferation of sensors in water resource systems is providing an opportunity to fill
persistent data gaps. For example, the introduction of smart water meters in homes is
providing much more precise information on the characteristics of water usage
(Cominola et al. 2015). Sensing is helping to understand agricultural water use and
monitor water quality as well as quantity. These innovations are having most impact in
water distribution systems, where sensing is helping to pinpoint leaks and target maintenance. We anticipate these technological innovations having increasing impacts on
water resource systems modelling and decision-making.

14.7.2   Economics

There is a long tradition of hydro-economics (Harou et al. 2009), in particular simulating the interplay between water resource systems and agricultural practices and production. Hydro-economic models simulate the market price for agricultural produce
and can seek to represent the dynamic response, for example in the practices (e.g. planting decisions) of farmers and their behaviour in water markets (in the few places where

water markets exist). Hydro-economic models have also taken an agent-based perspective to simulate the practices of, and sometimes competition between, different water users. There is a growing body of empirical research that seeks to quantify the interplay between water and the economy, in particular in economies that are highly dependent upon agriculture (usually, but not exclusively, poor countries) and are subject large hydrological variability (Borgomeo et al. 2018a; Brown and Lall 2006; Damania et al. 2017). Another strand of research examines the productivity of water and hence the wide economic effects of water shortages (Freire-González et al. 2017). This is challenging research because water is so pervasive in the economy, so its effects are difficult to isolate.

14.7.3 Finance

We have referred to investment in water resource systems, in particular in infrastructure and institutions, but so far have said little about how that investment may be financed. In fact, investment in water is a perennial challenge (WWC and OECD 2015) because water infrastructure has high capital costs, but the beneficiaries of water investment may not be willing or able to pay for the services that they receive, and they may not even be readily identified (see also Chapter 15, this volume). Water utilities rely heavily upon debt to finance their investments, but then become subject to vagaries of bond markets and credit ratings agencies who may not understand the nature of their business or the risks to which they are exposed. A growing number of financial instruments are becoming available to manage these risks, through insurance or credit default swaps (Brown et al. 2015). In developing country contexts where finance is particularly difficult to access because the perceived risks are so high, there is growing interest in blended finance instruments, which use development assistance from donors to leverage private finance for much-needed water infrastructure (Bender 2017).

14.7.4 Society

It is clear from everything that has been said above that we regard water as being deeply embedded in the functions of society, and by the same token river basins are almost all to some extent influenced by human activities, often profoundly so. Given that this observation has held true for generations, it is perhaps surprising that the study of the interplay between society and water has recently acquired the new title of socio-hydrology (Sivapalan et al. 2012). The emergence of socio-hydrology re-emphasizes a perennial need to better understand the complex human dimensions of water and incorporate these in the scientific analysis of water resource systems. These interactions operate at a very wide range of scales, from the choices made by individuals in households, through to the nature of water-related political conflicts in transboundary river basins.

14.7.5 The Environment

Throughout this chapter we have recognized the need to preserve water in the environment in order to conserve and enhance the resilience of aquatic ecosystems. We have noted that we depend upon these ecosystems for regulating water quantity and quality. For example, protecting watersheds can help to regulate flows (reducing the risk of flooding, promoting groundwater recharge), reduce polluting runoff and limit soil erosion. For the time being, policies for conservation of these ecosystem functions are mostly based upon regulating water quantity and quality, for example, through regulatory limits on water withdrawals in order to conserve environmental flows. However, environmental flow requirements are a rather blunt regulatory instrument, even if they are adjusted to represent seasonality (Pastor et al. 2014). We know that ecosystems can recover from quite severe droughts that have happened in the past in undisturbed catchments. The unanswered question in ecology is what range of stressor events (e.g. droughts) can ecosystems recover from? We know, for example, that the ecosystems in highly modified rivers are less able to cope with disturbance than unmodified systems (Dunbar et al. 2010). However, we do not know where the bounds on system resilience lie.

Looking to the future, we expect to see much more sophisticated understanding of the resilience of aquatic ecosystems. This understanding can hopefully be used in a more dynamic way to inform water resources management decisions. We might, for example, learn that restoring aquatic environments means that from time to time more water can be withdrawn for human use, provided that withdrawals do not persist beyond a certain length of time. For the time being we do not know that, and it would be risky to adopt policies based on this reasoning. An increasing emphasis needs to be placed upon learning about the resilience of aquatic ecosystems, especially under conditions of stress, whilst at the same time taking every possible step to preserve and restore healthy ecosystems, because we already recognize the many benefits they provide.

14.8 Conclusions

In this chapter we have set out the multiple, and mostly intensifying, challenges facing water resources around the world. These challenges come from increasing human demands for water in the context of a changing climate. They add to the complexity that water resources managers have always had to deal with, notably highly variable and unpredictable hydrological conditions, complex system responses, and the multiple values that we place upon water. Water resource planners face daunting decisions about how to sustainably manage the systems for which they are responsible.

Methods and models for water resource system simulation, risk analysis and decision analysis provide powerful tools for dealing with these challenges. They enable learning about the complex behaviour of river basins, testing of alternative scenarios, exploration of uncertainties and navigation of trade-offs. We have described a layered framework with deterministic simulation modelling at its core, surrounded by layers to simulate natural variability, test uncertainties and identify robust combinations of intervention options.

The framework that we have set out is now relatively mature, and is seeing increasing amounts of application, although there is still less uptake in practice than might be expected. We attribute that to the approaches being conceptually quite challenging and not always easy to align with existing decision-making processes. On the other hand, intensifying calls for 'outcomes-based' management of water resources, for reporting of climate-related risks and the legal duty in England to 'secure resilience' of water supplies, is now providing a powerful motivator for risk-based methods.

Acknowledgements

This chapter is an output of the MaRIUS project: Managing the Risks, Impacts and Uncertainties of droughts and water Scarcity, funded by the Natural Environment Research Council (NERC), and undertaken by researchers from the University of Oxford (NE/L010364/1).

References

Bender, K. (2017). *Introducing Commercial Finance into the Water Sector in Developing Countries*. Washington, DC: World Bank.

Borgomeo, E., Farmer, C.L., and Hall, J.W. (2015). Numerical rivers: a synthetic streamflow generator for water resources vulnerability assessments. *Water Resources Research* 51: 5382–5405.

Borgomeo, E., Vadheim, B., Woldeyes, F.B. et al. (2018a). The distributional and multi-sectoral impacts of rainfall shocks: evidence from computable general equilibrium modelling for the Awash basin, Ethiopia. *Ecological Economics* 146: 621–632.

Borgomeo, E., Mortazavi-Naeini, M., Hall, J.W., and Guillod, B.P. (2018b). Risk, robustness and water resources planning under uncertainty. *Earth's Future* 6 (3): 468–487.

Brown, C. and Lall, U. (2006). Water and economic development: the role of variability and a framework for resilience. *Natural Resources Forum* 30: 306–317. https://doi.org/10.1111/j.1477-8947.2006.00118.x.

Brown, C.M., Lund, J.R., Cai, X. et al. (2015). The future of water resources systems analysis: toward a scientific framework for sustainable water management. *Water Resources Resesarch* 51: 6110–6124. https://doi.org/10.1002/2015WR017114.

Cominola, A., Giuliani, M., Piga, D. et al. (2015). Benefits and challenges of using smart meters for advancing residential water demand modelling and management: a review. *Environmental Modelling & Software* 72: 198–214.

Damania, R., Desbureaux, S., Hyland, M. et al. (2017). *Uncharted Waters: The New Economics of Water Scarcity and Variability*. Washington, DC: World Bank Group.

Dinar, A., Rosegrant, M.W., and Meinzen-Dick, R. (1997). *Water Allocation Mechanisms: Principles and Examples* (English). Policy Research working Paper No. WPS 1779. Washington, DC: World Bank.

Dunbar, M.J., Pedersen, M.L., Cadman, D. et al. (2010). River discharge and local-scale physical habitat influence macroinvertebrate LIFE scores. *Freshwater Biology* 55 (1): 226–242.

Eden, J.M., Widmann, M., Grawe, D. et al. (2012). Skill, correction, and downscaling of GCM-simulated precipitation. *Journal of Climate* 25 (11): 3970–3984.

Fiering, M.B. and Jackson, B.B. (1971). *Synthetic Streamflows, Water Resources Monograph 1*. Washington, DC: American Geophysical Union.

Freire-González, J., Decker, C., and Hall, J.W. (2017). The economic impacts of droughts: a framework for analysis. *Ecological Economics* 132: 196–204.

Gallagher, L., Dalton, J., Bréthaut, C. et al. (2016). The critical role of risk in setting directions for water, food and energy policy and research. *Current Opinion in Environmental Sustainability* 23: 12–16. https://doi.org/10.1016/j.cosust.2016.10.002.

Garrick, D.E., Hall, J.W., Dobson, A. et al. (2017). Valuing water for sustainable development. *Science* 358: 1003–1005.

Haasnoot, M., Van Deursen, W.P.A., Guillaume, J.H.A. et al. (2014). Fit for purpose? Building and evaluating a fast, integrated model for exploring water policy pathways. *Environmental Modelling & Software* 60: 99–120.

Hall, J.W. and Borgomeo, E. (2013). Risk-based principles for managing and defining water security. *Philosophical Transactions of the Royal Society A* 371: 20120407.

Hall, J.W., Watts, G., Keil, M. et al. (2012). Towards risk-based water resources planning in England and Wales under a changing climate. *Water and Environment Journal* 26 (1): 118–129. https://doi.org/10.1111/j.1747-6593.2011.00271.x.

Hall, J.W., Grey, D., Garrick, D. et al. (2014). Coping with the curse of freshwater variability. *Science* 346: 429–430.

Hallegatte, S., Shah, A., Lempert, R. et al. (2012). *Investment Decision Making Under Deep Uncertainty: Application to Climate Change*, Policy Research Working Paper No. 6193. Washington, DC: World Bank.

Harou, J.J., Pulido-Velazquez, M., Rosenberg, D.E. et al. (2009). Hydro-economic models: concepts, design, applications, and future directions. *Journal of Hydrology* 375: 627–643. https://doi.org/10.1016/j.jhydrol.2009.06.037.

Knight, F. (1921). *Risk, Uncertainty and Profit*. Boston and New York: Houghton Mifflin Company.

Lempert, R.J., Groves, D.G., Popper, S.W., and Bankes, S.C. (2006). A general, analytic method for generating robust strategies and narrative scenarios. *Management Science* 52 (4): 514–528.

Matalas, N.C. (1967). Mathematical assessment of synthetic hydrology. *Water Resources Research* 3 (4): 937–945.

Murphy, J.M., Booth, B., Collins, M. et al. (2007). A methodology for probabilistic predictions of regional climate change from perturbed physics ensembles. *Philosophical Transactions of the Royal Society A* 365: 1993–2028.

Murphy, J.M., Sexton, D., Jenkins, G. et al. (2009). *UK Climate Projections Science Report: Climate Change Projections*. Exeter: Met. Office Hadley Centre.

Nicklow, J., Reed, P., Savic, D. et al. (2010). State of the art for genetic algorithm and beyond in water resources planning and management. *Journal of Water Resource Planning and Management* 136: 412–432.

Pastor, A.V., Ludwig, F., Biemans, H. et al. (2014). Accounting for environmental flow requirements in global water assessments. *Hydrology and Earth System Sciences* 18 (12): 5041–5059.

Pianosi, F., Beven, K., Freer, J. et al. (2016). Sensitivity analysis of environmental models: a systematic review with practical workflow. *Environmental Modelling & Software* 79: 214–232.

Reed, P., Hadka, D., Herman, J. et al. (2013). Evolutionary multiobjective optimization in water resources: the past, present, and future. *Advances in Water Resources* 51: 438–456.

Sadoff, C.W., Hall, J.W., Grey, D. et al. (2015). *Securing Water, Sustaining Growth: Report of the GWP/OECD Task Force on Water Security and Sustainable Growth*. Oxford: University of Oxford.

Simon, H. (1956). Rational choice and the structure of the environment. *Psychological Review* 63 (2): 81–97.

Simpson, M., James, R., Hall, J.W. et al. (2016). Decision analysis for management of natural hazards. *Annual Review of Environment and Resources* 41: 489–516. https://doi.org/10.1146/annurev-environ-110615-090011.

Sivapalan, M., Savenije, H., and Blöschl, G. (2012). Socio-hydrology: a new science of people and water. *Hydrological Processes* 26 (8): 1270–1276.

TCFD (2017). *Recommendations of the Task Force on Climate-related Financial Disclosures.* https://www.fsb-tcfd.org/wp-content/uploads/2017/06/FINAL-TCFD-Report-062817. pdf (accessed August 2018).

TEEB (2010). *The Economics of Ecosystems and Biodiversity: Mainstreaming the Economics of Nature: A Synthesis of the Approach, Conclusions and Recommendations of TEEB.* http://www.teebweb.org/publication/mainstreaming-the-economics-of-nature-a-synthesis-of-the-approach-conclusions-and-recommendations-of-teeb/.

Vörösmarty, C.J., Green, P., Salisbury, J. et al. (2000). Global water resources: vulnerability from climate change and population growth. *Science* 289 (5477): 284–288.

Vörösmarty, C.J., McIntyre, P.B., Gessner, M.O. et al. (2010). Global threats to human water security and river biodiversity. *Nature* 467: 555–561.

Wheeler, K.G., Hall, J.W., Abdo, G.M. et al. (2018a). Exploring cooperative transboundary river management strategies for the eastern Nile Basin. *Water Resources Research* 54 https://doi.org/10.1029/2017WR022149.

Wheeler, K.G., Robinson, C.J., and Bark, R.H. (2018b). Modelling to bridge many boundaries: the Colorado and Murray-Darling river basins. *Regional Environmental Change* 18 (16): 1607–1619. https://doi.org/10.1007/s10113-018-1304-z.

WWC and OECD (2015). *Water: Fit to Finance? Catalyzing national growth through investment in water security.* Report of the High Level Panel on Financing Infrastructure for a Water-Secure World. Marseille: World Water Council.

Yang, W., Seager, R., Cane, A. et al. (2014). The east African long rains in observations and models. *Journal of Climate* 27 (19): 7185–7202.

Zeff, H.B., Herman, J.D., Reed, P.M. et al. (2016). Cooperative drought adaptation: integrating infrastructure development, conservation, and water transfers into adaptive policy pathways. *Water Resources Research* 52: 7327–7346. https://doi.org/10.1002/2016WR018771.

15

Financing Water Infrastructure

Alex Money

Smith School of Enterprise and the Environment, University of Oxford, UK

15.1   Introduction

There is a large and growing body of evidence that underlines the strong causal relationship between water security, economic growth and sustainable development (Garrick et al. 2017; Sadoff et al. 2015). Achieving water security requires appropriate investment in infrastructure and capacity building, but there is a big gap between current investment in these key areas, and the amount that is required if meaningful progress is to be made against the UN Sustainable Development Goals (SDGs). Efforts to quantify the investment gap are more of an art than a science, but there is some consensus that it approximates to US$ 1 trillion per annum overall, of which water accounts for 15–30% of the total (Woetzel et al. 2017). By comparison, water attracts just 6% of the total investment that is being directed towards infrastructure assets. As things stand, the infrastructure that is necessary to make the SDGs achievable will not be financed; and what is more, changes to the status quo do not appear to be imminent.

There are water science, policy and management dimensions to the difficulties associated with insufficient or low-quality investment in infrastructure. Citizens of most countries in the world consider it the responsibility of their governments to ensure that core infrastructure is fit for purpose. In many industrialized countries, water infrastructure is perceived as a public good, although decades of relative underinvestment are in many cases degrading the quality of service that this infrastructure delivers. Meanwhile, in many of the less industrialized countries, the problems associated with underinvestment are being compounded by a perfect storm of population growth, increased urbanization and climate variability; that is likely to exert immense and potentially catastrophic pressure on existing systems.

Much has already been written about the problems of inadequate water infrastructure and the investment gap (Briscoe 1999; Camdessus 2003; Ward 2010). There have been long periods of relative underinvestment in the infrastructure of many countries, but public sector balance sheets came under particular strain during the financial crisis of 2008–2009 and the subsequent recession. Whilst those conditions have ameliorated somewhat in the decade since, the narrative around water infrastructure has also

Water Science, Policy, and Management: A Global Challenge, First Edition. Edited by Simon J. Dadson, Dustin E. Garrick, Edmund C. Penning-Rowsell, Jim W. Hall, Rob Hope, and Jocelyne Hughes.

evolved. There is now broad consensus that – at the global level, at least – the investment gap can be bridged only if public finance is augmented by capital flows from the private sector. The question as to how to attract private capital flows into water infrastructure projects has also been well explored in both the academic and practitioner literature (Estache et al. 2015; Gatti and Della Croce 2014; Woetzel et al. 2017), but to date remains inadequately answered. The reasons for this are both nuanced and context-sensitive, but one word – *bankability* – offers a synopsis of the problem. The term is commonly used by investors and project developers to describe the level of risk associated with achieving the anticipated financial return on a project. Where these risks are high, project bankability is low, and vice versa. The drivers of bankability vary from one project to another, but to generalize, the challenges can be grouped around securing a sufficiently attractive rate of return on a project to secure investment; and, around mitigating the many risks that a project might fail.

The rate of return that water infrastructure assets can sustainably generate depends on the income associated with those assets. Sources of income are frequently classified (Winpenny et al. 2009) as tariffs (user charges), taxes (government subsidies) and transfers (such as development assistance). There is extensive literature (see e.g. Bakker 2007; Gleick 1998 for an excellent overview) that discusses water pricing, the value of water, and the human right to water; and the objective here is not to engage in those debates. Rather, it is simply to highlight that for many reasons, the water supply and service tariffs levied in many municipalities around the world do not cover the full economic cost of provision, particularly when operating, maintenance and replacement expenses are considered. As a result, water infrastructure projects often rely on public or private subsidies, and this can be a constraint on the achievable rate of return – and by extension, project bankability.

Meanwhile, infrastructure risks can be grouped into political, economic and technical categories (OECD and World Bank 2015). Political and regulatory risks generally arise from government actions, the behaviour of government contracting agencies, or broader uncertainty associated with the policy environment. Macroeconomic and business risks arise from volatility in economic variables such as inflation, interest rates and exchange rates, or shifts in the business cycle. Technical risks are related to the competence and skill required to manage the strategic and operating complexities of a project. Risks can also be classified in terms of a project's lifecycle; from the development phase, through to the construction, operational and termination phases. The impact of these risks on project viability obviously varies markedly and is a core determinant of bankability.

15.2 The Infrastructure Financing Challenge

Over 15 years have elapsed since the World Panel on Financing Water Infrastructure, chaired by Michel Camdessus, published a report describing the challenge (Camdessus 2003). It set out some of the core reasons why water infrastructure projects struggle to raise finance. First, most projects require a high upfront investment, which is then repaid through small instalments over a long payback period. This profile is not attractive to many investors. Second, the water sector generally offers a low rate of return on investments, as the water tariffs charged to consumers are usually regulated. Third, international investors face foreign exchange risk, as the returns on their investment are usually generated in local currency. Fourth, there is an execution risk to the project, as local developers may lack the financial, technical, or managerial capacity to oversee a complex project. Fifth, there is a risk of political pressure being placed on contracts and tariffs, whilst the regulatory framework may be weak or inconsistent. Sixth, investors may face a contractual risk, where projects that are long-term in nature have to be entered into with limited initial information. To mitigate these risks, the Camdessus report identified seven categories of actors who needed to be engaged: central governments from both developed and developing countries, subsovereign bodies, community organizations and non-governmental organizations (NGOs), banks, and private investors, aid donors, multilateral financial institutions, and members of the UN system and other international organizations. The report also set out four areas that needed to be addressed as a priority: (i) a requirement for host governments to engage in strategic planning for the water sector; (ii) that existing financial facilities should be reused, replenished and enhanced; (iii) that the evaluation of new schemes and opportunities should be fast-tracked; and (iv) that necessary policy changes and reforms should be expedited.

The water financing landscape described by the Camdessus report in 2003 is instantly recognizable today. The risk attributes of water projects are much the same, whilst the actors identified as change agents remain highly salient. The four priority areas still feature consistently in policy recommendation documents. Local water authorities continue to rely on subsovereign entities to support the financing and implementation of improvements to collective water services. Their presence at the local level makes them well-placed to understand challenges in context and to make decisions quickly, but their capacity to act is often constrained by a lack of access to funds and limited management skills. In terms of funding, limits on subsovereign borrowing are often imposed by central governments, who may be competing for funds from the same lenders. The mitigation policies recommended by the Camdessus report include the development of domestic borrowing markets for subsovereigns; the introduction of local ratings agencies; the use of specialized financial institutions as intermediaries; and the creation of joint-liability credit pools for subsovereigns.

In the academic literature, questions around financing have typically been situated within broader debates on infrastructure, such the relative merits between open access versus private control. Economic arguments have been used to justify a commons management approach (Frischmann 2006) based on the value created for the demand-side (i.e. consumers). These arguments have gained traction amongst some practitioners, particularly in the aftermath of the financial crisis. Other arguments on consumer-oriented value from infrastructure have emphasized resource efficiency and service-focused models of delivery (Roelich et al. 2015), proposing the benefits of multi-utility service companies, for example. Meanwhile there is well developed vein of research on infrastructure asset management, such as work on mapping stakeholder requirements to the value provided by the asset (Srinivasan and Parlikad 2017), and developing system-of-systems analysis of national infrastructure (Hall et al. 2013). In terms of water infrastructure specifically, efficiency and optimization variously feature in the literature, for example on water metering (Arregui et al. 2012), appliances (Carragher et al. 2012), and municipal supply (Furlong and Bakker 2010). The policy and development implications of financing water infrastructure have also been of particular interest to scholars (Rouse 2014; Tortajada 2014, 2016). Meanwhile within the extensive literature on climate change, infrastructure and finance feature across many disciplines, with analysis that spans the built environment (Gill et al. 2007), water resource management (Vorosmarty et al. 2000), investment strategy (Hallegatte 2009) and natural capital (Helm 2015).

In summary, financing water infrastructure has been part of the academic and practitioner literature for more than two decades. For practitioners, attention has been directed towards how to improve creditworthiness and bankability, and on how to access and attract fresh sources of capital into the sector. Within academic research, the topic has been situated more broadly, to include the public good attributes of infrastructure. In addition, several policy-facing recommendations have been made by practitioners and academics alike, and many have emphasized improvements to the enabling environment: better governance, higher water tariffs, using public funds to mobilize private sector investment, and so on. And whilst there has been some progress, the financing gap remains daunting.

This should not be surprising – bankability involves some intractable challenges, as discussed – but nor should the current state of affairs necessarily be considered either inevitable or acceptable. The framework advanced in this chapter builds directly on the breadth of literature that precedes it and emphasizes an incremental contribution to closing the gap. It draws in part on collective knowledge produced from the experience of financing infrastructure assets in other sectors such as renewable energy and telecommunications. When it comes to water infrastructure, however, the value or otherwise of the arguments presented herein rest less in the conceptual discussion, and more in the capacity for practical implementation.

15.3   Bridging the Gap

It is difficult to make a confident estimate as to the aggregate amount of investment needed in water infrastructure in order to keep pace with projected growth. The most cited data comes from the grey literature, such as the McKinsey Global Institute (MGI), who estimated in 2017 that an annual investment of US$ 500 billion in water infrastructure is required from 2017 through to 2035, representing an aggregate spend of US$ 9.1 trillion (Woetzel et al. 2017) based on a 'business as usual scenario', that is, simply keeping pace with economic growth. Most of the investment is needed in the emerging markets, where the financing challenge is particularly acute. The MGI noted that whilst most G20 countries cut back their spending on infrastructure during and after the global financial crisis of 2008, investment rates have subsequently picked up. For many smaller, developing economies, the lack of domestic savings constrains the capacity for endogenous responses, rendering investment flows highly sensitive to changes in global sentiment.

Given the quantum of investment required, there has been an increased focus from various quarters on how the effectiveness and efficiency of infrastructure investment can be improved. The MGI estimates that up to 38% of spending is not efficient, due to bottlenecks, lack of innovation and market failures. They propose that required spending could be reduced by more than US$ 1 trillion per year, for effectively the same amount of infrastructure delivered. To derive these numbers, the MGI 'diagnostically measured' the efficiency of infrastructure systems in 12 countries and extrapolated their analysis. The diagnostic measurement was based on an assessment of each country's infrastructure balance sheet; the effectiveness of their delivery systems; and the performance outcomes as measured by productivity, benchmarked to costs and international comparators. In measuring effectiveness, the analysis evaluated five areas: project selection, funding and finance, delivery, asset utilization and maintenance, and governance. These were broken down into subcategories, such as whether a country's infrastructure strategy is closely linked to its socioeconomic objectives; or whether the procurement, tendering, and contracting processes are sufficiently transparent.

We concur that these five areas are highly salient to bridging the infrastructure investment gap. For the rest of this chapter, we focus on funding and finance, for three reasons. First, the innovation-based solutions in this area are perhaps the least intuitive, whilst the problems can appear to be the most difficult to address. Second, we present an innovation framework that includes, at its core, a fundamental change to the financing paradigm. Third, we anchor our discussion to the catalytic capacity for infrastructure to help deliver the SDGs – provided that appropriate and sustainable financing mechanisms are in place.

15.4 Stakeholder Collaboration and the Constructive Corporation

Of the US$ 2.5–3 trillion invested in infrastructure each year, the private sector accounts for US$ 1–1.5 trillion (Bielenberg et al. 2016). This is split between institutional investors, who commit capital as part of a broader portfolio, and corporations, who invest infrastructure as part of their strategic initiatives. Institutional investors account for an estimated 30–40% of the total, whilst corporations account for the balance. Notwithstanding elements of overlap and double-counting (for example, where institutional investors have equity and debt holdings in corporations, who use these proceeds to invest in infrastructure), it is evident that corporations – accounting for the majority of private sector investment in infrastructure – are fundamentally important actors in the financing landscape. However, the capacity and possible motivations for private corporations to invest beyond their own direct requirements and help to bridge the infrastructure gap has not been explored in detail. In common with development banks and philanthropic foundations – and in contrast to many subsovereign authorities such as municipalities – private corporations often have good access to low-cost capital for investment in infrastructure. Corporations may also have unique access to manufactured capital through their local operational presence.

Models of stakeholder collaboration that align the interests of different actors are more likely to deliver the desired outcomes on a sustainable basis – but these models generally require a catalyst. The economic and social rationale for corporate engagement is well-established in the literature on water stewardship (e.g. Hepworth and Orr 2013; Jones et al. 2015; Orr and Sarni 2015; Sojamo and Larson 2012). We highlight the institutional capacity of private corporations to participate in multi-stakeholder relationships, and the experience of many companies who are involved in various alliances around water stewardship. Corporations operating in sectors such as foods, beverages and apparel – for whom improved water infrastructure is highly desirable, but for whom delivering this improvement is not their *raison d'être* – are increasingly motivated to forge new collaborative relationships. Momentum has come from: the rising perception of water scarcity as a business risk (World Economic Forum 2018); variability and uncertainty associated with climate change (Vorosmarty et al. 2000); regulatory pressure (Kagan et al. 2003); and the rapid growth of consumer markets in water-stressed regions.

Concurrent with these developments, management scholars have highlighted a progressive change in how many corporations perceive their purpose, evolving from shareholder value maximization to something more aligned to responsible citizenship. For example, if the 'purposeful corporation' (Mayer 2016) is to prosper over time, every company needs to not only deliver financial performance, but must also show how it makes a positive contribution to society. Without a sense of purpose, a company risks losing its licence to operate from key stakeholders. Support for this argument can be found in the growing importance that investors place on the environmental, social and governance attributes of the companies that they own. And whilst forms of collective action – or 'corporate water stewardship' – have existed for some time, the widening infrastructure gap has highlighted the limitations of the status quo.

15.5 Hybridity and Blended Finance

Blended finance is defined by the OECD as the 'strategic use of development finance for the mobilization of additional finance towards sustainable development in developing countries' (OECD 2018). This is a useful definition as it introduces 'additional finance' as private finance that does not have an explicit development purpose; and 'development finance' as both public and private finance that is being deployed with a development mandate. As the OECD states, this framing distinguishes finance by purpose rather than by source, and highlights blending in terms of development and commercial finance, rather than public and private actors. Interest in blended finance appears to be growing strongly. According to the OECD, between 2000 and 2016 donor governments set up 167 dedicated facilities that pool public financing for blending, and the number of new facilities grew every year.

Meanwhile impact investment can be defined as investments made into companies, organizations and funds with the intention of generating an economic and social impact alongside a financial return. Impact investments can be made in both developing and developed countries, and can target a range of returns from below-market to market rate, depending on the investors' strategic objectives. An impact investor may be willing to accept a financial return that is lower than what they would expect to get from other investment opportunities in the market, because the economic and social impact associated with this investment is sufficient compensation for this. Impact investment is a growth area: a biennial review of investment strategies (Global Sustainable Investment Alliance 2016) indicates that funds with responsible investment strategies (a proxy for interest in impact investment) managed US$ 22.9 trillion of assets in 2016 – an increase of 25% from 2014.

By combining development finance with institutional investment, capital can be secured at a lower cost, making it a viable source of funding for projects that generate a lower financial return, but also produce a positive economic and social impact. However, for this funding to be unlocked, the appropriate enabling conditions need to be created. One solution could be the establishment of a special purpose vehicle (SPV), where the proceeds from any capital raising can be held, before being disbursed. The SPV could also facilitate the payment of investment returns, such as interest and capital repayment.

In order to access the capital markets, a borrower needs to demonstrate their creditworthiness. One of the biggest impediments at present to financing water infrastructure projects is that many of the project sponsors – such as municipalities or other subsovereign entities – are not deemed to be a good credit risk. This may be despite many of the projects themselves being intrinsically bankable. After all, the willingness to pay for improved water supply and services amongst even the least affluent communities, subject to affordability constraints, is established in the literature. However, projects rely on their sponsors for funding. The creditworthiness of a publicly owned water utility company in a country with a poor record of servicing sovereign debt may be higher if that utility was privately owned. By creating an SPV and injecting that vehicle with financial and other capitals secured through the stakeholder collaboration model described previously, we propose that a creditworthy entity could be established with direct access to the capital markets.

We identify three different types of fixed-income investors, but in practice there are a broad spectrum of investors in the market, targeting a range of risk and return objectives. We also distinguish between fixed-income (bonds and other debt instruments) and equity investments for simplicity. Here, we define impact investors as those most willing to accept a financial return that is below the market rate, provided the social or environmental impact associated with the investment meets their criteria. Opportunistic investors are defined as those who are interested in impact, but to a lesser extent, and therefore require a higher rate of return than pure impact investors, although still below the market rate. Mainstream investors are defined as those who are not explicitly focused on impact, but who target a market rate of return, adjusted for the associated risk. That is, they demand a higher return for investments that they deem high risk and will accept a lower return on investments that they consider to be low risk. Mainstream investors account for the vast majority of capital that is available for deployment. Their engagement is desirable in the short term, but vital in the long term. Whilst it may be possible to raise funds for a limited number of projects simply by relying on impact investors, the scale of the funding challenge means that unless mainstream investment capital is mobilized, then this approach can make a marginal contribution, at best, to bridging the funding gap.

By providing a differentiated proposition to impact, opportunist and mainstream investors, it is possible to raise investment funds from all three investor types at below-market rates, provided that there are investable projects available that meet the impact criteria. Investment commitments from impact investors reduce the financial risk for opportunistic investors, for any given level of return. Impact investors accept the risk of first loss – another way of describing a below market rate of return – providing some protection to opportunist investors. Similarly, investment is de-risked for mainstream investors, because opportunist investors have accepted the risk of 'second loss'; that is, losses that go beyond the capacity of impact investors. As further losses are less likely to be manifest, the risk of bearing them will be lower, provided first and second losses have been covered by other investors.

Structuring an investment proposal in this format requires time and preparation, as for each tranche of investors, contractual agreement needs to be reached on performance benchmarks and loss acceptance. However, in most financial markets there are various instruments in use that have been designed for similar purposes of risk attribution, and so some standards exist that can be adapted for purpose. Of note is that many developing countries with large infrastructure finance gaps also have well-established and sophisticated capital markets, with the technical and human capacity to introduce instruments of this type. We describe the tranches as senior, mezzanine and junior, consistent with the language of the debt capital markets. Senior debt is the lowest risk and must be repaid first. Mezzanine and junior debt are subordinate and incorporate a progressively higher rate of risk.

15.6 Blended Returns on Investments in Infrastructure

With funds available for investment, it is necessary to qualify which projects meet the criteria of investors. This may involve using established standards similar to those developed by the Climate Bonds Initiative (CBI), which consists of a certification process that is guided by a taxonomy of qualifying projects. To date, the most sophisticated standards have been developed for projects that deliver reductions in net greenhouse gas (GHG) emissions. However, this is a nascent space, and new standards are being developed that target a wider range of objectives than reduced emissions. In early 2018 the CBI set out the requirements that water infrastructure projects need to meet in order to be eligible for inclusion in a certified climate bond. The criteria cover both built and nature-based infrastructure, and are focused on GHGs, with mitigation, adaptation and resilience components. It is likely that further standards for water infrastructure assets will emerge that cover a broader set of impact criteria, including economic, social and environmental returns on investment (ROI), in addition to financial performance.

The point for emphasis is that different projects will have different return attributes, as benchmarked against these criteria. At the portfolio level (i.e. when assessing the overall impact of several projects), it is the aggregate return on investment that is the most salient data. So, it may be that one project has a high social return on investment – measured, for example against metrics of health, wellbeing, education or gender equality – but has a low financial return on investment. Meanwhile another project may offer high financial returns but contributes less to the other impact criteria. Both projects may be investable, if the aggregate ROI meets the threshold required by the investors. This model of diversification to capture multiple returns takes its inspiration from modern portfolio theory (Elton et al. 2014), which proposes that the risks associated from holding a single stock can be reduced by holding multiple stocks, whose performance are not closely correlated to each other. The analogy is illustrative rather than exact, not least because the methodologies to measure the impact performance of water infrastructure assets are still fairly undeveloped. However, it serves to highlight the benefits to investors of financing a diversified basket of projects, both from a reduced risk and optimal return basis.

The portfolio approach embeds flexibility in project selection at various levels. Both publicly and privately owned projects are in scope, as are built ('grey') or natural ('green') capital projects. Impact is measured in terms of outcomes, and emphasizes the service delivered by infrastructure, rather than the infrastructure asset itself; so projects that focus on models for maintenance or rehabilitation of existing infrastructure could also be eligible for investment. Project selection may be optimized to factor in the political, regulatory, macroeconomic or business environment of specific regions, countries or cities. The framework is designed to be agnostic as to which projects are investment candidates, provided they meet the criteria.

Financial ROI is highlighted because it is important to reiterate that this model for financing water infrastructure is not a charitable endeavour. Other, simpler platforms already exist for that purpose. Rather, this a model to generate financing for water infrastructure from a diversified set of investors, at the lowest sustainable cost of capital possible, across the broadest range of feasible projects. In addition to economic, social and environmental impact, investors will require the portfolio of projects to deliver some level of financial return, even if it is below market rates.

One way in which this could be structured involves the investors getting their principal – or initial capital – repaid to them over the duration of the bond, but the effective interest that is earned on that principal may be negligible or even negative. By way of illustration, consider a 'vanilla' (i.e. generic) bond for US$ 100 000 that matures in 20 years and pays a fixed coupon of US$8000 at the end of each year. If the principal was then repaid at the end of the period, then in nominal terms (i.e. not adjusted for inflation), the bond has an 8% yield.

An alternative arrangement could be where, rather than the principal being repaid in full at the end of the period, instead it was being paid down in regular instalments as part of the coupon. Such instruments, often called amortising bonds, are similar to the configuration of many residential mortgages, and reduce the credit risk of the loan because it is repaid over time, rather than as a lump sum on maturity.

Assuming the same amount of borrowing and coupon payments as described for the vanilla bond above, in nominal terms an amortising bond would result in the original investment being recovered and interest income of US$ 60 000 being received. This might appear to imply an interest rate of 3% over the period, which could still be an attractive return on a risk-adjusted basis. However, this is not an appropriate calculation, because it ignores the time value of money. The coupon received at the end of the first year is worth more to the investor than the same coupon received at the end of the 20th year, because it can be reinvested for the intervening 19 years. To adjust for this, we can calculate an internal rate of return (IRR), which is the discount rate that makes the net present value of all cash flows, whenever they were received, equal to zero. The calculation is iterative, and based on the illustration above, the IRR of this investment is in fact negative, at around −2.1%. To achieve an IRR of zero, the annual coupon would need to be more than $9000. To achieve an IRR of 3%, the annual coupon would be nearly $11 000. And to achieve an 8% IRR, the annual coupon on an amortising bond of US $100 000 would need to be $14 000.

As this simple example demonstrates, internal rates of return are highly sensitive to the amount and timing of cash flows. A core proposition of this model is that it is attractive to a heterogeneous set of investors who are willing to accept different rates of financial return, and to be compensated for this by the portfolio delivering impact performance.

15.7 Water Infrastructure Portfolio Management

We have described thus far the models of stakeholder collaboration required to mobilize the capitals to create a creditworthy SPV, that can raise finance from impact, opportunist and mainstream investors, and blend this commercial finance with development finance to unlock capital for investment in water infrastructure projects that deliver a combination of economic, social, environmental and financial returns, that are consistent with the requirements of those investors. To implement this framework in practice, a portfolio management layer is necessary.

Portfolio management in the context of multiple and discrete projects can be defined as the selection, prioritization and control of projects and programmes in line with the manager's strategic objectives and their capacity to deliver. The responsibility of the portfolio manager includes assessing whether the right projects are being selected to deliver the strategic investment objectives, subject to risk, resource constraints and affordability. Other considerations include assessing whether project managers are delivering these objectives effectively and efficiently; and whether the full potential benefits of the investment are being realised. The benefits of a portfolio approach include maintaining a balanced and strategically aligned portfolio in the context of changing conditions; and improving the returns from projects through a portfolio-wide view of risk, dependencies and scheduling. A clearly articulated strategy, along with a robust governance structure, helps to provide the capacity and commitment that is necessary for the portfolio manager to deliver against investment objectives.

The infrastructure portfolio manager is responsible for selecting the projects for investment, and performs four key roles. Firstly, projects need to be selected and then optimized against their risk and return attributes. The purpose of the optimization is to select a portfolio of projects that are suitably diversified both to lower risk (e.g. by spreading the investment across a number of projects) and to enhance return (e.g. by choosing projects that are expected to deliver a combination of economic, social and environmental impacts, as well as financial performance). After projects have been selected, their performance against these benchmarks of return need to be measured – the second functional role of the manager. Whilst financial return on investment is relatively straightforward, measuring economic, social or environmental returns on a consistent basis across different projects can be more problematic. Various methodological approaches are being developed to try and bring consistency and robustness to this process. Thirdly, the portfolio manager needs to ensure that projects maintain compliance with any guidelines or conditions of investment. Sanctions for non-compliance will vary between projects, but might include delaying disbursements, or even disinvesting from a project. Finally, the portfolio manager is required to intermediate the performance information from the range of projects into a consolidated format that is meaningful and relevant to investors, and then report this information on a regular basis.

15.8 Hybrid Income

To generate return sustainably over a multi-year investment timeframe, the portfolio must generate income, both to meet the operating and maintenance expenditures required, as well as to contribute to overall financial performance. The OECD defines three basic sources of revenue available to water supply and sanitation: tariffs, taxes and transfers (the 3Ts). It notes that most developing countries tend to draw heavily on transfers from overseas development assistance and philanthropy, whilst in developed countries revenues are more usually raised from tariffs, along with earmarked taxes. Developing a cost recovery strategy requires an appropriate combination of these sources of revenue. The analysis is typically conducted at a country-level scale. Our framework proposes combining these revenue sources at the scale of a diversified project portfolio that may span multiple countries, and embeds the capacity for operating and maintenance expenditure for some projects to be cross-subsidized at the portfolio management level, from the net income generated by other projects. In addition to tariffs, taxes and transfers, we introduce a fourth element of transactions.

Transactions become relevant in circumstances where assets in a portfolio are acquired by third parties. For example, a syndicate of private companies might acquire a wastewater treatment plant to direct further investment. A municipality might acquire grey (built) assets to meet rapid growth in user demand. A water fund might acquire natural capital assets such as wetlands as part of its development plans. There may be acquisition interest from financial investors seeking stable returns. It may also be that third parties acquire stakes in projects, rather than take full ownership. Various permutations are possible, and whilst it is impossible to be definitive on these outcomes in the context of this paper, there are many precedents in the sector where ownership has transferred during the life of the assets.

We suggest that the counterparty to such transactions would be the portfolio manager, who has a fiduciary duty to the investors in the fund. Any decision would need to consider the implications to the risk and return attributes of the entire portfolio, rather than being simply about the specific asset. The manager would need to evaluate how the transaction would affect both financial and non-financial performance of the residual portfolio. If this evaluation was effective in practice, it raises the prospect of a more integrated approach to the management of outcomes. That could help align decision-making more closely with the issues around project selection, delivery, asset utilization and maintenance, and governance, discussed earlier. From an income perspective, a transaction involves an injection of capital, which could be deployed in various ways. For instance, the portfolio manager could reinvest in other projects. Alternatively, the funds could be transferred back to the SPV, and used to enhance the returns paid to investors. Or, the funds could be used as shareholder capital within the SPV, allowing part of the existing capital to be returned to the original providers. Equally, all of these methods might be deployed, depending on conditions precedent. In summary, we propose an enhancement to the 3Ts framework that incorporates transaction activity at the portfolio level.

15.9 Synthesis

A representation of the integrated framework proposed herein comprises the five layers, as described in this chapter. A singular feature of this model is that the functional attributes of the underlying components are already well developed and applied in various contexts. For example, collaborations involving public–private partnerships are a staple of infrastructure finance, as are instruments that reallocate risk and return to suit investor preferences. The management layer is based on the basic attributes of project portfolio management, whilst diversification and divestment are widely used strategies to enhance risk-adjusted returns. The framework proposed is therefore less about execution at the component level, and more about hybridity through bringing together established practices in new ways. That said, we believe that there are areas in which the application of this model could yield fresh and important insights, both to the literature and to the field. For example, at the impact layer, there is still much work to be done around measuring and managing various forms of impact, particularly on a comparative basis.

15.10 Scaling the Model

The financing challenge will only be met by solutions that work at scale. In practice, this means that models need to be adaptive to different social, economic, political, institutional and regulatory frameworks.

The core elements required to deliver scale in this model are innovation, competition and diversification. Innovation at the financing level requires flexibility in terms of how different stakeholders collaborate; the capitals that are committed to the SPVs; the blended finance instruments used; the risk and return objectives that are established; the measurements of impact that are applied; and so on. There are many components and subcomponents to these relationships, and the different configurations are reflected through the range of SPVs that may be established. It may be that there are single, country-level SPVs; several SPVs established within a country; or a single SPV operating across a number of countries.

In terms of the competition layer, for this model to operate at scale, good execution at the project management level is critical. A broad range of skills and competencies is required to align different investor objectives with projects that deliver the necessary economic, social, environmental and financial ROI. Engaging a selection of portfolio managers with complementary domain expertise would significantly improve execution capability, particularly as the universe of potential projects expands. This approach also reduces the risk of rent seeking, as managers operate in a competitive environment where their performance against investment objectives can be measured and benchmarked. Managers that consistently perform better than the benchmark would probably receive a greater share of the investment pool to manage, whilst consistently underperforming managers would likely see their share of the pool being reduced. This approach to performance measurement, aligned with appropriate incentives, is commonly used in the investment management industry. It also offers the prospect of improved transparency and governance at the project level.

The diversification layer simply reflects the reality that water supply and sanitation is complex. The literature is replete with examples of this complexity, but it will suffice to note here that the financing challenges are local, context-dependent, and sensitive to the policy environment. For many reasons, including how water is priced, it is difficult to conceive of a templated set of water projects that can be repurposed for widely differing contexts. By contrast, in the renewable energy sector, it is exactly this sort of scalable replication that is contributing to lowered costs and accelerated rollout. However, by acknowledging from the outset that water infrastructure projects will likely be bespoke, we think that it should still be possible to develop diversified portfolios where the complementarity of projects can be optimized to generate strategically determined outcomes.

15.11 Conclusion

This chapter has framed the water infrastructure financing challenge within a historic context and has focused on efforts that have been made to bridge the gap between current rates of investment, and the investment needed to ensure that infrastructure is fit for purpose in the twenty-first century. A framework for financing water infrastructure that places the private corporation as a core stakeholder and change agent was presented here. Frameworks of this sort can help address the fundamental challenge of project bankability, by tapping into recent innovations in the financial markets, a resurgence of interest in blending finance, and the rapid growth of impact investment funds. In applying a blended approach to project selection as well as financing, we have described how a portfolio of projects can deliver economic, social, environmental and financial returns that are consistent with the requirements of mainstream and impact investors. The model relies on effective execution through portfolio management, and on developing sources of income that have not traditionally been associated with the sector. The functional attributes of the underlying components in the model are already well developed and applied in various contexts, and current research is focused on applying the overall framework at a country level. A key consideration is scalability, and this requires innovation, competition and diversification.

References

Arregui, F.J., Soriano, J., Cabrera, E. et al. (2012). Nine steps towards a better water meter management. *Water Science and Technology* 65 (7): 1273–1280. https://doi.org/10.2166/wst.2012.009.

Bakker, K. (2007). The "commons" versus the "commodity": alter-globalization, anti-privatization and the human right to water in the global south. *Antipode* 39 (3): 430–455. https://doi.org/10.1111/j.1467-8330.2007.00534.x.

Bielenberg, A., Kerlin, M., Oppenheim, J., and Roberts, M. (2016). *Financing change: How to mobilize private-sector financing for sustainable infrastructure*. Working Paper, McKinsey Center for Business and Environment (March), 29–35.

Briscoe, J. (1999). The changing face of water infrastructure financing in developing countries. *International Journal of Water Resources Development* 15 (3): 301–308. https://doi.org/10.1080/07900629948826.

Camdessus, M. (2003). Financing water for all excerpts from the executive summary report of the world panel on financing water infrastructure. *Water Resources* 5 (4): 17–19. https://www.jstor.org/stable/wateresoimpa.5.4.0017.

Carragher, B.J., Stewart, R.A., and Beal, C.D. (2012). Quantifying the influence of residential water appliance efficiency on average day diurnal demand patterns at an end use level: a precursor to optimised water service infrastructure planning. *Resources, Conservation and Recycling* 62: 81–90. https://doi.org/10.1016/j.resconrec.2012.02.008.

Elton, E.J., Gruber, M., Brown, S.J., and Goetzmann, N. (2014). *Modern Portfolio Theory and Investment Analysis*, 9e. Hoboken, NJ: Wiley.

Estache, A., Serebrisky, T., and Wren-Lewis, L. (2015). Financing infrastructure in developing countries. *Oxford Review of Economic Policy* 31 (3–4): 279–304. https://doi.org/10.1093/oxrep/grv037.

Frischmann, B.M. (2006). An economic theory of infrastructure and commons management. *Minnesota Law Review* 89: 917–1030. https://doi.org/10.1093/acprof:oso/9780199895656.003.0002.

Furlong, K. and Bakker, K. (2010). The contradictions in "alternative" service delivery: governance, business models, and sustainability in municipal water supply. *Environment and Planning C: Government and Policy* 28 (2): 349–368. https://doi.org/10.1068/c09122.

Garrick, D.E., Hall, J.W., Dobson, A. et al. (2017). Valuing water for sustainable development. *Science* 358 (6366): 1003–1005. https://doi.org/10.1126/science.aao4942.

Gatti, S. and Della Croce, R. (2014). Financing infrastructure – international trends. *OECD Journal: Financial Market Trends* 2014 (1): 123–138. https://doi.org/10.1787/19952872.

Gill, S.E., Handley, J.F., Ennos, A.R., and Pauleit, S. (2007). Adapting cities for climate change: the role of the green infrastructure. *Built Environment* 33 (1): 115–133. https://doi.org/10.2148/benv.33.1.115.

Gleick, P.H. (1998). The human right to water. *Water Policy* 1 (5): 487–503. https://doi.org/10.1016/S1366-7017(99)00008-2.

Global Sustainable Investment Alliance (2016). *2016 Global Sustainable Investment Review 1*. http://www.gsi-alliance.org/wp-content/uploads/2017/03/GSIR_Review2016.F.pdf (accessed 28 September 2018).

Hall, J.W., Henriques, J.J., Hickford, J., and Nicholls, R.J. (2013). Systems-of-systems analysis of national infrastructure. *Proceedings of the Institution of Civil Engineers – Engineering Sustainability* 166 (5): 249–257. https://doi.org/10.1680/ensu.12.00028.

Hallegatte, S. (2009). Strategies to adapt to an uncertain climate change. *Global Environmental Change* 19 (2): 240–247. https://doi.org/10.1016/j.gloenvcha.2008.12.003.

Helm, D. (2015). *Natural Capital: Valuing the Planet*. Yale: Yale University Press.

Hepworth, N. and Orr, S. (2013). Corporate water stewardship: exploring private sector engagement in water security. In: *Water Security: Principles, Perspectives, and Practices* (eds. B.A. Lankford, M. Bakker, M. Zeitoun and D. Conway), 220–238. London: Earthscan from Routledge.

Jones, P., Hillier, D., and Comfort, D. (2015). Corporate water stewardship. *Journal of Environmental Studies and Sciences* 5 (3): 272–276. https://doi.org/10.1007/s13412-015-0255-7.

Kagan, R.A., Gunningham, N., and Thornton, D. (2003). Explaining corporate environmental performance: how does regulation matter? *Law & Society Review* 37 (1): 51–90.

Mayer, C. (2016). Reinventing the corporation. *Journal of the British Academy* 4: 53–72. https://doi.org/10.5871/jba/004.053.

OECD (2018). *Making Blended Finance Work for the Sustainable Development Goals*. Paris: OECD Publishing. doi: 10.1787/9789264288768-en.

OECD and World Bank (2015). *Risk and Return Characteristics of Infrastructure Investment in Low Income Countries*. https://www.oecd.org/g20/topics/development/Report-on-Risk-and-Return-Characteristics-of-Infrastructure-Investment-in-Low-Income-Countries.pdf.

Orr, S. and Sarni, W. (2015). Does the concept of "creating shared value" hold water? *Journal of Business Strategy* 36 (3): 18–29. https://doi.org/10.1108/JBS-10-2013-0098.

Roelich, K., Knoeri, C., Steinberger, J.K. et al. (2015). Towards resource-efficient and service-oriented integrated infrastructure operation. *Technological Forecasting and Social Change* 92: 40–52. https://doi.org/10.1016/j.techfore.2014.11.008.

Rouse, M. (2014). The worldwide urban water and wastewater infrastructure challenge. *International Journal of Water Resources Development* 30 (1): 20–27. https://doi.org/10.1080/07900627.2014.882203.

Sadoff, C.W., Hall, J.W., Grey, D. et al. (2015). *Securing Water, Sustaining Growth: Report of the GWP/OECD Task Force on Water Security and Sustainable Growth*. Oxford: University of Oxford.

Sojamo, S. and Larson, E.A. (2012). Investigating food and agribusiness corporations as global water security, management and governance agents. *Water Alternatives* 5 (3): 619–635.

Srinivasan, R. and Parlikad, A.K. (2017). An approach to value-based infrastructure asset management. *Infrastructure Asset Management* 4 (3): 87–95. https://doi.org/10.1680/jinam.17.00003.

Tortajada, C. (2014). Water infrastructure as an essential element for human development. *International Journal of Water Resources Development* 30 (1): 8–19. https://doi.org/10.1080/07900627.2014.888636.

Tortajada, C. (2016). Policy dimensions of development and financing of water infrastructure: the cases of China and India. *Environmental Science and Policy* 64: 177–187. https://doi.org/10.1016/j.envsci.2016.07.001.

Vorosmarty, C.J., Green, P., Salisbury, J., and Lammers, R.B. (2000). Global water resources: vulnerability from climate change and population growth. *Science* 289 (5477): 284–288. https://doi.org/10.1126/science.289.5477.284.

Ward, F.A. (2010). Financing irrigation water management and infrastructure: a review. *International Journal of Water Resources Development* 26 (3): 321–349. https://doi.org/10.1080/07900627.2010.489308.

Winpenny, J., Jacobson, M., and Buhl-Nielsen, E. (2009). *Strategic Financial Planning for Water Supply and Sanitation*: *A report from the OECD Task Team on sustainable financing to ensure affordable access to water supply and sanitation*. Paris: OECD Publishing. https://www.oecd.org/env/resources/43949580.pdf.

Woetzel, J., Garemo, N., Mischke, J. et al. (2017). *Bridging Infrastructure Gaps: Has the World made Progress?* Discussion Paper. McKinsey Global Institute.

World Economic Forum (2018). *The Global Risks Report 2018*. Geneva: World Economic Forum.

16

Wastewater

From a Toxin to a Valuable Resource

David W.M. Johnstone[1], Saskia Nowicki[1], Abishek S. Narayan[2], and Ranu Sinha[1]

[1] *School of Geography and the Environment, University of Oxford, UK*
[2] *Aquatic Research, Swiss Federal Institute of Science and Technology (EAWAG), Zurich, Switzerland*

16.1   Introduction

This chapter discusses the development of wastewater treatment from the late 19th century when adequate treatment did not exist and wastewater was a major hazard to both public health and the environment. It uses Great Britain as a case study and charts the advancement of knowledge and process development up to the present day when all sorts of processes can be used to produce every possible quality of effluent including those for recycling. It traces the change in perception of wastewater from something toxic and nasty to one of a valuable resource from which materials can be recovered and energy generated. It also follows the rather tortuous development of the legal and institutional changes that made this development possible. It ends with a very brief discussion of the problems in developing countries where there are very few treatment facilities and gross pollution prevails.

16.2   The Early Formative Years

From the Middle Ages until the early part of the nineteenth century the streets of European cities were foul with excrement and filth to the extent that people often held a clove-studded orange to their nostrils in order to tolerate the atmosphere. The introduction of water-based sewerage systems merely transferred the filth from the streets to the rivers. The problem intensified, especially in Britain, by the coming of the Industrial Revolution and the establishment of factories along the banks of rivers where water was freely available for power, process manufacturing and disposal of effluents. This was accompanied by massive increases in urban populations needed to meet the demand of industrial labour, bringing with them large volumes of domestic sewage (Klein 1957). As a consequence, the quality of many rivers deteriorated to the extent that they were essentially no more than open sewers, aquatic life died, and water supplies and public health were placed in jeopardy. This is clearly illustrated by a series of cholera

Water Science, Policy, and Management: A Global Challenge, First Edition. Edited by Simon J. Dadson, Dustin E. Garrick, Edmund C. Penning-Rowsell, Jim W. Hall, Rob Hope, and Jocelyne Hughes.
© 2020 John Wiley & Sons Ltd. Published 2020 by John Wiley & Sons Ltd.

outbreaks, not only in Britain but all over Europe, the first in 1832–1833. That these were directly caused by contaminated drinking water was famously established in 1853/1854 in London by Dr John Snow who identified the water pump in Broad Street, Soho, as delivering sewage-contaminated water from the River Thames.

Dr William Budd described the situation in London during the hot summer of 1858, which has come to be known as the year of 'the Great Stink' (Klein 1957).

> For the first time in the history of man, the sewage of three million people had been brought to seethe and ferment under a burning sun, in one vast open cloaca lying in their midst. The result we all know… stench so foul had never ascended to pollute this lower air. For many weeks parliamentary committee rooms were rendered barely tolerable by suspension of blinds saturated with chloride of lime, and other disinfectants.

Even the construction of 132 km of interceptor sewers and 1800 km of trunk sewers by Joseph Bazelgette between 1858 and 1865 only transferred the waste outside the centre of the city; it did little at that time to improve the overall health risks. Conditions in other growing urban areas of the country and in parts of Europe were similar.

Finally, the government appointed two Royal Commissions on River Pollution, one in 1865 and one in 1868, to study and report on the problems. Both reported extensively on the shocking state of the country's rivers and a report of the second commission stated that 'of the many polluting liquids which now poison the rivers there is not one which cannot be either kept out of streams altogether, or so far purified before admission to deprive it of its noxious character'. The shocking reports of the two Commissions started to awaken public conscience and stir government into legislative action that resulted in the passage of the Public Health Act in 1875 and the Rivers Pollution Prevention Act 1876.

The 1875 Act is regarded as one of the most important sanitary reports of all time. It clearly recognized for the first time that care of public health was a national responsibility and established a system of local health administration setting down, amongst other things, the duties of local authorities with respect to the collection, disposal and treatment of sewage.

The 1876 Act formed the basis of all legal action connected with river pollution up until 1951. Two key offences were specified.

- Part I of the Act made it an offence to put solid matter into a stream, but it was necessary to prove that either pollution or interference with flow was caused.
- Part II of the Act prohibited the discharge of solid or liquid sewage matter into a river and it was no defence to argue that the river had already been polluted by sewage upstream.

Together these two Acts formed the legal and administrative basis of future developments. The 1875 Act resolved a major problem that beset progress in the later part of the nineteenth century. Up until then the only regular form of sewage treatment was by application to dedicated farms, which is very land-intensive, and there was considerable conflict between central and local government due to the reluctance of local government to occupy vast areas of development land with sewage farms; local government would not accept responsibility for sewage disposal. The 1875 Act solved that problem by specifying unambiguously that responsibility for treating sewage did indeed lie with local authorities, a situation that lasted up until 1973 with the passage of the 1973 Water Act that created ten Regional Water Authorities (RWAs) in England and Wales based on natural catchment boundaries and not political boundaries. The ten RWAs came into being on 1 April 1974. The RWA concept did not apply to Scotland where responsibilities were transferred to nine new Regional Councils in 1975.

Soon after the passage of the 1875/1876 Acts there was a particular legal problem that delayed any environmental improvement. The local governments that had become legally responsible for administrating the law were also the principal sources of sewage pollution. As a result, the government set up a number of river authorities and boards to administer the 1876 Act.

On the technical front, both Acts can be considered to have been before their time, as their ambitious targets were not achievable with the technology then available. However, they did stimulate research in treatment to add to the already established basic knowledge. As an alternative to land treatment there was a growing trend to use 'mechanical' methods. In 1898 Cameron and Cummins patented the 'septic tank', still often used in small treatment plants. Chemical precipitation methods were also developed during this period.

Much of the research was based on filtration aimed to simulate land treatment with a growing recognition that 'organisms' were responsible for degradation of the sewage. In 1882 Warrington wrote that 'sewage contains the organisms for its own destruction, and these may be so cultivated to effect the purpose'. And in 1887 William Dibdin stated that:

> In all probability the true way of purifying sewage will be first to separate out the sludge, and then turn into the neutral effluent a charge of the proper organism, *whatever that may be*, especially cultivated for the purpose, retain it for a sufficient period, during which time it should be fully aerated...

This is essentially the first statement of what would become the basic primary and secondary treatment method. However, despite this increasing knowledge, nothing much happened to put the knowledge into action, so the Government commissioned yet another Royal Commission which turned out to be the most critical, influential and effective study ever to be carried out on the subject. Many of the findings are still with us today, not only in the UK but internationally.

The Royal Commission on Sewage Disposal commenced work in May 1898 and completed the last of its nine reports 17 years later in 1915. The scope of the study was considerable, and the 8000 pages of the nine reports laid the foundations of much of today's wastewater practices. The eighth report (1912) introduced the BOD (biochemical oxygen demand) test which is still the most commonly used method for determining the strength of sewage and the effect that sewage would have on receiving watercourses. The fifth report (1908) introduced a rudimentary river quality classification based on oxygen absorption, but this was never passed into law. Perhaps the most famous outcome was the introduction of the 'Royal Commission Standards' of 20 mg/l BOD and 30 mg/l of suspended solids for treated sewage effluents based on a mass balance of river quality upstream and downstream of the discharge point, assuming an eight-fold dilution and an upstream BOD of 2 mg/l. These standards were used for decades as the norm for effluent quality and are still used today in many places. Even the basic European effluent standards for non-sensitive watercourses are not dissimilar; in fact, they are less strict.

In parallel to the work of the Royal Commission there was considerable research effort in various parts of the world. Ground-breaking research in Manchester resulted in the publication in April 1914 of the seminal paper on 'Activated Sludge' by Ardern and Lockett. Activated sludge and its variations form the bases for most modern suspended growth treatment processes, of which there are many, although there are other processes based on 'biological filtration' which had been applied in many towns and cities during the latter part of the nineteenth century.

16.3   Early Full-Scale Application and Process Development

Thus by 1914 there was, at least in Great Britain, a defined institutional and legal system, a rudimentary river quality classification with defined numerical effluent standards, and a basic knowledge of how to treat sewage. Under these circumstances, rapid application could have been expected, but that was not universally the case. Europe was to see the devastation of a major war that left little capital for sewage treatment well into the 1920s. The first British city to apply activated sludge was Sheffield in 1920, on a trial basis, but many municipalities had already invested in filtration plants and saw no reason to invest in the 'new-fangled activated sludge system'. As a consequence, the first very large treatment plants at Manchester and London (Mogden) were not completed until 1934 and 1935 (Cooper 2001). By contrast, development of activated sludge in the USA forged ahead in some places, but in others the municipalities decided to build filtration plants rather than pay steep royalties to UK companies that had taken out patents on activated sludge; hence the large-scale uptake was stalled until the 1940s when the patents expired (Schneider 2011). Although numerous activated sludge plants were built during the four or five decades following the initial discovery, much of the design was on a 'trial and error' basis and many of the early plants were unstable particularly with respect to secondary sludge settlement and nitrification.

16.4   The Age of Understanding

It was only from the early 1960s onwards that there were significant changes to our understanding of activated sludge using a more scientific bioengineering approach. The solution to the nitrification problem was discovered by Downing et al. (1964) with the elucidation of the kinetics of biological oxidation of ammoniacal nitrogen. Studies of aeration tank configuration using continuous flow reactor dynamics led to discoveries that different configurations produced different bacterial populations that had major effects on biomass settleability. The problem of poor settleability, known as 'bulking', had plagued the industry ever since the introduction of the continuous flow process. If uncontrolled, biomass solids can overflow into the receiving watercourse causing pollution and, at worst, can lead to the loss of the whole biomass and total system failure. Over the years 'bulking' had been attributed to overloading, underloading, over-aeration, under-aeration, short-circuiting, nutrient imbalance, high pH, low pH, high temperature, sewage septicity, and other causes (Tomlinson 1982). In fact, the real mechanisms were not understood until later when greater knowledge of the ecology of activated sludge and use of reactor dynamic studies showed that a 'completely-mixed' configuration produced poor settling sludge whilst 'plug-flow' configurations produced relatively good settling sludge. The simplistic answer is that each regime has a different effect on the generation of the filamentous bacteria that cause 'bulking'.

The biochemical engineering approach also led to the establishment of mathematical models that allowed process characteristics to be evaluated rapidly rather than waiting for the results of prolonged practical trials. This approach, along with the development of new monitoring devices, also led to the beginning of automatic control systems.

Not only had the understanding and knowledge of activated sludge systems been improved, but so also had that for so-called biological filters, a term which is a complete misnomer since these systems do not filter anything. They are biological treatment systems where the biomass is attached to the surface of a solid medium with a large surface area, originally coke or slag, although today there are many varieties that use plastic media. The biological reactions take place on a surface slime layer as the sewage percolates down through the bed, whilst air supplying oxygen travels upwards. These systems are the oldest of the constructed processes and pre-date activated sludge by decades. They are much in use today and are still referred to (wrongly) as biological filters. They have the advantage over activated sludge of being relatively low in energy consumption and are generally robust, but suffer the disadvantages of having a very large footprint, being rather inflexible, and performing poorly in very cold weather.

16.5   Some Important Legislative and Institutional Changes

At this juncture it is important to discuss some legal and institutional changes that influenced the development of wastewater treatment. Unfortunately, it is beyond the scope of this chapter to present the complete history of these changes, so only a few of the most significant are presented.

Paramount in establishment of all wastewater activities is the need to control industrial pollution, to which end the government passed the Public Health (Drainage of Trade Premises) Act 1937 that gave industry the right to discharge to a public sewer, subject to meeting conditions laid down by the appropriate public authority in a legal document giving 'consent' to discharge. This is essential to protect workers working in the sewer network, to protect the fabric of the sewer, and to protect downstream biological treatment plants and sludge disposal (see Johnstone 2003 for a more detailed account).

In 1948 a number of River Boards were formed to tackle the high levels of pollution that still permeated many rivers, with powers to set standards for rivers and effluent discharges. These powers were further strengthened under the Rivers (Prevention of Pollution) Act 1951, under which all new discharges had to have a 'consent' to discharge that specified both quantity and quality limits. This Act was further strengthened under the Rivers (Prevention of Pollution) Act 1961, and during the 1960s the River Boards were abolished and replaced with a smaller number of River Authorities with much wider powers. It should be pointed out that some of the legislation in Scotland was different from that in England and Wales, but the paths were similar.

These institutional and legal changes did drive improvements, but not enough, and sewage works of many Authorities failed to meet their designated 'consent' standards. Later the National Water Council introduced river water quality objectives (RQOs) based on the aquatic and anthropogenic requirements of a river, from which long-term targets were established. Effluent standards were then set to meet the relevant RQO, which in most cases led to a significant tightening of effluent quality. Another key legislative measure was the eventual enactment of the Control of Pollution Act 1974, with the establishment of Registers of Sewage Works Performance and easing of controls on prosecution. There then followed adoption of a number of European Directives such as those on Urban Wastewater (1991) and Bathing Water Quality (1992) that tightened regulation. More significantly, the EU introduced the Water Framework Directive (WFD) in 2000 that takes a more holistic approach to the aquatic environment by establishing basic management units within a river basin, which must address environmental, economic and social needs. A short but comprehensive history of the main long-term developments has been published by Woods (2003), whilst the Department for Environment, Food and Rural Affairs (DEFRA) presents an account of how the UK dealt with the European Directives (DEFRA 2002), and Chave (2001) gives an account of the WFD. All-in-all, this progressive drive to improve the aquatic environment with accompanying regulation did much to drive the development of better processes and promote enhanced resilience.

Two other institutional developments that led to significant increases in knowledge and understanding are worthy of mention. The first was the establishment of the Water Pollution Research Laboratory (WPRL), originally set up in 1927 by a Government that was 'increasingly concerned' with the problems of river pollution and its adverse effect on the supply of pure water for a growing population and industry. It moved to its own laboratory in 1947, and it was here that much of the fundamental research was carried out, including the aforementioned work on nitrification, mathematical modelling, 'bulking' and on the understanding of 'filter' beds.

The second event was the creation in 1974 of the ten RWAs in England and Wales based on managing the complete aquatic environment contained within natural catchment areas, and not bounded by political boundaries. These RWAs developed a much greater capacity to manage wastewater treatment than any of the previous disparate organizations, and developed large well-equipped centres of excellence devoted to the research, evaluation and application of wastewater processes, often in partnership with the WPRL. The result was arguably one of the most significant increases in wastewater process knowledge, with large positive influences on design and costs.

16.6   More Understanding and a Plethora of Processes

Up to the early 1970s the liquid stream at most treatment plants in Britain comprised preliminary treatment (removal of grit, coarse solids and rags, etc.); primary sedimentation (removal of settleable solids); secondary treatment (either based on activated sludge or 'biological filters') and occasionally a tertiary stage (usually rapid gravity sand filters) to improve effluent quality when necessary. Since then there have been very significant advances not only to the conventional process units, but also in the development of many alternative processes. Amongst the driving forces were:

- the need to meet increasing stringent effluent standards, particularly removal of nitrogen and phosphorus;
- a requirement to solve the problem of 'bulking';
- a desire to reduce the large 'footprint' of conventional systems;
- a need to improve the efficiency of aeration systems and reduce energy consumption and carbon footprint;
- a desire to minimize the production of sludge.

During this period many competing processes were introduced by contracting companies seeking a share of the large capital market behind the wastewater industry. Some were excellent, whilst others failed to live up to the claims. In fact, some of the claims turned out to be preposterous when examined against the rigours of scientific evaluation.

In addition to the introduction of many 'new secondary treatment processes', one of the most important but least heralded developments was the introduction of a new family of preliminary screens that continuously trap and remove particles greater than 6 mm from the incoming sewage. This made downstream operations much easier and helped the successful introduction of the 'new processes'; it also aided more effective control systems by minimizing the fouling of measurement probes.

Arguably the most important new processes were those aimed at the removal of nitrogen and phosphorus, especially from discharges to watercourses designated by the EU Directive (2000) as 'sensitive', which really means those with a potential for eutrophication. Following on from the discovery of the mechanism of nitrification, it became well established that subjecting a nitrified biomass to anoxic (very low dissolved oxygen) conditions would allow bacteria to use the oxygen atoms in nitrate for sustenance, liberating nitrogen in the process. Thus, incorporating 'anoxic zones' into aeration streams presented a way to remove nitrogen, and thus arose a number of processes with combinations of oxic (O) and anoxic (A) zones, some designed in a bespoke manner, and others proprietary processes with designations such as AO or A^2O.

The removal of phosphate can be achieved by adding chemicals such as ferric sulfate, but it can also be removed biologically by incorporating an anaerobic zone at the beginning of an activated sludge process which, if designed as a nitrogen removal system, yields a process for complete removal of the key nutrients, N and P. The key to biological phosphate removal was discovered by Barnard (1974) in South Africa. This is a biologically complex process beyond the scope of this chapter, but essentially when the biomass is subjected to anaerobic (An) conditions there is a release of phosphate from within the biomass cells into the surrounding liquor, which, when subjected to subsequent aerobic conditions, is taken up again: not only do the cells take up the phosphate released by the cells, but also the phosphate present in the incoming wastewater, with all the phosphate being incorporated into the biomass. And, as long as the biomass remains aerobic, the phosphate will stay there. However, there is a potential problem in that, if the biomass becomes anaerobic, the phosphate can be re-released. This gave rise to a number of modifications to the basic concept that seek to prevent reprecipitation and this produced even more 'new processes' with names such as Bardenpho, University of Cape Town (UCT), and many more. Most of these processes involve a series of (An), (O), and (A) zones in a plug-flow configuration with internal liquor recycling.

There are now many process options available that can achieve all standards of effluent and be located in any possible environment. It is beyond the scope of this chapter to discuss details, and the reader is directed to the many textbooks on the subject such as the massive 2018 pages in Metcalf and Eddy (2013). There is, however, one development that has completely changed the face of wastewater treatment and requires a more detailed description; that development is the introduction of membranes as a means of solid–liquid separation (see e.g. Faisal et al. 2013).

The possibility of using membrane technology was considered in the 1960s but was then thought too expensive. However, changes to manufacturing procedures in the 1970s and a subsequent competitive market brought down costs substantially to the extent that it is now an established, cost-effective system, especially when there is a need to produce very high-quality effluents. Asymmetric cellulose acetate membranes come in four size ranges with ever-decreasing pore size; microfiltration (MF), ultrafiltration (UF), nanofiltration (NF) and reverse osmosis (RO), although NF and RO tend not to be used in wastewater treatment. Between them MF and UF can remove most bacteria, colloidal particles, many viruses and large organic macromolecules, depending on the choice of membrane. Thus, when combined with biological treatment (usually activated sludge), very high-quality effluents can be achieved in a range of processes known as membrane biological reactors (MBRs).

Superficially MBRs are not cheap; the capital cost is high, they have relatively high energy consumption, and maintenance requirements are high compared with non-membrane systems, since the membrane has to be frequently backwashed, occasionally cleaned chemically and replaced every 12 years or so.

The use of MBRs has to be placed in context with other competing processes, but where space is limited and/or very high-quality effluent is required, they are ideal. They can also be engineered across the complete size range of wastewater treatment from the very large to the very small, including package plants and single building applications. However, their greatest niche is in re-use and recovery of water as discussed below.

16.7 The Question of Sludge

A common feature of all wastewater treatment processes is that they produce sludge, some more than others and, like the liquid stream, the knowledge and developments in sludge treatment and disposal have advanced greatly since the late nineteenth century.

The capital cost of sludge treatment is a significant proportion of the overall capital costs, often greater than the cost of the liquid stream. The treatment and ultimate disposal of sludge is usually the greatest operational burden. There is an old axiom amongst sewage treatment managers that 'if you take good care of the sludge the rest of the plant will take care of itself'. Yet, when it comes to deciding on new facilities, sludge often comes as an afterthought, whereas good counsel would suggest that the first question to be addressed at that stage is 'what are we going to do with the sludge?' rather than 'how do we achieve effluent quality?'.

There are, in essence, two types of sludge; those emanating from primary sedimentation and from secondary (biological) treatment. Primary sludge is generally obnoxious, odorous and a considerable hazard; secondary sludges are much less obnoxious and can have an 'earthy smell' depending on the nature of the secondary process.

The most common disposal routes are:

- Landfill; either monofill or combined with garbage – not significant in Britain
- Land: either agricultural, forestry, or sacrificial
- Incineration – not widely practiced in Britain but growing elsewhere
- Drying/palletization and used as fuel, particularly in power stations and cement factories – again, not widely practiced in Britain but growing elsewhere.

There are three main public health issues with sludge; so-called 'heavy metals'; pathogens; and chemical contaminants, with the hazards minimized by strict control and regulation. Regulations are geared towards the specific risks attached to each disposal route, which, in turn, determines the type and extent of treatment required. Most of the sludge in Britain is spread on agricultural land as governed by EU Directive 86/278/EEC (OJ No. L181/6) 1986 and implemented through a *Code of Practice for Agricultural Use of Sewage Sludge, 1996*.

16.7.1 Heavy Metals

Almost all metals found in sewage end up bound to the sludge; very little is discharged in the effluent stream. The metals of most concern are As, Cd, Cu, Pb, Hg, Mo, Ni, Se and Zn. Some, like cadmium and mercury, are directly toxic to humans and animals, whilst others like copper, nickel and zinc are phytotoxic, that is, they have a negative effect on crop yields. Many of these metals have their origin in industrial processes and hence the need for strong control of industrial wastewaters. To this end the 1937 Act was considerably strengthened by Part III of the Water Industry Act 1991. The current regulations on metals in soil are very strict, and water companies are obliged to develop sludge disposal safety plans including the need for storage.

16.7.2   Toxic Organic Chemicals

It is generally accepted that the risk to humans, animals and the environment from toxic organic compounds in sewage sludge is not as high as the risks from heavy metals, due to the very low concentrations usually found. The compounds of most potential concern are dioxins and several polychlorinated hydrocarbons. General opinion accepts that the risk from toxic organic compounds in domestic sewage sludge is very low, and that only places accepting large volumes of industrial wastewaters from certain industries would pose any risk, in which case such compounds should be eliminated at source.

16.7.3   Pathogens

The European Directive does not place limits on pathogens, but Britain controls pathogens by specifying the use of effective treatment processes with stringent conditions in terms of retention times of digesters, elevated temperatures, pH conditions, storage times and composting conditions.

Over the last century there has been a steady growth in sludge treatment technologies, with significant developments in equipment to 'thicken' (concentrate) sludges before main treatment and various types of presses and centrifuges after main treatment to dewater sludges, all ably assisted by dosing with cationic polyelectrolytes; thermal drying is also a feasible option. The choice of equipment depends on cost and the needs of the final disposal route.

Over the years the foremost process for sludge treatment has been mesophilic (35–37°C) anaerobic digestion (AD), but the digesters of today differ considerably from those found in the 1960s. Today's digesters are much more efficient and have been accompanied by other systems such as thermophilic (55°C) digestion, which is more effective in pathogen kill but can have stability issues. However, a development first implemented around 2010 is thermal hydrolysis (TH) applied upstream of AD. In this system, sludge is heated to around 160°C under pressure of about 7 bar where no pathogens can survive. The main benefit of TH, however, is that it changes the biodegradability and rheology of the sludge so that feed rate to the digester can be doubled and generation of methane greatly enhanced. Thus, energy recovery is greatly improved, but that is a subject for the next section.

16.8   A New Philosophy; A New Paradigm?

The developments so far discussed were driven initially by public health issues and later by ever-increasing environmental standards required to clean up rivers and improve aquatic life. This has been more or less very successful, with many river systems restored to good quality with considerable biodiversity, although the present should only be considered as a point in a continuing drive to further improvement, and there are still many issues to be resolved especially with sewerage networks and with the problems of residual pharmaceuticals and many other micro-chemical pollutants. The discussion also summarized very briefly the rather tortuous, but vital, institutional and legal developments that made these improvements possible.

The world of today is vastly different from that when this journey started, and so are our needs. The world now faces massive growth in urbanization caused by continuous population growth, and this is accompanied by a growing freshwater crisis in both quantity and quality. Increased demand for food and energy, along with reducing water resources, have resulted in complex food/energy/water nexuses that have to be managed in the future, all of which are further complicated by decaying assets, poor investment, and by the uncertainties of climate change. The future will not be easy and will require an entirely different philosophical approach to urban water issues and, in this respect, it will be essential that wastewater management is considered as an integral part of any management system that embraces all dimensions of sustainable development. It is no longer appropriate to take wastewater for granted.

Over the past decade or so, wastewater has been increasingly recognized as a valuable asset that can be re-used or recycled, and as a source of valuable materials that can be recovered. Indeed, the UN Sustainable Development Goal (SDG6a) advocates recycling and re-use. The most obvious recoverable materials are summarized below, but there are others.

16.8.1 Water

The most obvious material is water itself, since domestic sewage is at least 99% water. Of particular future importance will be wastewater re-use to supplement diminishing resources in procedures such as aquifer recharge, and in Indirect Potable Reuse (IPR). In the latter, highly treated effluents are discharged into river systems upstream of water intakes. More controversial is the use of treated wastewater for Direct Potable Reuse (DPR), which requires the application of much social science and public relations to convince populations of the safety of such an approach and to overcome the inherent 'yuck factor'. For comprehensive examples of the issues, research and application, visit www.WateReuse.org.

Today's technology allows re-use over a very wide size range. At one end, recycled wastewater arising from municipal networks can be re-used for either industrial or domestic (usually non-potable) purposes. At the other end, single building application with MBR/disinfection processes lodged in the basement are used to treat the building's wastewater and return treated water for toilet flushing or garden irrigation, often coloured to distinguish the recycled from the fresh.

16.8.2 Energy

Well-developed wastewater treatment is a large consumer of energy, most of which is used to keep aeration equipment constantly operational. It has been traditional practice to treat sludge by AD and use methane to generate energy for plant re-use to minimize other fuel purchases. More recently, efficient digesters and combined TH/AD means that much more energy is produced, and together there is a greater contribution to operating costs. In many places, energy production is supplemented by importing strong organic wastes and/or surplus food wastes for combined digestion. As a result, some treatment plants now generate an excess of biogas that can be sold directly to the national gas-grid; converted to electricity and sold to the national electricity grid; or used to fuel vehicles (for example, see http://geneco.uk.com). Not only does this practice produce an income for the operators, but it provides an environmentally sound solution to the problem of dumping food wastes.

16.8.3 Fertilisers

The disposal of sludge to agricultural land has been practised for decades in the UK, albeit that, in the past, farmers had to be persuaded to accept cost-free sludge on a regular basis. Today, with improved treatment processes and stringent regulations producing a better product, offsetting the high cost of inorganic fertilizers, farmers are willing to pay for sludge and for services to monitor and control application. This is a significant turn-around and highlights good recycling practice.

16.8.4 Phosphate

After recovery of water and energy, the recovery of phosphate is arguably the most important recoverable material, as described in Box 16.1.

16.8.5 Other Recoverable Materials

Over the last decade or so there has been much research and development on wastewater re-use and material recovery for a great variety of purposes, too many to be discussed here, but very well summarized by Lazarova (2013). In some parts of the world, materials such as building bricks and biofuel are regularly recovered and there is also current research on recovery of less obvious materials such as alginic acid and cellulose by, for example, Van der Hoek et al. (2016), to mention just one research group.

It is satisfying to say that wastewater is now recognized as a valuable resource. The philosophy has changed; now political and institutional paradigm shifts are required to implement the new philosophy on a very much wider scale.

16.9 The Uncollected and Untreated

The story reported in this chapter relates to Great Britain, and similar optimistic stories can be reported for most of the developed world, but it would be remiss not to mention the dire situation with wastewater in the developing world. Corcoran et al. (2010), in their report 'Sick Water', state that between 80% and 90% of all wastewater generated in developing countries receives no treatment whatsoever. They cite the situation in Jakarta by stating that 500 Olympic-sized swimming pools of wastewater are generated daily, but there is only capacity to treat 15 swimming pools' worth. Urban rivers in much of the developing world are effectively open sewers, and conditions are much the same as those found in Britain at the end of the nineteenth century.

Box 16.1 Phosphorus Recovery: Value and Security from Wastewater

Saskia Nowicki

Phosphorus recovery from wastewater is an emerging opportunity and a long-term necessity. Mining, fertilizer production, intensive agriculture, and inefficient waste management have exacerbated phosphorus losses, unbalancing the global phosphorus cycle (see Figure 16.1) and creating conditions for scarcity. Since pre-industrial times, the amount of phosphorus in terrestrial and freshwater ecosystems – where it is inaccessible to us – has increased by at least 75% (Bennett et al. 2001).

Phosphorus has no substitute. It is an essential component of genetic material, the building blocks of cell membranes (phospholipids), and the energy currency of cells (adenosine triphosphate). Most mined phosphorus is used in fertilizers or feed supplements for agriculture. Demand is increasing due to population growth, increased meat and dairy consumption, and cultivation of non-food crops for purposes like biofuel. Concurrently, phosphate rock reserves are depleting. As the best deposits are targeted first, mined phosphate rock quality is likely to worsen, meaning an increase in impurities (metals and radionuclides) and processing costs. Peak phosphorus is projected for as soon as 2025 or, more mercifully, around 2085 (Cordell and White 2014).

Although enhanced food system efficiency will be important in addressing the coming deficit, supply measures are also needed. Phosphorus in manure and human excreta, which account for 40% and 16% of lost phosphorus flows (Rittmann et al. 2011), must be re-used and recovered. The processes already used by wastewater treatment plants to control nutrient pollution are prerequisites for phosphorus recovery. Currently, the most economical phosphorus recovery methods work in combination with biological nutrient removal processes to precipitate either calcium phosphate or magnesium ammonium phosphate (struvite). Controlling struvite precipitation reduces treatment plant maintenance costs and produces a high-purity, odourless commodity that is easy to package and transport. Beyond good aesthetics, struvite is bioavailable and breaks down slowly, so it requires relatively low application rates and helps reduce wasteful phosphorus runoff.

Conventional fertilizer manufacturers are plagued by depleting reserves, increasing impurities and rising processing costs. In contrast, wastewater treatment plants can anticipate increasing demand for recovered phosphorus. Thus, they are faced with an economic opportunity, and an obligation, to help secure supplies of phosphorus and, therefore, food.

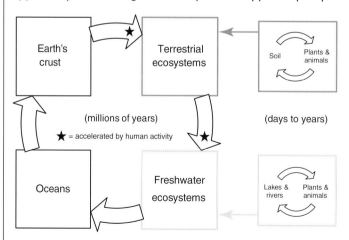

Figure 16.1 Global phosphorus cycle. *Source:* Authors. (*See color representation of this figure in color plate section*).

Why is this so? What are the barriers to wastewater development and what can be done to improve the situation? The problems are not treatment related; there are enough processes to suit all needs, including ones that are affordable and appropriate for developing world conditions. The problems are essentially related to governance, policy, regulation, institutions and finance. It is often said that the biggest barrier is that 'there are no votes in sewage', especially in countries facing other important investment needs. Two particular issues worth discussing are the problems with sewers and the lack of innovative institutional arrangements.

16.9.1 Sewers

The development of wastewater treatment is often inhibited by sewerage networks (or lack of) and problems in conveying wastewater to the treatment plant; the cost of supplying this infrastructure often far exceeds that for the treatment plant itself. In the developed world conventional networks are usually buried deep in the ground, making construction expensive, but there are many non-conventional sewerage systems that are much more affordable to the developing world, such as simplified sewerage and condominial systems (Reed 1995). Another option is to develop a decentralized network with a number of local treatment plants, which saves large network costs. One successful application in Addis Ababa is described in Box 16.2. There are many others.

16.9.2 Innovative Institutional Arrangements

It is axiomatic that development of wastewater treatment must be accompanied by robust institutional arrangements and the elements of good governance, including the legal and institutional framework for the control of industrial wastewaters. It is also vital to have some system that ensures adequate operation long after construction; the developing world is peppered with treatment plants that are either non-functional or abandoned for numerous reasons. In these cases, the costs have been high but the benefits zero. A system aimed at overcoming this has been developed in Brazil, and could easily find application elsewhere. It is described in Box 16.3.

16.10 Concluding Remarks

This chapter attempted to highlight some of the institutional and technical complexities ingrained in establishing sustainable wastewater treatment, and to show that wastewater itself can be considered a valuable asset. However, the discussion mostly concerned the developed parts of the world, whereas the bulk of wastewater in developing countries is neither collected nor treated. Dire though that statistic may be, it underestimates the problem: much of that wastewater is of industrial origin or highly polluted with industrial contaminants, many of which are toxic and banned elsewhere. Domestic wastewater is more-or-less biodegradable and, even if only partly treated, would eventually be incorporated into nature's self-purification processes. That is not the case with these toxic compounds, many of which are persistent, with devastating effects on human health and the environment. Of particular concern is that, if not treated (or

---

Box 16.2 Decentralized Wastewater Treatment in Addis Ababa, Ethiopia

Abishek S. Narayan

Rapid urbanization and a backlog of sanitation infrastructure provision has exacerbated the wastewater challenge in developing countries. Decentralized wastewater treatment (DWT) is an innovative strategy that Addis Ababa adopted for its affordable condominium housing schemes. This is a pro-poor and environmentally sustainable intervention that could be adopted in various high-density cities across the developing world.

Currently, over 40% of Addis Ababa does not have access to improved sanitation, and sewer connection exists for less than 10% of the population. The majority rely on vacuum truck emptying services. The wastewater infrastructure in the city is therefore limited, with insufficient capacity in central treatment plants to handle the wastewater produced. However, the ambitions of the federal government to reach the status of a middle-income country by 2025 places sanitation and urban housing as a priority (GTP-II 2016). To promote access to affordable housing as part of their poverty alleviation strategy, several condominiums were built in the capital city, Addis Ababa. The construction of condominiums without connections to the central wastewater systems put the local City Government (CG) under pressure to address this problem quickly.

In 2014, a decision was taken to introduce DWT systematically in 15 of the newly built condominiums at a cost of over 700 million Ethiopian BIRR (ETB) (US $25 million), by the CG, Water and Sewerage Authority, and the Addis Ababa Housing Development Project Office. Their combined treatment capacity is over 27 000 m³/day, serving 185 000 residents. Although high-end membrane technology was initially used, considering its cost and labour-capacity requirements, simpler activated sludge technologies are being piloted for subsequent projects.

The transformation of a potential sanitation crisis into a sustainable long-term sanitation solution addressing the challenges of urbanization came about through a combination of long-term national goals and short-term situational urgency. The autonomy and fiscal independence of the CG provided the flexibility to enable public finance for the project with no reliance on external donors or federal budget allocations. Further, the positive public perception of DWT and recognition of the importance of urban water bodies by government agencies were crucial to this successful pioneering of DWT by a municipal government in a developing country (Sankaranarayan and Charles 2017).

---

better, eliminated), they reach the oceans, causing inhibition of the primary oceanographic productivity that is so important in stabilizing global warming. The recent recognition of substantial pollution by microplastics adds further to the problem. In 2005 the Millennium Ecosystem Assessment (MEA 2005) reported that 60% of global ecosystem services had already been degraded by human activity. It will be disastrous if world communities make matters worse by not giving urgent and appropriate attention to their wastewaters, which grow in volume every day. It is a massive challenge whose solution will require application of every element of good Wastewater Science, Policy and Management. That, however, is another story.

Box 16.3 River Basin Clean-up in Brazil

Ranu Sinha

The Programa Despoluição de Bacias Hidrográficas (PRODES) or River Basin Clean-up programme was set up by the Brazilian National Water Agency (ANA) in 2001 to reduce water pollution caused by discharge of untreated sewage, particularly in urban areas. PRODES operates as an incentive payment to utilities that invest in the construction, enlargement or improvement of wastewater treatment plants (World Bank 2018). There are five critical steps in the PRODES design, as described below.

Registration	Utility presents proposal to ANA for registration. Must contain pollution reduction goals, approvals from Municipalities and river basin committees
Qualification	Proposals evaluated; compatibility of process and goals
Selection	Proposal selected by ANA – criteria include quality improvements and water resources
Contracting	Proposals contracted on priority basis depending on budget
Certification	Evaluation over three years with 12 evaluations

The first four steps take place within a year and occur before construction begins, whilst the final phase lasts for three years and can only begin once the plant is operational. PRODES subsidy levels vary and are determined using per capita cost reference values associated with two basic parameters: removal efficiency of specific pollutants (technology); and final treatment capacity in terms of effluent pollutant loads (population). Hence the amount of funds available can vary depending on the size of the municipality. Payment of the subsidy is made only when it is proven that there is a reduction of pollution loads over a three-year period in accordance with performance targets pre-established on each contract (Libanio 2015). At first, the operators monitor pollution loads and provide results to ANA. ANA then verifies whether the operational results have achieved the contractual goals, which include inflow and pollutant loads treated as well as pollutant removal efficiency. Once targets are achieved, reimbursements are made of full or partial capital costs.

Some of the key design functions of PRODES include: (i) utilities must get approval of their proposals from municipalities and river basin committees, forcing collaboration and transparency; (ii) resources are transferred to a specific escrow account related to the project, linked to a Fund, and can only be withdrawn after authorization from ANA; and, (iii) the requirement of output verification prior to disbursement of payments provides a critical fiduciary safeguard for the accurate targeting of funds as well as evidence that public funding was well spent. Since its implementation, PRODES has represented a paradigm shift in the water and sanitation sector in Brazil, with increased wastewater services and improvements to water quality indices of a number of rivers.

References

Ardern, E. and Lockett, W.T. (1914). Experiments on the oxidation of sewage without the aid of filters. *Journal of the Society of Chemical Industry* 33 (10): 523.

Barnard, J.L. (1974). Cut P and N without chemicals. *Water and Wastes Engineering* 11: 33.

Bennett, E.M., Carpenter, S.R., and Caraco, N.F. (2001). Human impact on erodible phosphorus and eutrophication: a global perspective. *Bioscience* 51: 227. https://doi.org/10.1641/0006-3568(2001)051[0227,HIOEPA]2.0.CO;2.

Chave, P. (2001). The EU Water Framework Directive: An Introduction. London: IWA Publishing.

Cooper, P.F. (2001). Historical aspects of sewage treatment. In: Decentralised Sanitation and Reuse: Concepts, Systems and Implementation (eds. P. Lens et al.), 11–38. London: IWA Publishing.

Corcoran, E., Nellemann, E., Baker, R. et al. (eds.) (2010). Sick Water? The Central Role of Wastewater Management in Sustainable Development – A Rapid Response Assessment. Nairobi: United Nations Environment Programme, UN HABITAT, GRID-Arendal.

Cordell, D. and White, S. (2014). Life's bottleneck: sustaining the world's phosphorus for a food secure future. *Annual Review of Environmental Resources* 39: 161–188. https://doi.org/10.1146/annurev-environ-010213-113300.

DEFRA (2002). Sewage Treatment in the UK: UK Implementation of the EC Urban Waste Water Directive. London: DEFRA Publications.

Dibdin, W.J. (1887). Sewage sludge and its disposal. *Proceedings of the Institution of Civil Engineers* 88: 155.

Downing, A.L., Painter, H.A., and Knowles, G. (1964). Nitrification in the activated sludge process. *Journal of the Institute of Sewage Purification* 2: 130.

Faisal, I.H., Yamamoto, K., and Lee, C. (eds.) (2013). Membrane Biological Reactors. London: IWA Publishing.

GTP-II (2016). Growth Transformation Plan II – Federal Democratic Republic of Ethiopia. Addis Ababa: National Planning Commission.

Johnstone, D.W.M. (2003). Effluent discharge standards. In: Handbook of Water and Wastewater Microbiology (eds. D. Mara and N. Horan), 299–313. London: Elsevier.

Klein, L. (1957). Aspects of River Pollution. London: Butterworth Publications.

Lazarova, V. (2013). Milestones in Water Reuse: The Best Success Stories. London: IWA Publishing.

Libanio, P.A.C. (2015). Pollution of inland waters in Brazil: the case for goal-oriented initiatives. *Water International* 40 (3): 513–533. https://doi.org/10.1080/02508060.2015.1010069.

Metcalf & Eddy, Inc (2013). Wastewater Engineering: Treatment and Reuse. Boston: McGraw-Hill.

Millennium Ecosystem Assessment (2005). Ecosystems and Human Well-Being: Synthesis. Washington, DC: Island Press.

Reed, R.A. (1995). Sustainable Sewerage. London: Intermediate Technology Publications.

Rittmann, B.E., Mayer, B., Westerhoff, P., and Edwards, M. (2011). Capturing the lost phosphorus. *Chemosphere* 84: 846–853. https://doi.org/10.1016/j.chemosphere.2011.02.001.

Royal Commission on Sewage Disposal (1898-1915). Nine Reports. London: HMSO.

Sankaranarayan, A. and Charles, K. (2017). *Decentralized Wastewater Treatment in Addis Ababa*. Research Brief 09/17. Oxford: REACH. www.reachwater.org.uk.

Schneider, D. (2011). Hybrid Nature: Sewage Treatment and the Contradictions of the Industrial Ecosystem. Cambridge: MIT Press.

Tomlinson, E.J. (1982). The emergence of the bulking problem and the current situation in the UK. In: Bulking of Activated Sludge – Preventative and Remedial Methods (eds. B. Chambers and E.J. Tomlinson), 17–23. Chichester: Ellis Horwood Ltd.

Van der Hoek, J.P., de Fooij, H., and Struker, A. (2016). Wastewater as a resource: strategies to recover resources from Amsterdam's wastewater. *Resources, Conservation and Recycling* 113: 53–56.

Warrington, R. (1882). Some practical aspects of recent investigations on nitrification. *Journal of the Royal Society of Arts*: 532–544.

Woods, D. (2003). Evolution of river basin management in England and Wales. In: FWR Guide FR/G0003. Marlow: Foundation for Water Research http://www.fwr.org/frg0003.pdf.

World Bank (2018). Wastewater: from waste to resource – the case of PRODES, Brazil (English). In: From Waste to Resource. Washington, DC: World Bank Group http://documents.worldbank.org/curated/en/885781521183256449/Wastewater-from-waste-to-resource-the-case-of-Prodes-Brazil.

17

A Road Map to Sustainable Urban Water Supply

Michael Rouse[1] and Nassim El Achi[2]

[1] *School of Geography and the Environment, University of Oxford, UK*
[2] *Global Health Institute, American University of Beirut, Lebanon*

17.1   Introduction

The Sustainable Development Goals (SDGs) are both ambitious and challenging. This chapter focuses on the goals for access to drinking water services (SDG 6.1). Goal 6.1 reads as follows: 'By 2030, achieve universal and equitable access to safe and affordable drinking water for all'. The goal can be seen as a stimulus for governments to take the actions necessary to provide affordable access with the many benefits related to health, quality of life and economic development. It is the latest in a series of stimuli, with earlier initiatives providing valuable information on the challenges and barriers to progress.

This chapter addresses two questions. What does SDG 6.1 mean in practice, and how is it different from earlier international stimuli? What steps need to be taken to achieve sustainable water supply services for all? An outline road map is proposed.

17.2   International Stimuli – What Has Been Achieved?

17.2.1   A Brief History Before the Water Decade of 1981–1990

The beginning of serious aid programmes (Black 1998) took place in the 1960s with the aim of tackling poverty, but the approach was largely that of promoting developed countries' technologies on both water and wastewater, with sewers and water mains and associated treatment works, for application in developing countries. Such an approach was very costly and had limited success in urban areas due to inadequate local capability and lack of training for operations, and little consideration for future maintenance. These approaches were wholly inappropriate in rural areas. To a large extent the approach was driven by donors, including the World Bank, as monies for large-scale infrastructure were the only form of investment available. Public health professionals with knowledge of developing world conditions were pushing for more appropriate technology. Amongst others, Robert McNamara, President of the World Bank from

Water Science, Policy, and Management: A Global Challenge, First Edition. Edited by Simon J. Dadson, Dustin E. Garrick, Edmund C. Penning-Rowsell, Jim W. Hall, Rob Hope, and Jocelyne Hughes.
© 2020 John Wiley & Sons Ltd. Published 2020 by John Wiley & Sons Ltd.

1968 to 1981, recognized that this approach to reducing poverty was not working. Boreholes and handpumps for water, and improved pit latrines for sanitation, were being proposed. There was a requirement for demonstration applications. The World Bank did not have a remit to finance such projects, but teamed up with UNDP to sponsor the first generation of projects which became known as the UNDP–World Bank Water and Sanitation Program. The amount of annual investment was low, around US$15 million, but it stimulated other international and national donors, governments and NGOs to adopt the 'appropriate technology' approach. Although this was a technology push, both UNDP and the World Bank recognized the need for not just hardware, but improved governance issues such as community support and involvement of women. In parallel with this development, the World Health Organization recognized the need for safe drinking water, and sanitation and hygiene to tackle immense and growing waterborne disease, particularly diarrhoea, which was claiming the lives of millions of children, particularly those under five years old.

The World Water Conference in 1977 adopted the Declaration of Mar del Plata, which established the Water Decade. International support for the Decade was in part due to the UNDP–World Bank Water and Sanitation Program with its poverty-reduction focus. The background to the decision in 1977 was 1.8 billion people living in rural areas in developing countries, with only one in five having access to safe water. It was estimated (GDRC 2008) that 590 million children under 15 years old (around 40%) did not have safe drinking water. In developing countries, due to polluted water, perhaps one hospital patient in four suffered from waterborne illness.

17.2.2 The Water Decade 1981–1990

The objective of the Water Decade was to make access to clean drinking water available across the world, effectively universal access, with safe water and sanitation for everybody by 1990. It became recognized that there were significant obstacles to achieving the objective, some of which were whether developing countries would give sufficient priority to the aim, whether effective delivery organizations could be established, availability of finance, sufficient manpower training, and whether appropriate technology would be used.

According to the Global Development Research Centre (GDRC 2008), the decade ending in 1990 was estimated to have brought water (of undefined quality) to around an additional 1.2 billion people and sanitation to around 770 million people. However, the population growth during the decade was around 1.2 billion people (Figure 17.1).

The reported outcome of the decade for all developing countries is given in Table 17.1 (World Health Organization 1991).

The above figures are inconsistent with those reported by GDRC. The WHO figures may be covering a different component of the world population, but indicate that the number of people in developing countries without water services had reduced to around 1 billion people. This is less than the 1.3 billion estimated by the WHO and UNICEF Joint Monitoring Programme (JMP) in its report on the outcome of the Millennium Development Goals (MDGs), which is covered later. This highlights the importance of clarity of definitions and robustness of reporting systems.

Towards the end of the decade the annual investment in the developing world was around US$10 billion. It was recognized that the requirement was many times this

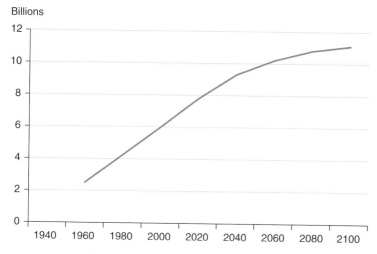

Billions

Figure 17.1 Estimated global population 1960–2015 and predicted population 2020–2100. *Source:* UN Population Division (2015 Revision).

Table 17.1 Levels of service coverage, all developing countries, 1980 and 1990 (in millions).

Population with services (millions)	Urban		Rural	
	1980	1990	1980	1990
Water supply	710	1159	676	1817
Sanitation	627	953	847	1274

Population without services (millions)	Urban		Rural	
	1980	1990	1980	1990
Water supply	226	173	1627	842
Sanitation	309	379	1456	1385

Source: World Health Organization (1991). Reproduced with permission of the World Health Organization.

amount, and that even the poor would have to make a contribution to the costs. This triggered a focus towards community input and the involvement of women. The UNDP–World Bank Water and Sanitation Program was continued until 1998, so in a sense the Decade became two decades. Generally, although the ambitious aim was not achieved, it made progress and was applauded for its international focus on approaches appropriate to country situations. This is perhaps summed up best by quoting from a 1993 publication (Choquill et al. 1993): 'Despite the failure to meet the quantitative goals, much was learnt from the experience of the water and sanitation decade… There was further realisation of the importance of comprehensive and balanced country-specific approaches to the water and sanitation problem… Most importantly, perhaps, was the realisation that the achievement of this goal that was set at the beginning of the decade would take far more time and cost far more money than was originally thought'.

17.2.3   Millennium Development Goals (MDGs)

In September 2000, world leaders met in the UN Headquarters in New York and adopted the UN Millennium Declaration. This established eight goals to be achieved by 2015. One of the targets was to halve the number of people without access to 'improved sources of water'. Sanitation was omitted initially, but this was corrected at the World Summit on Sustainable Development in 2002 in Johannesburg with the inclusion of a target of halving the number of people without access to 'basic sanitation' by 2015. In 1990 UNICEF and WHO set up the JMP (Bartram et al. 2014) to monitor progress on the MDGs on water and sanitation. There had been some monitoring from 1930 by the League of Nations, the forerunner to the United Nations which was formed in October 1945 in the aftermath of the Second World War. Monitoring results were of doubtful quality until the JMP was established later, but the experience of the MDGs was that much more attention still needed to be given to definitions, for example, what is meant by safe drinking water. This was not addressed effectively until after the SDGs came into force in January 2016.

Whatever the limitations of monitoring, good progress was made towards achievement of the MDG target for water. The JMP reported that by 2015 an additional 2.8 billion people had gained access to improved water sources since 1990. There was less progress on sanitation. The MDG target called for halving the proportion of the population without continuous access to basic sanitation between 1990 and 2015. This target translated into achieving basic access for 77% of the population. The achievement was 68%, 9% short of the target, but with an additional 2.1 billion people getting access to basic sanitation.

In 2008 the cost of meeting the MDG for water in developing countries was estimated to be US$42 billion and for sanitation the estimate was US$142 billion (Hutton and Bartram 2008). The higher cost for sanitation would be expected to be a significant factor in less progress being made on sanitation. An added consideration is that it is more difficult to implement cost recovery measures for sanitation.

17.3   Sustainable Development Goals (SDGs)

17.3.1   Formation and Definitions

The United Nations Conference on Sustainable Development held in Rio de Janeiro in June 2012 began consideration of what would follow the UN's MDGs. The new SDGs were ratified by UN member states at the UN General Assembly in September 2015. The Drinking Water Goal (SDG 6.1) is to achieve universal access to water services by 2030.

In July 2017 the JMP issued a report which provided an update on progress on drinking water, sanitation and hygiene and established the baselines for the SDGs (WHO and UNICEF 2017). Based on experience with the MDGs, specific attention was given to definitions on 'quality and safety' of water with a monitoring ladder as shown below.

Service level	Definition
Safely managed	Drinking water from an improved water source that is located on premises, available when needed and free from faecal and priority chemical contamination
Basic	Drinking water from an improved source, provided collection time is not more than 30 minutes for a round trip, including queuing
Limited	Drinking water from an improved source for which collection time exceeds 30 minutes for a round trip including queuing
Unimproved	Drinking water from an unprotected dug well or unprotected spring
Surface water	Drinking water direct from a river, dam, lake, pond, stream, canal, or irrigation canal

Note: Improved sources include piped water, boreholes or tubewells, protected dug wells, protected springs and packaged or delivered water.

The definitions in the ladder are important. The report estimates that 844 million people in 2015 lacked access to a basic level, whereas the number without access to a 'safely managed source' (the highest rung in the ladder) was estimated to be around 2.1 billion people. In addition, by 2030 the world's population could have increased by 1 billion, with Africa expected to have the highest growth. This chapter is being written in 2018 with less than 13 years left before the 2030 target date. The above numbers would suggest that access for around an additional 170 000 people per day from now until the end of 2030 is required to achieve universal access to basic services. That is the task.

The 2017 report shows that faster progress than that made during the MDGs will be required in a majority of countries in order to achieve universal access to basic services by 2030. In some countries, largely in sub-Saharan Africa, there was a reduction in people receiving a basic level of service over the period 2000–2015 (Figure 17.2). To prevent further reduction, attention to supply system maintenance and refurbishment will be required, this being one of the major challenges discussed later in the chapter. The report refers to major data gaps, and states that effective monitoring will require significant improvements in the availability and quality of data underpinning national, regional and global estimates of progress.

17.3.2   Water and Sanitation as a Human Right

It is important to include a brief description of UN Resolution 64/292, 3 August 2010, which recognizes water and sanitation as a human right, because it is complementary to the MDGs. A discussion of the resolution is given in Rouse (2013). Two key parts of the resolution are reproduced below:

1) *Recognizes* the right to safe and clean drinking water and sanitation as a human right that is essential for the full enjoyment of life and all human rights;
2) *Calls upon* States and international organizations to provide financial resources, capacity-building and technology transfer, through international assistance and cooperation, in particular to developing countries, in order to scale up efforts to provide safe, clean, accessible and affordable drinking water and sanitation for all.

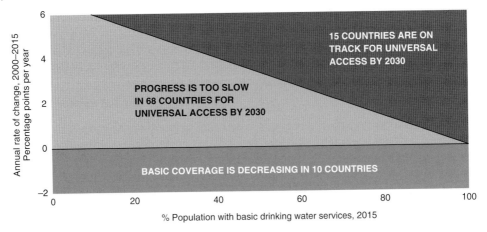

Figure 17.2 Rate of progress over period 2000–2015 on basic coverage of drinking water, illustrating the need for effort to be increased to achieve the 2030 goal. *Source:* World Health Organization (WHO) and the United Nations Children's Fund (UNICEF) (2017). Reproduced with permission of the World Health Organization.

For drinking water, there are elements that provide the framework for implementation, namely availability, quality/safety, acceptability, accessibility and affordability. To these are added non-discrimination, participation, accountability and sustainability. Non-discrimination reinforces the universal access aim of SDG 6.1. Quality/safety is defined as 'water should be of a quality that it does not pose a threat to health'. The basic access from an improved source (as shown in Figure 17.3) may be equivalent, but not as clearly defined in safety terms as the 'safely managed' level. Sustainability states the need for effectively maintained supply systems, as loss of access is loss of a human right and not acceptable. The authors of this chapter suggest that there should be joint monitoring of SDG 6.1 and the elements that make up achievement of human rights on water.

17.4 Challenges to be Faced

17.4.1 Sustained Political Commitment to Goal

The water sector is capital-intensive with long asset lives. This is generally at odds with a political environment that encourages short-term decision-making. For sustainability, policies will have to make provision not just for achieving the goal of universal coverage of a basic service by 2030, but in maintaining that service beyond that date, and achieving the ultimate goal of universal access to safely managed service level. This requires joined-up government, particularly integrated planning (see Figure 17.4), but also all political factions agreeing to continuity of commitment, direction and policies. Water should not be a political football. At each stage there should be a lead or coordinating Ministry, ideally the one responsible for water. Finance and health must be involved for obvious reasons. The social affairs function is vital in relation to equality of access and affordability. Responsibility for reporting to the JMP in any particular country could rest with the Water Ministry, but this role might be handled by other ministries under the remit of international affairs as defined in Figure 17.4.

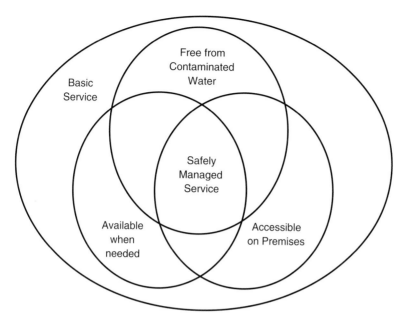

Figure 17.3 JMP monitoring levels for drinking water. *Source:* World Health Organization (WHO) and the United Nations Children's Fund (UNICEF) (2017). Reproduced with permission of the World Health Organization.

17.4.2   Reliable Data

In many countries there is a lack of reliable data on supply systems and their condition, on water abstracted and treated, and on population served. In some local authorities water services are part of other local authority operations without a clear separation of operating costs. Reliable data are essential not just for accurate reporting on progress towards goal 6.1, but also to determine what has to be done, how it will be financed, human resource requirements and developments, and for realistic improvement delivery planning. Achieving reliable baseline data, and putting in place effective management and operational database systems, requires planning and investment.

Where there are combined water, sanitation and wastewater service organizations, all aspects could be included in the database. Equally, such a database is just as important for rural borehole or handpump supply systems as it is for urban piped systems, particularly in relation to investment and maintenance as discussed later in the chapter. Included in the data system would be responsibilities for reporting to the JMP on progress and the processes for collecting, analysing and verifying the performance data.

17.4.3   Effective Planning

In many countries planning takes place on a one-year cycle as part of annual budgeting, with independent decisions on large infrastructure projects. In some countries there are detailed design plans for infrastructure for 20 years or more. Neither approach is

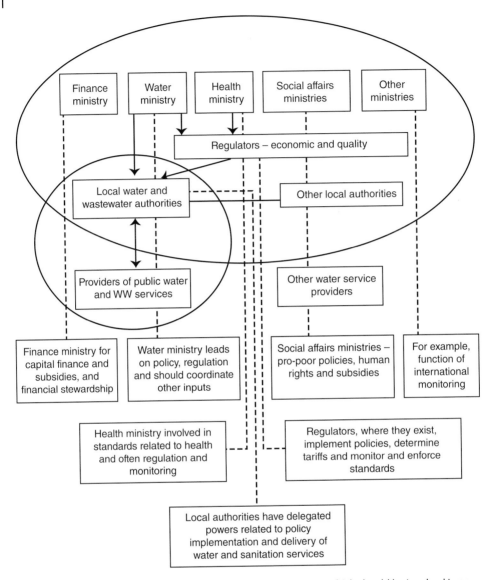

Figure 17.4 Diagram showing the various functions of government which should be involved in an integrated way. *Source: Routledge Handbook of Water and Health* (2015). Reproduced with permission of Taylor and Francis.

effective. The former is far too short-term and does not allow for integrated investment decisions, and water service managers can spend a large part of their time obtaining budget approval and associated government subsidies. The latter, so-called master plans, are expensive to produce, are not water-consumer service driven, and do not provide flexibility to address unpredictability; nor do they allow for the benefits of future improved technology.

The requirement is for strategic plans and delivery plans. A strategic plan should be long-term, perhaps looking 20–25 years ahead. With the SDG 6.1, a strategic plan leading to the achievement of that goal in 2030 would be appropriate. Even then, there would be benefits in considering the requirements to achieve sustainability of service provision beyond that date. A strategic plan is not detailed. It is about the 'what' and the 'when'. It should set down a long-term vision, expressed in terms of service delivery standards. It should discuss the starting situation and the challenges of meeting the long-term goals. It will reflect water service reform laws, stated government policies and the responsibilities of the various stakeholders. It will give an indication of the level of investment required and how it will be funded, and will be realistic about the time-frame with political commitment to achievement of the goals.

The strategic plan will provide the basis for delivery plans. Typically in five-year stages, delivery plans take the 'what' and decide on the 'how'. Stress is on the word 'deliverable'. The stated actions must be technically and operationally capable of being implemented according to timetable, with the human resources and financial resources available to carry out the work. For this reason, adequate time, perhaps a year, must be devoted ahead of each five-year period to obtain the data and understanding necessary for firm plans to be established. The plans will include monitoring and reporting requirements with procedures for any required corrective action.

Following the discussion on Water Resources and Water Distribution, see Section 17.4.6 on City Planning, which requires a wider perspective on how water supply systems are designed.

17.4.4   Water Resources

The plans should include the development of water resources, whether surface water and water treatment, boreholes, wells, springs, and perhaps a contribution from rainwater harvesting, to meet the specified levels of service delivery. In areas where there are already piped supply systems, the contribution that can be obtained from reducing leakage must be an important consideration (see the discussion on leakage in Section 17.4.5).

Systems have to be designed to provide for supply to those currently unserved, and provide for the rapid growth in urban population. Water is critical to economic development (Sanctuary et al. 2005) and it might be expected that economic development would increase demand in per-capita water use, thus adding to the water resource requirement. Average per-capita figures worldwide are reported to have increased by only 10% in the decade from around 1900 to 2000, with increases in total consumption dominated by the population increase over that period (Kummu et al. 2016). In the authors' view, as there is little reliable per-capita data available, there must be doubts about the validity of the 10% increase figure.

With apparent greater uncertainty for rainfall, water storage becomes even more important. In addition to increased lake and reservoir capacity, the storage capacity of aquifers requires greater attention. Over-abstraction is causing water quality problems. The lower and lower groundwater levels suggest that there is much unused underground storage capacity. Controlled runoff through creating 'contour bunding' has benefits of groundwater recharge, reduced soil erosion, improved irrigation and improved water quality. This can be done on a large scale (Rouse 2013; Everard 2015). It costs very little in areas where people are pleased to provide labour. Rainwater harvesting from roofs and other hard surfaces can add to the overall storage capacity. Storage should be considered in an integrated way.

17.4.5 Water Distribution

Water distribution tends to be the 'Cinderella' of water supply. Governments, donors and academia give it insufficient attention. Perhaps this is because it is not pictorial. There are no attractive scenic pictures to illustrate reports, and it is not seen to be part of the environment. Compared with a water treatment works, it is difficult to recognize water distribution infrastructure with a plaque on the wall, and with a politician having a photo opportunity at an opening ceremony. Equally, the timescales of construction and infrastructure life tend to be much longer than the tenure of governments. Perhaps distribution research is seen to be trivial – it is not!

Water is heavy and moving it around is costly. Unlike telecomms data, there are no prospects of better physics to transmit water. Unlike power, where the major costs are in energy production, water costs are dominated by water distribution, with infrastructure construction and life costs being around 65% of total supply costs. Those urban inhabitants without a piped supply rely on informal water vendors for their water supplies, often for the amount required for potable use. However, there is no alternative to piped distribution for an affordable, satisfactory urban service, as other methods are much more expensive. Distribution by road tankers typically costs 14 times as much as piped water. Bottled or sachet water ends up being retailed at around 1000 times the cost of piped distribution per litre. The question is often asked: 'why treat all water to drinking water standard, when only a small amount needs to be potable?' The answer is quite simple – having two distribution systems is even more expensive. This is because distribution systems contribute around 65% to the total cost of water supply (Rouse 2013). There is also the issue of the already high level of 'pipe congestion' under the streets.

The required investment in piped distribution systems will be the dominant cost of achieving SDG 6.1. In addition to new systems there is a large backlog of investment required for existing systems. In most parts of the world, both developed and developing, there has been a neglect of distribution systems, partly because they are 'out of sight, out of mind', but also because utility income has been insufficient for effective maintenance. In the UK prior to 1990, every year, in many of the water service providers, water managers would propose a sum for system refurbishment, with the response being that it can wait for another year, and so on. Over the last 25 years there has been investment of £116 billion, much of that sum on underground system refurbishment, and much more investment is still required to catch up on perhaps 40 years of pre-1990 neglect (see Section 17.4.7 on finance, which discusses the investment challenge).

The neglect of the distribution systems has led to high levels of leakage. The figures quoted for NRW (non-revenue water, which consists of both commercial and physical losses) are typically in the range 15–30%. This unmeasured level is quoted as it is seen to be acceptable. In most cases in the developing world, leakage is not measured. Supply meters, which measure the amount of water going into supply, have long since stopped working and have not been replaced. Generally there are no distribution zone meters and often there are no consumer meters, with water charges levied as a fixed sum. Where leakage has been measured, it can be as high as 70% of water put into supply. More typically, it is around 50%. By reducing it to 30% the water available for consumers is increased by 40%. This can be a substantial water supply enhancement which so often is ignored, with additional water resources pursued instead.

The importance of leakage management is most evident as part of achieving the service delivery requirement of achieving continuous supply. In many parts of the developing world (Reiff 2015) the population receives only a few hours of supply a day, or in a week: so-called intermittent supplies. The argument is that as there is insufficient water to supply everyone all of the time, water has to be rationed through sequencing groups of people for a few hours at a time. At one time in India (A.K. Jalakam, personal communication, 2018) some systems were even designed to facilitate this mode of operation. That this approach requires less supply water is an illusion. It also puts the safety of drinking water at great risk. Frequent pressurization and depressurization of the pipes results in deterioration of the system, there is increased leakage at joints, and an increased number of burst mains. Leakage losses increase, resulting in a reduction in pressure and loss of supply to the more distant points. Pumping pressure is increased to compensate, resulting in yet higher leakage, and a cycle of decline. Often, water service suppliers are unable to keep to intermittent supply schedules, such that consumers lose confidence in the timing of supplies. They compensate by filling all available containers when the supply is on, and leave taps open in case they miss the next 'delivery'. Also, they tend to dump any 'old' water when the new supply arrives. The consequence of all these factors is that more water is required than with continuous (so-called 24/7 supplies). This was demonstrated by the Karnataka Urban Sector Improvement Project (World Bank WSP 2010). It is now the policy of the Government of India to work towards universal 24/7 supplies. This requirement is also recognized in the UN Handbook on water and sanitation as a human right (De Albuquerque 2014), because unsustainable systems result in a loss of supply and thus a loss of a human right.

The question is how, with existing systems, to get from intermittent to 24/7. Attempts are made to carry out leakage control across whole-city or part-city distribution network systems. Generally this has not proved to be possible for two reasons. First, it requires a significant proportion of available investment money, so has to be phased over time. Second, the system is deteriorating with new leakage as existing leaks are fixed, so it is not possible to achieve a sufficiently leakproof system to maintain pressure. The solution is to form distribution metering areas (DMAs), which are small enough for affordable rapid rehabilitation, and which can be maintained as 24/7 as the rehabilitation progresses from one DMA to another until the whole urban area is completed. There is the added advantage that the population in the new 24/7 areas are much more willing to pay water bills (World Bank WSP 2010). This improves the financial situation in the utility to be able to make progress across the remaining parts of the system.

The DMA process requires a good network analysis of the distribution system which often does not exist, and moreover, the data for building models does not exist. But to achieve the sustainable aim there is no option but to get the data and build the models. (see Section 17.4.2). This is likely to require some outside assistance, the investment for which would be included in the delivery plans. Investment sums would have to be included for the installation of distribution zone meters and valves, and provision made for some investment in pipes. Once the DMA zones had been established it would be possible to systematically measure leakage and identify the worst parts of the system. Whilst the work on the network was proceeding, work could begin on locating the large leaks by walking the relevant streets looking for wet patches and helped by residents' observations. For a comprehensive manual on leakage control see Farley and Trow (2001). The box below shows a major advance in DMA development for sustainable 24/7 water supplies in Amman, Jordan.

Box 17.1 Amman Case Study

Jordan is one of the world's driest countries, with a daily water supply of 120 l per capita. In the densely populated capital, Amman, the population is rising rapidly, and with it water demand, due to refugee influx from surrounding war-torn countries. To cope, Amman water authorities deliver piped water for an average of 48 hours per week. The city was ideal for assessing a new hydraulic modelling approach to achieving continuous supply, with its critical water status and available updated water network maps.

A gradual transition from intermittent to continuous water supplies appears to be feasible, but it is proving difficult in practice, both technically and financially, to refurbish whole networks at once. Moreover, the previously reported gradual models focus exclusively on upgraded areas, not impacts on service to the surrounding zones during the transition. A hybrid hydraulic model was therefore designed using WaterGEMSV8i modelling software applied to Amman's 'Tariq' residential area (population 120 000), as a pilot. The area's water network has recently been refurbished, transforming its four large and interlinked District Metering Areas (DMAs) into seven smaller and completely independent DMAs. Such re-zoning is crucial since gradual transition to 24/7 is not achievable without each DMA being able to supply water independent of the others.

For a given model simulation, the demand pattern of the upgraded DMA was set as that of continuous supply having dual spike demands. The remaining DMAs were given demand patterns having flat and constant supply throughout the usual 48 hours.

Seven cumulative simulations were carried out, to attain zone-by-zone 24/7 supply of the seven Tariq DMAs, by considering the water demand of rationing supply areas as the baseline for the diurnal pattern of the upgraded regions. For each simulation, the system was 'designed' to guarantee water pressure of 3–9 bars, even during peak times, by adding pressure-reducing valves. The model results met the objective of stage-by-stage achievement of 24/7 without service degradation in the other DMAs.

The approach allows for progressive refurbishment of each zone at an affordable rate. 24/7 supplies could be achieved with only a 15% increase in water supply relative to the 48 hours arrangement. However, with the refurbishment saving a proportion of the current NRW value of 43%, 24/7 could be possible with a reduction in water resources. The model can be applied widely as an affordable method, and as an important component in meeting SDG 6.1 in urban areas.

All data and maps, part of USAID's NRW reduction project, were kindly provided by Jordan Water Company Miyahuna and Engicon Consulting Group.

17.4.6 City Planning

Uncertainties in city development require a change in thinking from large-scale water resources and treatment, to a more flexible modular approach. A full discussion of this would require another chapter. China has now adopted the concept of 'Sponge Cities' (Zevenbergen et al. 2018) to promote an integrated approach to city and water supply development, incorporating water re-use and links with energy (see also Novotny and Brown 2007 and Hao et al. 2010, which discuss the various aspects of water and sanitation development and the need for these to be considered in the context of city planning).

17.4.7 Finance

Water is a capital-intensive industry, and the capital charges from investment in infrastructure make up a large proportion of the supply costs. Chapter 15 considers the three 'ts' of tariffs, taxes and transfers, and adds a fourth 't', that of transactions. It discusses innovations in private finance and blended solutions, and concludes that every initiative deserves attention. There are a number of estimates of the level of investment needed that include the cost of refurbishing neglected existing infrastructure: US$23 trillion over 20 years (Doshi et al. 2007); US$22.6 trillion by 2050 (OECD 2016). Whatever the actual number, the concern is that investment needs are much greater than current investment levels (OECD 2018). Even the much lower estimate of providing a service to the unserved in connection with SDG 6.1 and 6.2 of US$1.7 trillion (Hutton and Varughese 2016) is three times the current investment level.

The large cost of refurbishing existing distribution networks deserves particular attention. Whereas new infrastructure can be considered to be associated with 'new customers', there are no new payees on existing infrastructure, so convincing private investors is more difficult. Consideration has to be given to regard such investment as equivalent to new water resources, as reduced water distribution leakage can be considered as the most accessible increase in water resources for urban supplies.

Wherever the investment comes from (other than grants) it has to be paid for, either through tariffs or taxes. There are many challenges here, including achieving a service for which people are willing to pay, provisions for the poor, and long-term political commitment.

An important factor is how to spread the costs over the generations who will benefit from the investment. The Government of Malaysia established a government water infrastructure funding agency able to obtain long-term low-interest loans, such that costs could be recovered by leasing charges over 40 years.

17.5 Reform Requirements

Achieving SDG 6.1 will require substantial reform of water services, in particular urban services. Substantial investment and policy/operational development are required and there are no short cuts. It is instructive to consider some successful water service reforms and to identify the success factors.

17.5.1 Phnom Penh

In 1993, following the fall of the Khmer Rouge, Cambodia had to rebuild from scratch. In the capital, Phnom Penh water services were poor, almost non-existent, but 13 years later in 2006 were regarded as the best in South Asia (Biswas and Tortajada 2010). Table 17.2 gives the performance data for 1993 and 2006 (Chan 2009).

Table 17.2 Phnom Penh urban water performance indicators, 1993–2006.

Performance indicator	1993	2006
Coverage area (%)	25	90
Number of connections ('000s)	27	147
Supply duration (hours/day)	10	24
Revenue collection (%)	48	99.9
Non-revenue water (%)	72	8
Staff per 1000 connections	22	4
Financial situation	Subsidy	Full cost recovery

Source: Courtesy of Ek Sonn Chan.

Since 2006 there has been ongoing progress, with there being virtually 100% coverage in 2018 and NRW less than 6%. The success factors are:

a) A remarkable water manager, Ek Sonn Chan
b) Consistent government policy and support
c) Initial financial and technical support (from the French government)
d) Effective staff training to become technically self-sufficient
e) Elimination of corruption; all staff involved in corrupt practices were fired
f) Staff bonus incentives for improved performance
g) No interference by government in water service operations
h) Recognition that there was no shortage of resource water and that leakage control would provide water for the whole population and provide a 24/7 service
i) Public consultation: Ek Sonn Chan went from street to street discussing the reform programme and getting public support
j) Recognition that for affordability, there has to be efficient service operations
k) Provisions for the poor with pro-poor subsidies, but within a non-general subsidy culture. Full cost recovery achieved by 2006.

17.5.2 National Water and Sewerage Corporation (NWSC), Uganda

Following a government decree in 1972, the National Water and Sewerage Corporation (NWSC) became operational in 1973. There followed a period of political unrest, including during the Idi Amin regime, with a series of NWSC Boards and Managing Directors (MDs) but with political interference and failures to improve water services. Although various initiatives and injections of donor investment had achieved some improvements, the NWSC was virtually bankrupt, inefficient, and service was poor. In 1988 a new NWSC Board was established with the authority to appoint a new MD and to function without political interference. The appointed MD was William Muhairwe who had a business management background. The turnaround story is told in detail in Muhairwe's book (2009). Essentially Muhairwe took a business performance approach with areas competing to be the best and with associated rewards. The 1998–2008 results in Table 17.3 show the performance improvements.

Table 17.3 Urban water performance indicators, Uganda.

Performance indicator	1998 performance	2008 performance
Service coverage (% of population served)	47	72
Total piped connections	50 826	202 559
New connections per year	3317	24 384
Staff per 1000 connections	36	7
Non-revenue water	51%	33.5%
Metered accounts	65%	99%
Income (UGX billion)	21.9	84.0
Operating profit after depreciation (UGX billion)	−2	3.8

Source: Muhairwe (2009). Reproduced with permission of IWA Publishing.

Since 2008 the NWSC has expanded its operation from 22 towns to 218 towns, now covering most of Uganda. What have been the success factors?

a) Consistent government policy and support
b) Effective leadership with a business approach with inclusive area management involvement and encouragement to take initiative
c) No interference by government in water service operations
d) Staff bonus incentives for improved performance with effective staff training
e) Comparative competition through benchmarking with publicized rewards to areas for 'top' performance
f) Elimination of corruption; all staff involved in corrupt practices were fired
g) Recognition that for affordability, there has to be efficient service operations
h) Recognition of the importance of reducing NRW, both reducing leakage and increasing revenue collection, for sustainability.

17.5.3 Chile

Water services reform in Chile began after the Allende communist period of 1970–1973, during which water charges were just 5% of costs, such that consumers wasted water and there was insufficient income for maintenance of the systems. There were a number of governance changes (Rouse 2013) which made improvements, but the most effective step was the Tariff Law of 1998 and the establishment of the regulator, Superentendencia de Services Sanitarios, SISS. This was accompanied by the formation of state-owned water companies. SISS drove service improvements and efficiencies through a performance-based delivery plan process. Chile now has the most developed urban water service system in Latin and South America, with 99.9% coverage. What have been the success factors?

a) Consistent government policy and support even during a change from a conservative to a socialist government
b) A corruption-free governance system

c) Commercially regulated water companies. The major improvements were achieved with public sector operations. Privatization (partial) took place later to raise capital for wastewater treatment, this being necessary to achieve the environmental standards to satisfy a free-trade agreement with the EU

d) Full cost recovery with tariffs determined by the regulator based on need

e) Efficiency improvements through regulator-driven benchmarking. In 2013 there were 2.4 staff per 1000 service connections

f) Provisions for the poor through incremental charges for connections and direct subsidies for those on a means-tested register

g) Extension of the distribution systems to serve poor areas through capital subsidies

h) 24/7 services, although NRW is an average of 30% against the SISS benchmark of 15%.

17.5.4 Singapore

Singapore has probably the most developed water service operation in the world, but that was not the situation back in the 1970s when Singapore was a relatively poor country. The story of water and wastewater services reform began around 1971 (Tortajada et al. 2013). What have been the success factors?

a) A strong political objective to be independent of Malaysia for water resources

b) Integration of water and wastewater with integrated and consistent planning over 45 years

c) Attention not only to multiple water resource developments, but equal attention to the demand side, with investment in water distribution and NRW down to less than 5%

d) Full cost recovery including an environmental charge

e) Extensive professional and technical staff development and training with the ability to learn about and incorporate new technology and systems

f) A culture of never being satisfied with progress and of always looking for improvements.

17.5.5 Conclusions

Based on the discussion in Section 17.4 and supported by the four country case examples, there are the following key requirements for achieving universal access to safe drinking water:

a) Reform takes time, covering several decades, with a requirement for government policies to be consistent across jurisdictions.

b) Water cannot be considered in isolation of city planning. There is a requirement for integrated city and water planning.

c) The major cost of achieving SDG 6.1 will be in the development of the distribution systems. Sustainable water services cannot be achieved without continuous 24/7 supplies. This is rarely recognized but there is a requirement for investment to obtain good data, to develop network models, to extend the systems to unserved areas, and to deal with leakage.

d) The governance structure must make provision for delivery plans (typically five years). These are hard-nosed business plans to deliver defined customer-service objectives stage by stage towards the long-term strategic directives. These plans must be fully costed, with committed finance, and must be verifiable, monitored and enforced.

e) Recruitment of skilled and committed staff is vital. Currently most public workers are paid poorly compared with the private sector, so the best people leave for better remuneration. Staff remuneration and conditions, which are comparable with the private sector, and which have incentives to reward improved performance, should be introduced. In the authors' view this does not require privatization and is most acceptably done by converting public sector providers from a municipal model to a so-called corporatized public company one. These companies are required to adopt private-sector approaches, including on remuneration, whilst remaining publicly owned. This should be accompanied by zero tolerance of corrupt practices.

f) General government subsidies to reduce the cost of water services exist in the belief that they help the poor. In practice it is a disincentive for water service management to be more efficient, and does not allow for increased investment in poor areas. Changing to targeted subsidies to help the poor, as in Chile, increases access for the poor, and achieves greater efficiency leading to sustainable 24/7 services.

g) Although there will be high investment costs in infrastructure, which will require grants and subsidies for affordability, there must be sustainable cost recovery for operating costs and maintenance. This is essential to achieve sufficient income to maintain systems, and to gain the level of creditworthiness necessary to attract private finance. The success in Phnom Penh has shown that it can be achieved in a very poor country.

17.6   Achieving Awareness of What Needs to Be Done

Governments, which have to deal with many problems competing for attention and investment, require some incentive to give priority to water service provision. Generally, the individual 'cabinets' will not be in office long enough to see the full success from reforms and investment, but there should be opportunities to recognize and reward work in progress. Perhaps each year, the UN could give recognition to the ten countries making the greatest relative progress through a badge of honour presented by the Secretary General, with due international publicity.

This could be accompanied by donor banks funding the transfer of knowledge and experience of the 'best in class' countries, or perhaps more likely cities, to hold training workshops, and provide hands-on capacity-building in countries with similar circumstances.

17.7   An Outline Road Map to the Sustainable Development Goal (SDG) on Water

Previous international stimuli made progress in the delivery of water services, but fell well short of safe sustainable services for all. SDG 6.1 will be no different unless the challenges are addressed effectively. The following road map provides a suggested sequence of actions leading to the achievement of SDG 6.1.

1) Government commitment across political factions, leading to a review of the government structure to cover all the various functions related to achieving the goal, and to provide an effective governance structure.
2) The lead government department to take the lead in preparing the strategic plan as discussed in Section 17.4.3.
3) The lead government department consults other departments including the treasury to discuss likely investment costs and funding. Early meetings with potential donors are advisable at this time.
4) Water service providers to be required to carry out a review of population numbers and predicted changes, infrastructure and availability and accuracy of operational data, including cost and revenue information.
5) Based on the results of 4, investment in data and data systems to determine baseline information to allow calculation of the investment options leading up to 2030. The required actions would most likely be in the first year of the delivery plan (6 and 7 below).
6) The lead government department (or regulator if there is one) to develop a template for five-year delivery plans. A timetable should be set for development of the first delivery plan.
7) Water service providers to prepare first delivery plan, assisted as necessary. These must be realistic and deliverable. They should include a financial model to facilitate discussion of options with the Finance Department and Donors.
8) The lead government department, with the Finance Department, holds detailed discussion with funding agencies.
9) The lead government department, or regulator if there is one, specifies the reporting requirements, and monitors progress.
10) Parallel with the delivery plans there has to be a human resource development programme and associated training.
11) A review of the process is carried out, say at the end of year 4, desirable changes are made to the planning process, and work begins to plan the next five-year delivery plans. This cycle is repeated not just up to the SDG target year of 2030, but beyond, to sustain the achieved level of customer service and move (as affordable) to higher levels of quality of delivered water related to Figure 17.3.

References

Bartram, J., Brocklehurst, C., Fisher, M. et al. (2014). Global monitoring of water supply and sanitation: history, methods and future challenges. *International Journal of Environmental Research and Public Health* 11 (8): 8137–8165.

Biswas, A. and Tortajada, C. (2010). Water supply of Phnom Penh: an example of good governance. *International Journal of Water Resources Development* 26 (2): 157–172.

Black, M. (1998). Learning what works: a 20 year retrospective view on international water and sanitation cooperation (English). In: *Water and Sanitation Program*. Washington, DC: World Bank Group.

Chan, E. (2009). Bringing safe water to Phnom Penh's City. *International Journal of Water Resources Development* 25 (4): 597–609.

Choquill, C., Franceys, R., and Cotton, C. (1993). *Planning for Water and Sanitation.* Loughborough, UK: Loughborough University Centre for Development Planning Studies.

De Albuquerque, C. (2014). *Realising the Human Rights to Water and Sanitation: A Handbook by the UN Special Rapporteur Catarina de Alberquerque.* https://www.ohchr.org/en/issues/waterandsanitation/srwater/pages/handbook.aspx.

Doshi, V., Schulman, G., and Gabaldon, D. (2007). Lights! Water! Motion! In: *Global Perspective: Strategy + Business Issue.* Washington, DC: Booz, Allen, Hamilton.

Everard, M. (2015). Community-based groundwater and ecosystem restoration in semi-arid North Rajasthan (1): socio-economic progress and lessons for groundwater-dependent areas. *Ecosystem Services* 16: 125–135.

Farley, M. and Trow, S. (2001). *Losses in Water Distribution Networks.* London: IWA Publishing.

GDRC (2008). *The First Water Decade, Urban Environment Management.* https://www.gdrc.org/uem/water/decade_05-15/first-decade.html (accessed July 2018).

Hao, X., Novotny, V., and Nelson, V. (2010). *Water Infrastructure for Sustainable Communities – China and the World, Cities of the Future Series.* London: IWA Publishing.

Hutton, G. and Bartram, J. (2008). Global costs of attaining the millennium development goal for water supply and sanitation. *Bulletin of the World Health Organization* 86 (1): 13–19.

Hutton, G. and Varughese, M. (2016). *The Costs of Meeting the 2030 Sustainable Development Goal Targets on Drinking Water, Sanitation, and Hygiene.* Washington, DC: World Bank.

Kummu, M., Guillaume, J., de Moel, H. et al. (2016). The world's road to water scarcity: shortage and stress in the 20th century and pathways towards sustainability. *Scientific Reports* 6 (1).

Muhairwe, W.T. (2009). *Making Public Enterprises Work; From Despair to Promise: A Turn Around Account.* London: IWA Publishing.

Novotny, N. and Brown, P.e. (2007). *Cities of the Future: Towards Integrated Sustainable Water and Landscape Management.* London: IWA Publishing.

OECD (2016). *Policy Perspectives on Water Growth and Finance.* Paris: OECD Publishing.

OECD (2018). *Policy Perspectives on Financing Water – Investing in Sustainable Growth,* Environment Policy Paper No 11. Paris: OECD Publishing.

Reiff, R. (2015). *The Problem with Intermittent Water Delivery in Developing Countries.* Water Quality and Health Council.

Rouse, M.J. (2013). *Institutional Governance and Regulation of Water Services: The Essential Elements,* 2e. London: IWA Publishing.

Sanctuary, M., Tropp, H., and Haller, L. (2005). *Making Water Part of Economic Development: The Economic Benefit of Improved Water Management and Services.* Stockholm: Stockholm International Water Institute (SIWI).

Tortajada, C., Joshi, Y., and Biswas, A. (2013). *The Singapore Water Story.* Hoboken, NJ: Taylor and Francis.

UN Population Division (2015). *World Population Prospects: The 2015 Revision, Key Findings and Advance Tables. Working Paper* ESA/P/W.P.241.

World Bank WSP (2010). *The Karnataka Urban Water Sector Improvement Project. 24/7 Water Supply is Achievable.* Water and Sanitation Program, Field Note, September 2010. New Delhi: Water and Sanitation Program.

World Health Organization (1991). *Evaluation of the International Drinking Water Supply and Sanitation Decade, 1981–1990*. Report by the Director-General. Geneva: World Health Organization.

World Health Organization (WHO) and the United Nations Children's Fund (UNICEF) (2017). *Progress on Drinking Water, Sanitation and Hygiene: 2017 Update and SDG Baselines*. United Nations Children's Fund (UNICEF), World Health Organization Joint Monitoring Programme.

Zevenbergen, C., Fu, D., and Pathirana, A. (2018). Transitioning to sponge cities: challenges and opportunities to address urban water problems in China. *Water* 10 (9): 1230.

18

Equity and Urban Water Security

Katrina J. Charles[1], Thanti Octavianti[1], Erin Hylton[2], and Grace Remmington[3]

[1] *School of Geography and the Environment, University of Oxford, UK*
[2] *Concurrent Technologies Corporation, Washington, DC, USA*
[3] *Cranfield University, UK*

18.1 Introduction

Urbanization is a critical development pathway. Managing water security effectively for urban spaces, particularly during the transitions of development, is vital to support that pathway. Disruption to the pathway can come in multiple water-related forms: extremes such as floods and droughts, difficulties securing sufficient water, or marginalization from access for certain urban dwellers. Without adequate management, the potential disruptions to people and the economy are huge: in Jakarta in 2007, a catastrophic flood killed 79 people and displaced 500 000. In São Paulo in 2015, an historic drought threatened the city's water supply, forcing the utility to implement pressure reductions that limited access to peripheral and elevated communities, and the army to be readied to respond to potential civil unrest. Whilst these impacts are most notable in megacities, similar issues are playing out in small urban centres as people migrate for economic opportunities and better access to services, but find opportunities constrained by water insecurity.

Whilst meeting water security needs in urban areas has always been a critical challenge, the unprecedented levels of urbanization and concentration of world's population in cities means that the issue is becoming even more critical to a sustainable global future. The urban population globally increased from 30% in 1950 to 55% in 2018 (United Nations 2018). This is projected to increase further to 68% by 2050. Even more critically, urban areas are estimated to create over 70% of GDP globally. Therefore, urbanization and related industrialization are expected to drive this statistic significantly upwards in the coming decades, placing enormous pressure on limited services and resources. As populations in urban spaces include diverse water users, from the poor attracted by economic opportunities or better access to services, to international industries, the impacts of water insecurity are diffuse across urban spaces.

The challenge of urban water security is to balance the water security trade-offs for these heterogeneous populations and interests to achieve sustainable development. But is this a new challenge? Many developed cities have stories of their own pathways to

Water Science, Policy, and Management: A Global Challenge, First Edition. Edited by Simon J. Dadson, Dustin E. Garrick, Edmund C. Penning-Rowsell, Jim W. Hall, Rob Hope, and Jocelyne Hughes.

improved water security. Two pathways commonly taught to students relate to large infrastructure development in the mid-1800s: London's Great Stink was a focusing event that lead to sewering the city, and New York's basin transfer scheme brought safe water to Manhattan. These are useful illustrations of decision-making processes, but offer limited transferability to many modern-day scenarios, as we discuss later on in this chapter. There is much to learn from more recent developments in these water systems, such as the adoption of ecological classification of water bodies and the benefits of catchment management; however, there is also a need for new pathways to be developed that allow for developing urban areas to meet their water security needs. Hence, we frame urban water security here not as an engineering challenge of new pipework and large infrastructure, but as a management challenge of how to innovate for better water outcomes, through policies, practices and technologies.

In this chapter we focus on improving water security in the urban space under the dynamic conditions therein. These dynamic conditions of urbanization and industrialization are a major challenge to delivering the Sustainable Development Goals (SDGs) by 2030. The explicit overlap between two goals highlights the necessity of treating urbanization and urban water security as interrelated issues. SDG 11 aims to make cities and human settlements inclusive, safe, resilient and sustainable, which includes providing basic services such as drinking water and sanitation and reducing the impact of water-related disasters, whilst SDG 6 seeks to 'ensure availability and sustainable management of water and sanitation for all'. The key challenges for the water sector related to achieving SDG 11 are explored through this chapter. To do this, we draw from our research in megacities (Jakarta, Indonesia, and São Paulo, Brazil); these are the most intractable challenges in urban water security due to multiple dimensions of lock-in that have occurred through their historical development pathways. We start with an international perspective on the challenge of urban water security, considering the continuum of urban spaces and the challenges from nascent urban spaces through to more established cities. Secondly, we explore the trade-offs in urban water security through the case study of Jakarta, with particular reference to the role of politics in water security, and how the marginalized can bear the greatest negative impacts of development. We counter more utilitarian perspectives by presenting a positive example of how urban water security can be universal and equitable from São Paulo.

18.2   Urban Water Security: Framing the Global Challenge

Water security has been defined in a variety of ways and these definitions continue to evolve. As a concept, it has been operationalized in a range of ways in urban areas. Regardless of definition, improving water security requires recognizing that water management must consider the interactions between water users via the water cycle, necessitating trade-offs. In this section we explore the definitions of urban water security and the urban space before addressing the global challenge of urban water security.

18.2.1   Urban Water Security

One of the earliest and encompassing definitions of water security is 'the availability of an acceptable quantity and quality of water for health, livelihoods, ecosystems and production, coupled with an acceptable level of water-related risks to people, environments and economies' (Grey and Sadoff 2007). Rooted in risk science, water security has also been defined as 'the tolerable water-related risk to society' (Grey et al. 2013). However, water security means different things to different people, as it depends upon one's perspective and particular geographical setting. From a legal perspective, for example, the key determinant of water security is water allocation to secure certain water quantities, whilst protection from floods and droughts is considered as the main component of water security in agriculture.

To define urban water security, it is important to recognize the variety of conceptual domains constituting the term. Within the growing literature, water security is often framed as an integrative concept. The core dimensions of water security represented in the literature include meeting basic human needs, environmental needs and productive uses (particularly on agricultural productivity), as well as resilience to water-related disasters. Governance has also been considered as a separate dimension by some studies (e.g. Jepson 2014; Mason and Calow 2012).

Urban water security is further complicated by fuzzy analytical boundaries. Urban areas cannot be divorced from their relationship with rural areas. Indeed, their fate is intertwined: urban populations rely on the rural hinterlands to grow food and raise livestock to support urban development and to produce raw materials for industries. Additionally, rural land is managed to reduce flooding in downstream cities and to provide reliable water supplies of good quality for human health and industrial use. Urban water security is thus dependent on actions outside the urban centre; however, for the purposes of this chapter we will focus on the urban only.

Urban water security is a distinct component of national water security and faces particular challenges (Hoekstra et al. 2018); therefore, a specific definition is critical. We define urban water security as a tolerable level of water-related risks to urban society, that facilitate equitable human and economic development within the urban space, whilst protecting the environment, without restricting development in rural areas. In practice, urban water security is highly political as trade-offs are managed between wealthy corporations and elites, and marginalized groups with limited agency, as we highlight in our case study on Jakarta.

18.2.2 The Importance of the Urban Space

Whilst the concept of 'urban' may conjure very definite images to the reader, the definitions in practice are not always clear or able to inform operational decisions. Above we defined urban water security without specifically defining the concept of an urban space. The term 'urban' is nebulous and needs further unpacking. Urban areas are defined in terms of density, size, employment, or administrative arrangements, with definitions varying between countries. These factors will also influence appropriate water security pathways. There is a need to transcend traditional dichotomies of rural and urban 'towards a more complex settlement differentiation' (McGranahan and Satterthwaite 2014, p. 20) that facilitates planning which is reflective of these complexities. For the purposes of this chapter, where we are considering the water security pathways that are associated with dynamism in these areas, we use the term 'urban space' to collapse the narrative around different types of urban settlement, incorporating a spectrum from small towns to megacities, formal to informal areas, and going beyond administrative boundaries to peri-urban areas.

SDG 11 incorporates cities and human settlements in the goal. This stand-alone goal for urban areas was developed to reflect the importance of their role in meeting the overarching aim for sustainable development. Despite cities and urbanization present-ing challenges for environmental sustainability and poverty reduction, research has also demonstrated that cities can be a driver of growth and development (Glaeser 2011). Cities offer economies of scale, both for economic activity and the provision of basic infrastructure, including for water services. They offer reduced transport and transac-tion costs, opportunities for knowledge sharing and networks, and access to markets, hospitals and universities (Satterthwaite 2017; Turok and McGranahan 2013). Cities, it is argued, provide a transformative function to development which engenders a con-centration of resources, human capital and services which, in turn, accelerate economic activity and drive innovation (Revi and Rosenzweig 2013). As a result, urbanization can be positively correlated with economic growth, where water security (and other chal-lenges) is adequately managed.

18.2.3   The Challenge of Water Security for Urban Spaces

Urbanization often has a dual identity. In addition to being viewed as an economic benefit as described above, it is also viewed as a 'challenge' to be overcome which cre-ates issues of marginality and exclusion, as well as creating environmental degradation, unmanaged expansion and the concentration of risk (Hoekstra et al. 2018). These are key water security challenges, creating specific impacts for water services and environ-mental protection because water management needs to balance multiple, diverse trade-offs between different actors and users. Particularly, there is a need to balance these trade-offs whilst safeguarding human health and the environment, and simultaneously promoting economic growth. Here we focus on three facets of urban dynamism that impact water security, firstly population growth, secondly geographies of governance, and finally the temporality of urban water security pathways.

At the core is population growth: too often, urbanization and population growth is inadequately planned. Smaller urban spaces have different institutions, resources and needs to larger urban areas, which require different approaches. As population density increases, the potential for epidemics increases if public health is not adequately man-aged through clean drinking water and sanitation provision. 'Urban' approaches to water services have been designed for established cities, without considering how to adapt them to smaller centres or how to transition them as the population grows. Smaller settlements on the continuum between urban and rural face significant issues from having weaker institutional and infrastructural capacities, lacking the economies of scale to meet physical and hydrological challenges and regional distribution of goods and services (Adank et al. 2016). In small towns, these mismatched approaches create significant stresses for small and intermediate urban settlements whose institutions and infrastructure, as well as inability to meet costs and increased staffing demands, present a challenge to meeting water supply needs following 'urban' models (Rouse 2013; Satterthwaite and Tacoli 2003). In the case of water and sanitation, some of these inter-mediate settlements have generally low institutional capacity, coupled with a weaker enabling environment, which results in greater risks to the provision of water supply, and less ability to plan for a growing population.

Long-term changes in population are commonly addressed through strategic plan-ning, but rapid changes in population are rarely considered. Rapid changes occur when populations, particularly from rural areas, move to urban areas in periods of drought, conflict or economic downturn. Whilst these movements may be temporary, the impact on the urban space can be lasting as land is developed to make housing, water services are rapidly expanded, and the urban space has to cope with additional sanitation needs. A recent example is the Syrian conflict, which has resulted in high migration into neighbouring countries, notably into urban areas in Lebanon, placing high stress on existing water systems. In these periods of rapid change, much of this response is unplanned, resulting in occupation of flood plains or land reserved for future service development. Urban water security is often locked-in by the planned and unplanned developments in the space. Particularly, this rapid development challenges the abilities to create and implement long-term and sustainable plans for managing water services.

Another facet of the urban water security challenge is the geographies of governance. Administrative delineations can create barriers to water security: spatial boundaries create geographical inequalities and policy development in silos hinders the integration of water management. Administrative boundaries between urban and rural or peri-urban, and between formal and informal areas, often prescribe the level of water security afforded to residents. Definitions of what constitutes 'rural' or 'urban', and what constitutes 'formal' or 'informal' within an urban space is both contentious and context-specific. This complexity, which constitutes the definition of urban spaces, creates difficulties in understanding the different needs and capacities of water supply provision.

'Slums' and informal settlements, including peri-urban areas, have often been excluded from accessing the services afforded to the city, including water provision and sanitation, but also other water security interventions such as flood protection. In some cases, urban planning has further entrenched marginality and increased risk to these settlements, such as through measures that see more informal settlers building homes on flood plains (Braun and Aßhauser 2011). Although there have been some efforts to mainstream or formalize service provision to these areas, the occupants of informal settlements and peri-urban areas still often bear disproportionate water risks compared with formalized cities. These risks include public health risks (e.g. waterborne disease) but also loss of assets, housing and social networks (e.g. during flooding events) (Braun and Aßhauser 2011). In order to realise SDG 11 by making 'cities and human settlements inclusive, safe, resilient and sustainable' by 2030, much more critical attention needs to be paid to the exclusion from water-related services within the cityscape (Mitlin et al. 2011).

At a policy level, the components of water security are commonly managed by different ministries of a government. Flood protection, water abstraction, water supply, sanitation and environmental water quality are commonly considered separately. Whilst not unique to urban settings, the scale of each aspect can encourage a gulf between the broadly defined 'water' policies.

Despite these challenges, there are many positive international examples of managing water security through urbanization. For example, China's 'sponge cities' create approaches to managing storm water that reduce flood risk, and improve environmental conditions and social wellbeing. However, caution has to be applied to translating pathways for urban development to different social, political and climatic contexts. Urbanization has progressed at varying rates in different continents, providing different challenges. In Europe, urbanization occurred with industrialization over a period of more than 100 years, achieving over half the population living in urban areas by the 1950s; this milestone was reached even earlier in North America. There are many examples of how river water quality challenges, drinking water quality, sanitation and industrialization were managed, but with limitations on their transferability to present-day conditions. The Thames' clean-up in London in the nineteenth and twentieth centuries relied on significant capital investment to construct a sewer scheme, and on the ability to export polluting industries overseas. In New York, a cholera outbreak in 1832 created the impetus to develop an aqueduct to bring clean water to Manhattan from the rural Croton watershed. Yet protected rural areas are not always readily available in countries with higher rural populations.

Even where positive urban water security examples exist, inequalities in water security still remain an issue. For example, the lead contamination case in Flint, Michigan, highlights the wealth inequalities in water security in a developed country. The development trajectories in other regions have differed from Europe and North America, and have resulted in different experiences of urbanization. In Latin America, urbanization occurred later, but more rapidly, over the twentieth century. Asia is only now achieving 50% urbanized, whilst Africa has 43% of its population living in urban areas. These later, and faster, trajectories of urbanization impact upon development pathways as international policies, technologies and priorities change.

Crucially, there is a geographical and temporal component to pathways to improved urban water security. Due to physical geographical factors (e.g. hydrology and climate) and human geographical factors (political context, development plans, technological and policy shifts), the trajectories of urban development will not be the same in Africa or Southeast Asia as in Europe and North America. Current urbanization processes are taking place during a period of international policy that focuses on equality and dignity through the SDGs, which will promote different pathways to improved water security than those that have driven European and North American urban water security. The climates in which they are developing mean a concerted international effort will be needed to eradicate diseases readily wiped out of Europe and North America. The development of other countries internationally mean that they are recipients of waste and toxic industries, rather than being able to export these issues. These global structures mean that countries in the Global South are more likely to need to deal with pollution issues from industry, whilst also supplying water services that protect health.

This nexus of urbanization and urban water security creates specific challenges and opportunities for delivering secure water services in urban spaces, and for meeting the SDGs (particularly dealing with the complexities of different boundaries and classifications of 'urban' and its implications for service delivery). Dealing with rapid urbanization processes requires agile and equally rapid responses to ensure the protection of cities against shocks, to support economic growth and development, and to 'ensure availability and sustainable management of water and sanitation for all' (SDG 6).

18.3   Trade-offs in Urban Water Security

The need to manage competing interests, making trade-offs between them, makes water security an inherently political process. In this section we explore the competing priorities for water security in the case study of Jakarta, Indonesia, where large infrastructure has been touted as the solution to a complex water security situation. We find that trade-offs between both water sectors and users are ignored by a project designed to address part of the biggest risk, sea-level rise and coastal flooding. We first explore the dimensions of the water security challenge, then analyse the trade-offs in achieving water security for different sectors from the perspective of politics, and then examine users' water security from the perspective of inequalities.

18.3.1 The Water Security Challenge

Jakarta has multiple water security challenges. At the core is land subsidence: 40% of the Indonesian capital is already below mean sea-level, and it is settling at a rate estimated in some areas to be as high as 15 cm/year (Abidin et al. 2011). As sea-levels continue to rise, subsidence is increasing the coastal flood risk in this delta city already burdened by pluvial and fluvial flooding. In 2007, the worst flood in recent memory inundated 60% (400 km^2) of the city, killing 79 people, forcing 500 000 people to evacuate, and causing almost US$680 million in damage (Bappenas 2007). Whilst harrowing, these statistics do not include the loss of productivity during the flood events or the associated public health impacts, and they mask the disproportionate burden on poor and marginalized people. The flood was attributed to a combination of high rainfall intensity within the city, runoff of water from poorly managed upland areas, and high tides. Despite other serious problems of poor water and sanitation coverage and polluted waterways, it is flooding that has developed as Jakarta's primary water security concern.

Processes of urbanization have contributed to increasing water insecurity in Jakarta. After Indonesia declared its independence in 1945, Jakarta's population significantly increased from 600 000 people in 1945 to 1.3 million people in 1949. This exponential growth generated uncontrolled development within and beyond Jakarta's administrative boundaries. Dedicated green areas for stormwater infiltration were converted to settlements and commercial spaces. As people choose to live outside the city, development beyond (particularly upstream of) Jakarta continues to grow. Today, Jakarta's metropolitan area is one of the largest urban agglomerations on the planet with 30 million inhabitants. As the capital, Jakarta has inherently more power and wealth compared with other cities in the region, but decentralized policy, which gives authority and financial independence to local governments to manage their own jurisdictions, has limited Jakarta in imposing its power on other local leaders. Urbanization has also impacted the city's water supply and wastewater provision, as the government was not able to cope with rapid population growth. In the early independence era, the central government was busy creating a modern image in the city by building monumental symbols (Abeyasekere 1989). As a result, resources were less available to provide basic services, including water. As the city was busy growing, the importance of urban ecosystem was not a priority on the government's agenda. Only recently did the poor quality of waterbodies started to become a serious issue.

18.3.2 One Solution for a Complex Issue

After the 2007 flood, the government proposed to take action to mitigate floods by constructing an ambitious 32-km offshore seawall project, part of the NCICD plan (National Capital Integrated Coastal Development), to close Jakarta Bay and reduce sea-level in the created coastal lagoon within the wall. A privately led land reclamation project was planned to fund a significant portion of this US$40 billion project. This proposal would address some aspects of flooding, but the potential trade-offs between water risks are not addressed, with the potential to cause unintended effects or externalities of the megaproject. Some of the key trade-offs that are not fully addressed include:

- Land subsidence: this has been purported to be caused by over-extraction of groundwater due to insufficient piped water services. NCICD has been narrated as the solution to solve the sinking problem, without containing any concrete policy action to solve the subsidence. Resources are focused to pursue NCICD, and therefore provision of water supply and a ban on groundwater use have become secondary objectives of Jakarta's water policy (Octavianti and Charles 2018).
- Water quality: Initially, the 1 million m^3 retention lake was projected to serve as both a drinking water reservoir and a means to reduce mean sea-level. Policy-makers were keen to develop their own water source to slake the thirst of the megacity, yet the idea was criticized from a water quality perspective. Given that all 13 rivers flowing into Jakarta Bay have poor water quality, the cost to treat the water to an acceptable standard would be prohibitive.
- Wastewater treatment: Proponents claim that the integrated approach of NCICD would be able to improve the poor environmental condition of Jakarta Bay, which was caused by wastewater discharge from the city. Whilst NCICD was able to reinvigorate the city's 15-year-old sewerage plan, there has been limited progress in realising that plan. The goal is to increase sewerage coverage from the current 7% (one of the lowest rates in urban areas in Asia) to 75% by 2020 in line with the seawall timeline (PAL Jaya 2018). However, neither the Jakarta provincial government nor the central governments have made any concrete commitments. PAL Jaya, the city's wastewater institution, is struggling to find investors for the sewerage project estimated to cost almost US$5 billion. Without timely measures to stop water contamination, the retention lake will turn into a large septic (anoxic) lagoon.
- Runoff from the catchment: The sea wall project has a myopic focus on the urban centre, without considering urban–rural linkages to areas that are not contained in the administrative boundaries of Jakarta. The programme does not address the risk from these regions and the role in flood management in upland catchments.

Flood control is the main purpose of NCICD, but it is worth questioning its ability to achieve that aim. One key weakness that has not been addressed is how maintenance of the seawall and dykes will be ensured, as this an area in which Indonesia does not have a strong track record. In 2016 alone, dyke failures caused five flood incidents (Ramadhiani 2016), including the collapse of a dyke protecting Pantai Mutiara, an exclusive area in the north of Jakarta, resulting in the inundation of homes in up to a metre of water, causing substantial damage for the residents. Without proper maintenance, NCICD may actually pose a higher risk of flooding whilst encouraging more people and assets to move into vulnerable areas of the new waterfront city.

Overall, will this seawall support the city's economic growth? The reclamation element of the NCICD has been promoted to create employment and add enormous value for the city. According to the master plan, it will structurally employ over 550 000 people and temporarily employ 4250 persons per year during construction of the sea wall. This added value is estimated at US$64 billion (present value) through to 2040 (NCICD 2013). Significant development on reclaimed land was initially proposed which would offset the costs and attract investment, but a large-scale corruption case related to these developments has made their future uncertain. If it does go ahead, the land reclamation policy indirectly encourages people to live in the capital and consequently concentrates development there. Moreover, these people who reside at the forefront of the bay are at high level of exposure from water-related disasters. By putting people and assets at the forefront of a defence mechanism (Nicholls 2011), land reclamation increases the risk of damage or loss if disasters occur.

In Jakarta, water security has been framed as an issue of flood control because floods have caused massive economic and human damage. However, the project proposed only partly addresses flood risk and ignores interrelationships between flooding and other critical water sectors. This narrative has been created deliberately to package the seawall as an instant solution to the flooding problem (Octavianti and Charles 2018). In reality, key causes of flooding including land subsidence due to over-extraction of groundwater, and upstream catchment management are not being addressed, whilst a lack of attention to the severe pollution in coastal areas due to poor sanitation service threatens to create a toxic lagoon that would undermine potential economic benefits. Therefore, collective water security is unlikely to be achieved if the project's externalities are excluded.

18.3.3 Universal and Equitable Development

One of the key principles of the SDGs is that development should be universal and equitable. In cities, the urban poor are often marginalized when it comes to water security, as in so many other aspects. They are more likely to live on marginal land, subject to flooding. They are more likely to lack access to adequate drinking water and sanitation services, to find water harder to access in a drought, and to suffer the health burden from environmental pollution. Despite better physical access to hospitals and healthcare, the urban poor often have higher rates of stunting and mortality in children under five than the rural poor. Moreover, it can be hard to identify these hardships, as the urban poor are often poorly visible in the metrics used to track water security, such as access to drinking water. The large-scale statistics used for tracking progress of the SDGs mask more granular inequalities.

In Jakarta, the proposed seawall development focuses on safeguarding and developing economic interests, at the expense of marginalized groups. The narratives around the project have excluded the needs of the poorest, ignoring the lack of access to water and sanitation, and focusing direct negative impacts onto one marginalized group. Here a significant trade-off is made that imposes the burden onto those with the least power.

In the discourses around the NCICD project, the equity issue boiled down to the claim that relocating 27 000 poor fishers is necessary in order to save 3.5 million Jakarta people from sinking. This line has undergone serious simplification by confronting the interest of fishers as a minority compared with the general public.

Traditional fishers are socially and economically marginalized communities in Jakarta. They are well-established communities who have been contributing to the local economy and provide affordable sources of protein for people living in Jakarta. Their settlements often equate to slums and they are considered to be squatters (Padawangi and Douglass 2015). Despite the fact that the current state of marine resources is declining due to pollution and overexploitation, Jakarta Bay still provides livelihoods for millions of people in the coastal areas (Baum et al. 2016). The estimated 27 000 fishers in

Jakarta Bay, 56% citizens of Jakarta, does not include other professions related to the fishing industry, such as salted fish workers and green mussel cleaners (BPS 2015). Their resistance to the project is primarily because land reclamation activities will eventually destroy their only source of livelihood.

The narrative that the project will be 'saving 3.5 million people from sinking' is a powerful storyline, but this is a flawed statement and should not be taken for granted, as discussed above. Our concern here is that the negative impacts of development are easily focused on the most vulnerable people who do not have adequate resources and power to fight back.

Different philosophies view this trade-off from different perspectives. An economic egalitarian, who believes people are equal and is committed to reducing gaps in the economic means of people, would tend to support the fisher community. A utilitarian would give weight to the priority of the greater good, which is the public safety in this case. A no-nonsense libertarian would support the project in general, considering the fact that this project uses private investments to fund the seawall project, and consequently the relocation of the fishing community would not be an issue.

There are at least two main forms of injustices/inequities to fisher communities affected by the NCICD project. First, the NCICD may cause fishers to lose their only source of income. They fear that the seawall construction will put them out of business permanently. A closure of Jakarta Bay for fishing activities could amount to a total annual production loss of approximately US$53 million (Bakker et al. 2017). Many fishers have been complaining of their decreased fish catch due to poor water quality; now they need to fish in an area further off to sea. Despite locks added in the NCICD masterplan to accommodate fishers, the fuel cost of travelling to fishing grounds far from the shores will reduce their income significantly. A dedicated rumah susun or rusun (low-cost apartment) in the reclaimed land for ex-fishers who want to work as blue-collar workers for the islands will be provided. The government believe that this is the best solution they can offer to improve the living quality of the poor. However, in the NCICD business case, no clear financial provisions are made for this plan.

The second form of inequity is that the relocation of traditional fishers to *rusun* (low-cost apartments) will destroy this local community and traditions. The integrated approach to coastal development in NCICD is promised to improve equity in north Jakarta where both wealthy and poor communities live on the shores. New homes offered to the fishing families are far away from the bay, which makes it impossible for this low-income group to commute from the suburbs and maintain their livelihood (Padawangi and Douglass 2015). Existing eviction of poor communities is criticized. People of the Muara Angke fishing settlement already face eviction from their homes due to the ongoing reinforcement of dykes (Bakker et al. 2017). Although they agreed that dyke reinforcement is needed, the relocation procedures do not provide adequate information on the process and location of the *rusun*.

Those two interrelated forms of inequity are perceived differently by the government and the fishers. The government believes that they provide justice via solutions to accommodate the fishers: *rusun* housing, locks to facilitate fishing activities, as well as opportunities to work as blue-collar workers in the new land. Meanwhile fishers, whose lives have been lived in the *kampung* as fishers, are embedded in their way of life. The government understands this social process as linear; they think that moving them to *rusun* can open a new path for a better life. These different perceptions are partly caused by ineffective communication between government and the fishers. There is a possibility that information received by the affected people was distorted or they simply have not been informed about the project and the relocation plan. The affected people want to be given opportunities for dialogue in search of a win-win solution to minimize the social consequences of this infrastructure project (Bakker et al. 2017). In this case, the cause of injustice can be viewed as both subtle and complex: subtle in that the injustice can potentially be reduced with open communication between stakeholders, but also complex because people are being restricted from opportunities to achieve their fullest potential.

Such situations where inequalities are exacerbated, rather than reduced, are not uncommon in urban water security improvements. Major infrastructure projects assume utilitarian principles of justice, enabling them to ignore inequalities. The Jakarta example highlights that to achieve universal and equitable development the marginalized need a voice. The section below presents a contrasting perspective to Jakarta, offering an example of where the traditionally marginalized have been targeted for water security improvements to the benefit of the city's water security.

18.4 Inclusive Water Security: A Case Study of São Paulo's Water

A bottom-up approach to addressing the needs of the urban poor can reduce inequalities and have benefits beyond serving the underserved, as is highlighted in the case of the megacity of São Paulo, Brazil.

In São Paulo, a major drought in 2014–2015 reduced the city's water reservoirs to 3% capacity. Worried about the dwindling supply but simultaneously trying to avoid outright rationing, the water and sanitation company (SABESP) chose to engage in a public education campaign and offer financial incentives for users to reduce their consumption. However, they also chose to reduce the system pressure, leading to *de facto* rationing for many users. The impacts were disproportionately felt by marginalized residents living on the urban periphery or in elevated areas where pressure reductions significantly limited, and in some cases terminated, access to piped supply. As the drought intensified, the army was readied to respond to any civil unrest that accompanied a shift to standpipes. Fortunately, rain began to refill the city's reservoirs before the situation became worse.

A one-in-250-year event, the 2014–2015 drought was twice as severe as SABESP's worst-case scenario planning. Part of SABESP's response has been to expand its service to informal areas through the *Água Legal* programme, launched in March 2017 (SABESP 2016). Roughly 30% of São Paulo's 12 million residents live in slum-like conditions, lacking access to basic infrastructure such as water or sewage (Marques and Saraiva 2017). These informal areas are quite diverse, ranging from newly constructed shacks on the margins of polluted waterways to two-storey, cinder block homes that have been occupied for decades in pockets of the urban centre. What they have in common is their uncertain land tenure, which under Brazilian law means that it is illegal for utilities to provide services to residents. Faced with this reality, residents commonly create *gatos* (unsanctioned water connections) by tapping into SABESP's water mains. Although the mains contain potable water, shoddy materials, untrained installation of above-ground pipes, and the lack of sanitation services lead to contamination of the water supply before reaching the point of delivery. Additionally, these *gatos* fail to maintain sufficient pressure for 24/7 service, even when SABESP is not engaging in pressure reductions. When the water is flowing, residents are not incentivized to conserve because they do not pay for their consumption.

Responding to the dual pressures of the drought and community demands for formal water provision, SABESP has worked with municipal authorities to secure a path to service extension for some communities under *Água Legal*. The utility has budgeted US$51.5 million to upgrade 160 000 connections and avoid 3.3 billion litres of physical and commercial losses per month through 'regularization' of water connections. The programme includes four parts: upgrading physical infrastructure, installing water meters, administratively recognizing connections, and engaging in social outreach to reduce consumption in informal areas (SABESP 2017). The programme has not been without its challenges; SABESP must obtain the necessary permissions, work with residents who prefer the clandestine supply, and overcome the physical constraints associated with working in unplanned settlements on marginal land.

For those that the programme has reached, *Água Legal* has achieved high levels of resident satisfaction (Hylton and Charles 2018). In addition to providing water, regularizing the water supply has provided residents with a water bill, which serves as a proof of address that empowers residents to access other formal services, such as registering for employment or opening a bank account. SABESP is looking to incorporate other services as well; eventually piped sewage and water supply will be installed together. They are also looking at installing Wi-Fi to facilitate mobile payments, to incentivize bill payment, and to encourage broader community development, which will contribute to beneficial educational and economic outcomes.

From the perspective of informal residents, *Água Legal* improves water security by providing a connection to a reliable supply of water, by reducing the water quality risks associated with clandestine connections, and facilitating engagement with formal markets. For SABESP, the programme significantly reduces non-revenue water, provides a new source of funding, eliminates the infrastructure deterioration associated with *gatos*, and reduces the risk of infiltration or inflow. Taken together, these outcomes will improve São Paulo's urban water security by building resilience to future droughts and reducing water quality risks.

18.5 Conclusions

Urbanization is a critical development pathway that presents a challenge for water managers. Achieving SDG 11 by making cities and human settlements inclusive, safe, resilient and sustainable requires trade-offs in water security to be managed in complex technical, social and political environments. Here we summarize the challenge to water managers working in urban spaces in this SDG period and beyond.

First, achieving integrated water management in socially and administratively complex urban spaces presents a challenge. A water security approach can help ensure that the interconnectedness of water systems and water risks are addressed, in the urban space as well as in the upstream and downstream communities. Flood management, sanitation, water supply, groundwater and ecosystems are interrelated systems that are managed through complex, often disparate, governance arrangements. The associated water-related risks are rarely viewed equally, giving power to some interests at the expense of others. In Jakarta, flooding has been perceived as the dominant risk, with the push to deliver flood protection stealing focus from the impacts on other water sectors. Managing different agendas will be key to delivering an integrated system.

Second, urban water security is not apolitical. Water management is often assumed to be the domain of water engineers, solving technical challenges. However, urban water security is heavily influenced by politics, and has to be responsive to all populations, regardless of their political or economic clout. Water managers will require inclusive, interdisciplinary approaches to be effective.

And finally, water management in urban spaces has to consider explicitly who wins and who loses, and the benefits of pro-poor approaches. In São Paulo, a drought stimulated a programme that empowered residents through service extension and administrative inclusion in the formal economy, whilst simultaneously improving the physical and financial health of the water utility, and thus the water security of the city. Investing in equitable water services for the poor can provide additive benefits for the broader urban space.

Acknowledgements

This work is partially funded by the REACH programme (www.reachwater.org.uk) funded by UK Aid from the UK Department for International Development (DFID) for the benefit of developing countries (Aries Code 201880). However, the views expressed, and information contained in it are not necessarily those of or endorsed by DFID, which can accept no responsibility for such views or information or for any reliance placed on them.

References

Abeyasekere, S. (1989). *Jakarta: A History*. Singapore: Oxford University Press.

Abidin, H.Z., Andreas, H., Gumilar, I. et al. (2011). Land subsidence of Jakarta (Indonesia) and its relation with urban development. *Natural Hazards* 59 (3): 1753–1771. https://doi.org/10.1007/s11069-011-9866-9.

Adank et al. (2016). WASH Services in Small Towns. OneWASH Plus Programme report. The Hague: IRC.

Bakker, M., Kishimoto, S., and Nooy, C. (2017). *Social justice at bay: The Dutch role in Jakarta's coastal defence*. Amsterdam, Netherlands: SOMO, Both ENDS, and TNI https://www.somo.nl/wp-content/uploads/2017/04/Social-justice-at-bay.pdf.

Bappenas (2007). Hasil penilaian kerusakan dan kerugian pascabencana banjir awal februari 2007 di wilayah Jabodetabek (Damage and loss assessment report in Jabodetabek area caused by the early 2007 flood). http://www.bappenas.go.id/files/8913/5441/6576/hasil-penilaian-kerusakan-dan-kerugian-pasca-bencana-banjir-awal-februari-2007-di-wilayah-jabodetabek__20081123211335__1300__0.pdf (accessed 25 January 2017).

Baum, G., Kusumanti, I., Breckwoldt, A. et al. (2016). Under pressure: investigating marine resource-based livelihoods in Jakarta Bay and the Thousand Islands. *Marine Pollution Bulletin* 110 (2): 778–789. https://doi.org/10.1016/j.marpolbul.2016.05.032.

BPS (2015). *Statistik Daerah Provinsi DKI Jakarta 2015*. Jakarta: Badan Pusat Statistik Provinsi DKI Jakarta.

Braun, B. and Aβhauser, T. (2011). Floods in megacity environments: vulnerability and coping strategies of slum dwellers in Dhaka/Bangladesh. *Natural Hazards* 58: 771–787.

Glaeser, E. (2011). *Triumph of the City: How Our Greatest Invention Makes Us Richer, Smarter, Greener, Healthier and Happier*. London: Macmillan.

Grey, D. and Sadoff, C.W. (2007). Sink or swim? Water security for growth and development. *Water Policy* 9 (6): 545–571.

Grey, D., Garrick, D., Blackmore, D. et al. (2013). Water security in one blue planet: twenty-first century policy challenges for science. *Philosophical Transactions of the Royal Society* 371 https://doi.org/10.1098/rsta.2012.0406.

Hoekstra, A.Y., Buurman, J., and van Ginkel, K.C. (2018). Urban water security: a review. *Environmental Research Letters* 13 (5): 053002.

Hylton, E. and Charles, K.J. (2018). Informal mechanisms to regularize informal settlements: water services in São Paulo's favelas. *Habitat International* 80: 41–48.

Jepson, W. (2014). Measuring "no-win" waterscapes: experience-based scales and classification approaches to assess household water security in colonias on the US-Mexico border. *Geoforum* 51: 107–120. https://doi.org/10.1016/j.geoforum.2013.10.002.

Marques, E. and Saraiva, C. (2017). Urban integration or reconfigured inequalities? Analyzing housing precarity in São Paulo, Brazil. *Habitat International* 69: 18–26. https://doi.org/10.1016/j.habitatint.2017.08.004.

Mason, N. and Calow, R. (2012). *Water Security: From Abstract Concept to Meaningful Metrics (Executive Summary)*. London: Overseas Development Institute.

McGranahan, G. and Satterthwaite, D. (2014). *Urbanisation Concepts and Trends*. IIED Working Paper. London: IIED.

Mitlin, D., Satterthwaite, D., and Bartlett, S. (2011). *Capital, Capacities and Collaboration: the multiple roles of community saving in addressing urban poverty*. IIED Poverty Reduction in Urban Areas Working Paper 34. London: IIED.

NCICD (2013). *Pengembangan Terpadu Pesisir Ibukota Negara (National Capital Integrated Coastal Development)*. Jakarta: The Coordinating Ministry of Economic Affairs http://en.ncicd.com/ncicd/downloads.

Nicholls, R.J. (2011). Planning for the impacts of sea level rise. *Oceanography* 24 (2): 144–157. https://doi.org/10.5670/oceanog.2011.34.

Octavianti, T. and Charles, K. (2018). Disaster capitalism? Examining the politicisation of land subsidence crisis in pushing Jakarta's seawall megaproject. *Water Alternatives* 11 (2): 394–420.

Padawangi, R. and Douglass, M. (2015). Water, water everywhere: toward participatory solutions to chronic urban flooding in Jakarta. *Pacific Affairs* 88 (3): 517–550.

PAL Jaya (2018). Cakupan layanan air limbah (Wastewater service coverage). http://www.paljaya.com/jasa-layanan-kami/air-limbah-sistem-perpipaan (accessed 1 August 2018).

Ramadhiani, A. (2016). Tanggul Laut Raksasa Dinilai Bukan Solusi Atasi Penurunan Tanah Jakarta (Giant Sea Wall is not a Solution for Jakarta's Land Subsidence). *Kompas*. 14 May 2016. http://properti.kompas.com/read/2016/05/14/154112221/Tanggul.Laut.Raksasa. Dinilai.Bukan.Solusi.Atasi.Penurunan.Tanah.Jakarta (accessed 9 June 2017).

Revi, A. and Rosenzweig, C. (2013). *The Urban Opportunity: Enabling Transformative and Sustainable Development*. Research Report. New York: Sustainable Development Solutions Network. http://unsdsn.org/wp-content/uploads/2014/02/Final-052013-SDSN-TG09-The-Urban-Opportunity1.pdf.

Rouse, M. (2013). *Institutional Governance and Regulation of Water Services*, 2e. London: IWA Publishing.

SABESP (2016). *SABESP Sustainability Report – 2016*. http://site.sabesp.com.br/site/uploads/file/asabesp_doctos/RS2016_ing.pdf (accessed 15 December 2018).

SABESP (2017). *Programa Água Legal vai evitar a perda de 3,3 bilhões de litros em ligações clandestinas*. http://site.sabesp.com.br/site/imprensa/noticias-detalhe.aspx?secaoId=65&id=7404 (accessed 17 June 2017).

Satterthwaite, D. (2017). The impact of urban development on risk in sub-Saharan Africa's cities with a focus on small and intermediate urban centres. *International Journal of Disaster Risk Reduction* 26: 16–23. https://doi.org/10.1016/j.ijdrr.2017.09.025.

Satterthwaite, D. and Tacoli, C. (2003). *The urban part of rural development: the role of small and intermediate urban centres in rural and regional development and poverty reduction*. IIED Working Paper. London: IIED.

Turok, I. and McGranahan, G. (2013). Urbanization and economic growth: the arguments and evidence for Africa and Asia. *Environment and Urbanization* 25 (2): 465–482. https://doi.org/10.1177/0956247813490908.

United Nations (2018). *World Urbanization Prospects: The 2018 Revision, Key Facts*. https://population.un.org/wup/Publications.

19

Reflections on Water Security and Humanity

David Grey

School of Geography and the Environment, University of Oxford, UK

> *We are at a unique stage in our history. Never before have we had such an awareness of what we are doing to our planet. And never before have we had the power to do something about it. Surely we have a responsibility to care for our blue planet. The future of humanity, and indeed all life on Earth, now depends on us.*
>
> (David Attenborough 2017)

19.1 Introduction

The focus of this chapter on water security is the relationship between humanity and hydrology, with an underlying hypothesis that there is a correlation between hydrological complexity and economic status, other factors being equal (Grey and Sadoff 2007; Hall et al. 2014).

There is a particular focus on those parts of the world where water and climate systems are exceptionally variable and complex, and, at the same time, people and governments are relatively poor, with limited resources and capacity to manage such complexity. Herein lies the challenge of achieving and sustaining 'water security', an enduring human goal since the origins of humanity and civilization.

In our time, this goal is becoming increasingly difficult to achieve and sustain, due to the impacts of rapid changes in population, economic growth, industrialization, and – we now know for certain – climate. But it is a goal we must achieve in this century, if humanity as we know it is to survive and flourish.

All known life on Earth requires liquid water, without exception (Ball 1999). Temperatures on Earth are favourable for the presence of water in all its forms, solid, liquid and gas, and, driven by the energy of the sun, a relatively small proportion of the Earth's water is constantly moving in the 'water cycle' (Chapter 2), which supports life throughout its journey. The impacts of climate change on the water cycle are uncertain. However, significant loss of ice from the poles, leading to sea-level rise, will increase the 97.5% of the Earth's water that is in the oceans and reduce the already only 2.5% of the Earth's water that is fresh.

Water Science, Policy, and Management: A Global Challenge, First Edition. Edited by Simon J. Dadson, Dustin E. Garrick, Edmund C. Penning-Rowsell, Jim W. Hall, Rob Hope, and Jocelyne Hughes.
© 2020 John Wiley & Sons Ltd. Published 2020 by John Wiley & Sons Ltd.

19.2   Human Origins and Water: Then and Now

It is useful to reflect on the past as context to the present and our future. We can therefore perhaps see in sharper focus the needs of water management today and for tomorrow.

19.2.1   African Beginnings

Water has played and continues to play a central role in human development (Finlayson 2014). During the last century, much work has been done to understand the rise and fall of great lakes in Africa and the relationship with early hominids, including the existing lakes of east and central Africa, the dying lakes, such as Lake Chad, and the fossil lakes of the Sahara and Kalahari deserts (Maslin 2017).

Investigations in the early 1970s on the margins of the great Makgadikgadi salt pans in Botswana's Kalahari Desert mapped gravel ridges that could only have been shorelines, and identified Paleolithic stone tools that had been rolled by wave action in a wet period, and later, Mesolithic, finely-worked tools which had not been rolled, in a drier period (Grey and Cooke 1977). Much research in the Kalahari has since been undertaken, providing evidence for the relatively rapid fluctuation and disappearance of a great lake, impacting the settlement of our human ancestors since the Stone Age (Burrough 2016).

Recent theories propose that *Homo sapiens* evolved and moved in response to complex and changing climate systems in equatorial Africa. Evidence suggests that *H. sapiens* first moved out of Africa about 100 000 years ago, some moving along the lower Nile valley into the Middle East, continuing eastwards about 70 000 years ago into south Asia and on to east Asia (Shultz and Maslin 2013). It is along this route that we find early evidence of the Neolithic revolution that began about 5000 years ago, with irrigated agriculture in the four 'great river valley civilizations' of the Nile, the Tigris and Euphrates, the Indus, and the Yellow rivers. The first three of these are discussed below. The Yellow River story is well told by Ball (2016).

19.2.2   The Nile

The Nile is the best known and understood of the great river civilizations, as its 5000 year old story is first told in cuneiform script deciphered from the Rosetta Stone, and in the surviving pyramids and tombs along its banks in Egypt and northern Sudan. The Nile's role has been interwoven with human history since then and it continues to be (Collins 2002).

The longest river in the world today, at 6500 km from its most distant source to the Mediterranean Sea, the Nile waters and the silt that the river brings down created the conditions for civilization to emerge in a desert landscape. Water was harnessed in the lower floodplains for extensive irrigation systems, requiring strong institutions to manage water allocations, creating great wealth and enabling widespread trade. Variable rainfall patterns in the headwaters and both droughts and floods in Egypt resulted in the development of mathematical tools and measuring devices ('nilometers') to predict flows, and hence crop planting times and potential yields. Longer-term variability resulted in the rise and fall of the pharaonic dynasties. About 2400 years ago, Herodotus wrote that 'Egypt is the gift of the River' (Griffiths 1966).

Today, the Nile is shared by 11 countries. Burundi, Rwanda, Tanzania, Kenya, Uganda, Democratic Republic of Congo, southwest Ethiopia and South Sudan are the sources of the 'White Nile', which provides a low but steady flow, buffered by the vast Sudd wetlands of South Sudan. The Ethiopian highlands are the primary source of the Blue Nile, which has a highly variable flood flow, and of the Atbara River, whose Tekeze tributary is shared with Eritrea. The Blue Nile joins the White Nile at Khartoum (Sudan), picking up the Atbara River further north and flowing on to Egypt.

The population of the Nile Basin countries is about 500 million and increasing rapidly, with great and growing demand for water, food and energy to support expanding cities and industries across the basin. There has been, and continues to be, tension between Nile riparian states, and much has been written on Nile 'hydropolitics' and the associated risks (Kimenvi and Mbaku 2015; Waterbury 2002). Egypt is particularly concerned, due to millennia of complete dependence on Nile flows from upstream. The Nile countries have worked hard together for many years to build trust and cooperation, establishing the Nile Basin Initiative in 1999 (see: www.nilebasin.org), to promote collaborative endeavours. Regional diplomacy and international support continue today through the Nile Basin Initiative and its associated investments, laying the foundations of trust and capacity that will be essential for the goal of long-term, cooperative Nile management and sustainable regional development.

19.2.3 The Tigris and Euphrates

The Tigris and Euphrates river basin, characterized by upland headwaters, highly variable flows and an extensive floodplain, is also one of the world's great international river basins, shared by Turkey, Syria, Iran and Iraq. The opportunity for irrigated food production, coupled with the imperative of flood regulation, gave birth in Mesopotamia (land 'between two rivers') to one of the oldest known civilizations about 5500 years ago, including the Sumerian, Babylonian and Akkadian empires. Here we find the first evidence of writing, the wheel, and other recorded 'firsts' (Kramer 1981).

Population and economic growth over the past century within today's basin states have seen major development of the river, including hydropower, primarily in the headwaters, and extensive irrigation. This relatively recent development in the basin is creating new hydrological, environmental, economic, legal and political challenges.

Downstream, Iraq's water challenges are amongst the most serious globally, with increasing water and soil salinity and decreasing water quantity becoming a crisis that needs urgent action (Rahi and Halihan 2018). The basin's extensive marshlands are of regional and international significance and are now under severe threat. All the riparian states are signatories to the Ramsar Convention, and the Mesopotamian Marsh is a Ramsar site, with all signatories under an international obligation to protect it, requiring international cooperation. Uncertain climate futures, coupled with growing water demand and very limited international cooperation, are likely to make both the quantity and quality crises even more severe in the future.

19.2.4 The Indus

The Indus River rises as snow melt from the high Himalayas flowing across a great floodplain with highly variable monsoonal rainfall, leading to both serious floods and droughts. Major tributaries include the Kabul, the Swat and the Kurram from the west, and the Jhelum, the Chenab, the Ravi, the Beas and the Sutlej from the east.

The Indus Valley civilization lay almost hidden until the 1920s, when ancient mounds near the city of Harappa and the Ravi River in the Punjab were excavated (Wheeler 1953). Soon after, similar mounds, called Mohenjo-daro, were excavated 600 km south-west along the Indus main stem. The remains of two major cities emerged, and soon after many other smaller cities were discovered, all with evidence of urban institutions and architecture. These findings revealed a great civilization, established along the Indus about 4500 years ago, in which water development was pioneering, not only for extensive irrigation and industry, but also for public baths, household water connections and early 'sewers' for waste water (Jansen 1989).

Today the Indus River basin, with its interlinked tributaries, supplies water to the world's largest contiguous area of irrigated land. The Indus tributary headwaters pass through four nations, Afghanistan, China, India and Pakistan, to form the Indus main stem flowing down Pakistan's Indus valley to the Arabian Sea near Karachi. Most of the basin lies within India and Pakistan, with about 13% situated upstream in China and Afghanistan. With the Partition of India in 1947, the new India–Pakistan border cuts across the upper Indus and its five eastern tributaries, giving India the upstream water-rich headwaters and the potential opportunity to prevent downstream flows. Pakistan was left as the water-short downstream riparian, with 95% of its population and most of the Indus basin's irrigable and irrigated area within its new borders.

A Treaty was needed. After eight years of World Bank-brokered negotiations, the Indus Waters Treaty was signed in September 1960 (Gulhati 1973). The Treaty's key points are the allocation of the eastern rivers (Ravi, Beas, Sutlej) to India, and the western Rivers (Indus, Jhelum, Chenab) to Pakistan, and the establishment of the Indus Commission to implement and monitor that treaty. Major new storage infrastructure was needed in Pakistan, to feed link canals needed to move water into the eastern rivers whose flows across the border were to cease.

The Treaty's success is evidenced by the fact that it has withstood repeated tensions, including wars of 1965, 1971 and 1999 (McKinney 2011; Sarfraz 2012). However, with India now developing hydropower on the western (i.e. Pakistan's) tributaries, allowed under the Treaty with specific conditions, there have been two recent cases, (the Baglihar difference and the Kishanganga dispute) requiring external arbitration. The division of the Indus waters remains tense, significantly contributing to already tense relationships between India and Pakistan (Haines 2017).

19.2.5 What Might We Learn from These Reflections?

These stories are of the central role that water played in early human origins, in the development of the earliest known large civilizations, and continuing today amongst nation states. They are instructive for many reasons, not least in leading to important questions. What was the historical relationship between human development and water management, and what, if any, are the implications for today? What was the direction of causation, and was it the same everywhere? As to causation, Mithen differs somewhat from Wittfogel's view that powerful institutions were needed to build irrigation systems (Mithen 2012; Wittfogel 1975). Instead, Mithen argues that power grew from small communities controlling access to water, and that building irrigation and other water systems created powerful leaders. In ancient Egypt, irrigation created great institutions, whilst major flood events and long periods of drought brought down dynasties (Zhang et al. 2008). In ancient China, the leaders that managed the floods are still revered, and those that did not were reviled (Ball 2016).

A lesson that appears to emerge from early human history is that societies and their leaders that managed water well became strong and successful and that societies that did not manage water well suffered and even collapsed (Cullen et al. 2000; Kaniewski et al. 2015; Mithen 2012; Yancheva et al. 2007). This lays the foundation for the concept of *water security* as an enduring human goal of individuals, leaders and societies, encompassing the complex and evolving relationships between humans and water. The lesson of history to be drawn here is that society and its leaders today must develop the knowledge, skills and institutions to manage water wisely if they are to survive and flourish.

These stories of three of today's great international river basins provide examples of how water security and national and international security are interwoven, and of the need for interdisciplinary skills, from hydrology to politics, economics and negotiation, to address the challenges of dispute resolution, international cooperation and benefit-sharing (Sadoff and Grey 2002, 2005; Wheeler et al. 2018).

19.3 Water Security and Risk

Water is critical for human life, healthy ecosystems and food production and it is important for energy production and navigation. As a consequence, water has great cultural (including religious) value in most societies and can foster cooperation within and between communities. Through its presence or absence, water can also be a cause of death, destruction, poverty, dispute, and even conflict. There is an ancient Chinese proverb:

水能载舟，亦能覆舟 *Not only can water float a boat, it can sink it also.*

At any time, we can scan the news and find water-related crises around the world. For example, Japan, one of the world's wealthiest and most technologically advanced nations, suffered from its worst rainfall-generated floods in 36 years in July 2018, with 177 people killed and 70 missing. Eight million people were ordered to evacuate their homes, 270 000 homes were without water supply services, and 1000 more homes without electricity.

Water security is a relatively recent, still evolving, debated and contested concept that addresses this dichotomy (e.g. Beck and Villaroel-Walker 2013; Cook and Bakker 2012; Grey and Sadoff 2007; Mukhtarov and Gerlak 2015; Zeitoun et al. 2016). There is also a debate about whether the concept of 'integrated water resources management' (IWRM) is in competition with the concept of 'water security'. IWRM has been successfully promoted by the Global Water Partnership as 'a process which promotes the coordinated development and management of water, land and related resources, in order to maximize the resultant economic and social welfare in an equitable manner without compromising the sustainability of vital ecosystems' (Global Water Partnership 2017). IWRM is thus a 'process' or a 'journey'.

In contrast, as with food security and energy security, water security is an 'outcome' or a 'destination'. Food security and energy security both mean adequate access to food and energy (with 'adequate' encompassing factors such as reliability, affordability and quality). The concept of water security is broader, as it needs to include two elements in the 'destination': first, adequate access to water for multiple uses, and, second, managed impacts of water-related shocks, including floods, droughts and contamination. One of the most cited definitions is that 'water security is the availability of an acceptable quantity and quality of water for health, livelihoods, ecosystems and production, coupled with an acceptable level of water-related risks to people, environments and economies' (Grey and Sadoff 2007, p. 547).

Both the productive and destructive impacts of water can be described in terms of risk – the risks of inadequate access and the risks of unpredictable and/or unmanageable events. A complementary and more reductionist definition has been proposed: 'water security is a tolerable level of water-related risk to society' (Grey et al. 2013, p. 4). 'Security' here means freedom from want, doubt, anxiety, fear and danger. In risk science, 'tolerable' means risk that is as low as reasonably practicable – a zone that is between acceptable and unacceptable, which will be very context-specific and dependent upon social, economic and cultural factors. 'Society' means an aggregate of people within a community, including needs and values.

This construct is controversial, as rich societies will (and can afford to) tolerate lower risk – for example through investing in flood, drought and pollution risk reduction measures. Poor societies may not have the financial and institutional capacity to reduce such risks to a desirable level, and must therefore tolerate higher risks. This argument reinforces the value of a risk-based approach, underscoring the need for financial transfers to support water security investment in poorer parts of the world where society carries a much higher burden of risk. The concept of tolerable risk is also dynamic. Water risks may become less tolerable with economic growth and increasing wealth, and water risks may increase with exogenous change (such as climate) and endogenous change (such as forest and land use) and become less tolerable.

Water risks can inhibit economic growth, and mitigating these risks can expand economic growth opportunities, beyond water (Khan et al. 2017). Water security can thus be considered to be an outcome providing two freedoms: freedom from intolerable water-related risks and freedom to pursue otherwise constrained social and economic opportunities, drawing upon Amartya Sen's concept of 'development as freedom' (Sen 1999). It is important, therefore, to make the economic case that development finance transfers from richer to poorer nations to reduce water risks may both reduce direct water impacts and catalyse economic investment in other sectors, such as education, infrastructure and industry, all of which are necessary conditions for national political and social stability, with regional and global benefits.

Our planet will see dramatic changes in this, the twenty-first century. From a global population of about 1.6 billion in 1900, the UN estimates that it is 7.6 billion today and it will climb to 9.7 billion by 2050 and 11.2 billion by 2100 (UNDESA 2017). Water security, and associated food and energy security, will be an existential challenge for all of humanity. But several billion people are water insecure today – and this is an existential challenge for them now.

One measure of the scale of water-related risk is captured in the World Economic Forum's Global Risk Perception Survey (GRPS), which is undertaken annually to capture the views of the Forum's 'network of business, government, civil society and thought leaders', with 684 respondents contributing to the 2018 Survey (World Economic Forum 2018). Participants rank a wide range of global risks in terms of both their likelihood and their impact. Over the past five years, 'water crises' feature every year amongst the global risks perceived both to be most likely to occur and to have the highest impact.

In the 2018 report, 'water crises' feature amongst several water-related global risks associated with climate change, biodiversity, ecosystems, food, and involuntary migration, together considered to be the most likely to occur and to have the highest impact. This deep and growing concern about water risks and their impacts on society is thus not just the voice of water scientists and practitioners, but, increasingly, the voice of a wide network of opinion-formers around the world. Given this, the challenge for water specialists is to analyse challenges, define potential solutions, make their voices heard by leaders across the world, and monitor actions and results.

As the lessons of the distant past have suggested, achieving water security made leaders and societies strong and successful. Imagine a place where 98% of households have no electricity, 41% have no toilets, and 78% go 300 m for water. There is low literacy and life expectancy, with people living on subsistence farming on exhausted soil, suffering from frequent, catastrophic flooding. This could describe the poorest communities in Africa or Asia today (Kummu et al. 2011; World Bank 2018a, b). But this was the Tennessee Valley, in southern USA, in 1935. Franklin D. Roosevelt stood for the Democratic nomination in 1932, promising 'a new deal for the American people'. Elected and then inaugurated in 1933, he announced his New Deal in his first 'hundred days', including the launch of the Tennessee Valley Authority. Within a generation: $224 m/year of flood damage was prevented; almost 100% literacy was achieved; life expectancy increased to over 70 years; smallpox, malaria and typhoid were eradicated; industrial production was up by over 500%; 700 miles of navigable waterways to the sea were developed; and median incomes in the Valley matched the national level. With flood management came large reservoir storage and hydropower generation, supporting large-scale production of aluminium (ALCOA) for aircraft manufacture in time for the war effort. This is a tale of water security challenges and solutions, of leadership and of economic transformation (J. Delli Priscoli, personal communication).

19.4 Eight Major Global Water Security Challenges

There are many water security challenges faced by society across the world. Here I identify eight major challenges, based on a career spent working on these challenges in many countries and basins across the world, seeking elusive solutions.

19.4.1 The Dynamic Challenge of Water Security Risks in Changing Climates

A first and overarching water security challenge is managing water security risks associated with the climate regime within which different societies have developed and live now, coupled with the way in which these climate regimes may be changing.

History tells us that there has always been a complex relationship between water and human development and stability (Vorosmarty et al. 2000). But is this true today? Why are some parts of the world poor and water insecure and others wealthy and water secure? Is poverty the cause of water insecurity? Or is water insecurity the cause of poverty? Or is there no relationship?

Scientists and field practitioners have postulated for some years that climate, in particular rainfall variability, has a significant impact on economic growth (World Bank 2006). Recent analysis has demonstrated a relationship between runoff variability and the affected economy, with runoff being a useful parameter, as it incorporates several factors including rainfall, temperature, topography, soil and land use. The wealthiest parts of the world have the lowest runoff variability, yet have had the resources to invest the most to mitigate that variability and ensure water security, whilst the poorest parts of the world have the greatest runoff variability and, unsurprisingly, have invested the least and remain water insecure (Hall et al. 2014).

Climate change could increase runoff variability significantly (Banze et al. 2018), potentially deepening water insecurity in the poorest regions of the world and increasing the difficulty and costs of achieving and sustaining water security for the several billion people that are water insecure there today (Collins et al. 2013).

19.4.2 The Challenge of Water Supply and Sanitation

A second – and the most urgent – water security challenge for humanity is to remove health and dignity risks to every human by guaranteeing water supply and sanitation (WSS) services at all scales, starting at the household level and including schools, villages, cities and nations.

Despite the 1980s International Drinking WSS Decade, a global initiative seeking to deliver 'water and sanitation for all', and progress under, first, the Millenium Development Goals (MDGs) and now under the Sustainable Development Goals (SDGs), ensuring reliable and affordable WSS services remains a massive challenge.

An estimated 2.1 billion people are without access to clean water at home and 4.5 billion people are without safe sanitation (WHO 2017). In some poorer countries, schools do not have toilets, and even if they do they are inadequate and often not segregated, with evidence from Pakistan that this results in girls in rural areas not going to school (Asian Human Rights Commission 2012). By 2050, the world's urban population of 4.2 billion in 2018 is expected to reach 6.7 billion, with the rural population declining slightly (UNDESA 2018). With urban centres growing rapidly, particularly in Africa and Asia, innovative and affordable WSS solutions are essential to meet the needs of these large, densely populated and rapidly growing cities.

19.4.3 The Challenge of Hunger

A third critically important water security challenge is to remove the risks of hunger by ensuring water availability at the farm level in many parts of the world, to increase the crop yields needed to feed the world's growing population.

The number of malnourished people has been increasing, to 815 million in 2016, with stunted growth of 155 million children, and 3.1 million child deaths annually due to starvation (FAO et al. 2018; World Hunger Education Service 2018). In Africa and Asia, where precipitation is highly variable, irrigation is particularly important to buffer rainfall variability and ensure higher and reliable crop yields.

However, irrigation consumes more water than any other human activity, and inefficient irrigation is widespread, wasting water that could provide higher social and economic returns elsewhere (FAO Aquastat 2016). Due to its institutional and policy regime, India's rice yield per hectare is only a third of that of China and half of those of Vietnam and Indonesia (World Bank 2012a, b). If the world's population in 2050 is to be fed, irrigated areas will need to be greatly expanded and irrigation efficiencies greatly increased. Current research suggests that crop water requirements can be reduced significantly, making the available water stretch further (e.g. Ragab et al. 2017).

19.4.4 The Challenge of Floods

A fourth and particularly difficult water security challenge is coping with the episodic risks of floods, a natural consequence of a wide range of events, including storms, earthquakes, tsunami, ice, and snowmelt, often aggravated by land-use change and man-made infrastructure (whose failure may cause floods).

Floods can be managed for a purpose, such as recession agriculture. If unmanaged, floods can be uniquely destructive, as they take lives, reshape landscapes and destroy infrastructure and property. There are many recent and catastrophic examples, with the most vulnerable being the large and growing populations of south and southeast Asia at the mercy of the annual monsoon, where both too much rain and too little rain can have great impacts.

The exceptional 2007 South Asia monsoon resulted in widespread flooding with major impacts across India, Bangladesh, Nepal and Pakistan, with over 4000 lives lost and massive crop losses (UNICEF 2007). A year later, on 18 August 2008, the Kosi river (at a time of low river flows) breached its left embankment 13 km upstream of the Kosi Barrage inside Nepal (constructed and maintained by India, under the 1966 Kosi Treaty). Most of the river flowed out through the 10 km wide breach, crossing the national border into Bihar and flowing onward along former river courses. Within a few days, an unknown number of people were killed, 50 000 people in Nepal and 3.5 million people in India were affected and large areas of crops destroyed. The flood and its widespread impacts were the result of inadequate maintenance and infrastructure failure, aggravated by the absence of early warning (Dixit 2009).

The 2010 monsoon in Pakistan brought the worst floods in its history and catastrophic impacts, with 2000 deaths, 20 million people displaced, 1.7 million homes destroyed and an estimated economic loss of about US$15 billion (Relief Web 2011). The 2011 monsoon brought more floods, with many still in camps from the previous year and many more millions affected. There was further flooding in the 2012 monsoon.

In Thailand in 2011, exceptionally heavy rain from May onwards led to extensive flooding. By October, the floods were close to Bangkok. Whilst the city itself was protected, surrounding areas were flooded, including many large industrial plants. The numbers are stark: 880 people were killed, 1.5 million homes damaged, 25% of the rice crop destroyed and 7500 industrial plants flooded (World Bank 2012b).

19.4.5 The Challenge of Drought

A fifth water security challenge is coping with local risks of drought, which occur in most parts of the world and are relatively easily managed in rich countries, through grain storage and trade and bulk water supply restrictions. However, in poor countries, commonly characterized by high rainfall and runoff variability, drought can have massive humanitarian impacts and social costs.

The rains failed in the Horn of Africa in 2010 and 2011, resulting in severe drought and crop failure, coupled with internal conflict. The consequences were catastrophic, with about 15 million people in Somalia needing food aid, with 350 000 refugees moving across the border into Kenya's Dadab camp. About 260 000 people died of starvation between April 2010 and July 2012, of whom half, about 130 000, were children under five years of age (Checchi and Robinson 2013).

Syria suffered a severe drought between 2006 and 2011, affecting two to three million people in half of the country, with 75% of crops failing and 85% of its livestock dying (Kelley et al. 2015). Farmers moved to major towns and cities to seek paid work.

In 2010, Russia suffered a severe drought with loss of life, extensive forest wildfires, the wheat harvest down by 40% and exports of grain banned in August of that year. This extreme drought in Russia in the 2010 summer almost exactly coincided with the extreme floods in Pakistan, to the south across the Tibetan plateau, with evidence that the two events were 'physically connected' (Lau and Kim 2012).

Much of southern Africa (including Angola, Namibia, Botswana, Zimbabwe, Lesotho, Malawi, Madagascar and South Africa) suffered severe drought in 2015 and 2016 (and locally into 2017), with food insecurity impacting about 30 million people (Relief Web 2017). Early in 2017, the drought broke and heavy precipitation occurred across much of southern Africa, resulting in widespread flooding, with many deaths, hundreds of thousands of people displaced and road and rail infrastructure damaged (Relief Web 2017; World Meteorological Organization 2018). Taken together, this cycle of drought and flood is a very difficult water security challenge, constraining social and economic stability and restricting economic growth in some of the poorest communities in the world.

19.4.6 The Challenge of International and Transboundary Waters

A sixth water security challenge is managing the local and regional risks associated with transboundary waters. There are an estimated 260 international transboundary river and lake basins and about 600 transboundary groundwater basins (Wada and Heinrich 2012). China, whose territory includes much of the Himalayan 'third pole' (with the largest body of ice outside the two poles), is an upstream riparian state sharing 110 rivers and lakes with 17 neighbouring countries, with about 60% of the global population living in the countries that share these basins (He et al. 2014). China has one basin agreement (with Russia) but is now exploring ways to cooperate with neighbouring states, such as the establishment of the Lancang-Mekong Cooperation agenda in 2015, establishing broad economic cooperation with downstream states Myanmar, Laos, Thailand, Cambodia and Vietnam.

There has always been, and there will continue to be, some tensions between nation states sharing transboundary basins, exacerbated where basin knowledge and national capacity is limited, resulting in uncertainty and mistrust. There have been efforts to reduce tensions in many of the great river basins of the world, by investing in shared information systems, transboundary institutions, and joint infrastructure (and other) investment. Considerable efforts over many years to move from dispute to cooperation are increasingly in national leaders, where they need to be if real cooperation is to be achieved (Sadoff and Grey 2005; Sadoff et al. 2003).

19.4.7 The Challenge of 'Spillovers': From Local to Global

A seventh major challenge is to predict and mitigate the 'spillover' of the impacts of local water insecurity on social, economic and political stability in other parts of the world – in unexpected and often unpredictable ways. Three examples illustrate this.

The 2006–2010 Syrian drought, briefly described above, is an important example. There is strong evidence that climate change caused by human activity across the world played a part in the most severe drought ever recorded in Syria, and that this drought played a part in the civil war in Syria that started in 2011 (Kelley et al. 2015). This conflict has escalated and continues, resulting in the large-scale movement of migrants within Syria, into neighbouring countries and beyond the region, and engaging other regional powers in the conflict, as well as western and eastern superpowers. Whilst the drought was not the cause, it was a local trigger for dangerous regional and even global disputes that continue.

A different situation arose from droughts in Russia, Argentina and China, and floods in Pakistan and Australia in 2010 and 2011. These events resulted in reduced food stocks and increased international food prices, including in North Africa, with food and other riots, as the 'Arab Spring' took hold. A 2011 IMF Working Paper concluded that: 'In low income countries increases in the international food prices lead to a significant deterioration of democratic institutions and a significant increase in the incidence of anti-government demonstrations, riots and civil conflict' (Arezki and Bruckner 2011).

The 2011 Thailand floods, whose local impacts, including the damage to industrial sites, are described above, resulted in global supply-chain disruptions, with major global impacts. The latter included loss of vehicle component manufacture for vehicle assembly, such as US and Japanese production of Toyota and Honda vehicles, and electronics industries, such as mechanical hard drives for Seagate and Western Digital (Chongvilaiyan 2012; Kim 2011). With economic costs of US$45 billion, this flood was then the fifth costliest natural disaster on record (World Bank 2012a).

19.4.8 The Challenge for the World's 'Low Latitude' Regions

The eighth and overarching water security challenge is regional in scale but of global importance. This challenge is to understand the complex and little-understood hydrological endowments of 'low-latitude' continental regions of the world, and to identify and implement effective water security solutions there.

Whilst these regions include monsoonal south and southeast Asia, where the challenge is great, nowhere is this challenge greater than in Africa. A very large continent (equal in area to the USA, Europe, India, and Australia combined) straddling the Equator, Africa's current population is 1.2 billion, and is projected to grow to 4.5 billion by 2100 (UNDESA 2017).

An Economic Commission for Africa working paper indicated that: 'Scientific understanding of the African climate system is very low, and the level of understanding varies significantly from region to region. Though it is improving, our understanding of the drivers of the African climate and their complex interactions is still very poor. This lack of knowledge limits our ability to analyze and understand African climate variability, detect and attribute climate change, and predict the climate with an appreciable degree of accuracy' (UNECA ACPC 2011, p. 9).

Africa's inter-annual and intra-annual rainfall variability is very high and there is evidence of a correlation between this variability and economic growth (Odusola and Abidoye 2015). Africa has also inherited complex 'geopolitical legacies', with many international borders originally negotiated and drawn in European capitals during the colonial era as lines on maps, taking limited account of ethnic and natural boundaries. As one consequence, Africa has about 60 international river basins, with a greater of number of basins shared by 3 or more countries than any other continent, with every country sharing at least one basin, 41 countries sharing 2 or more basins, 15 countries sharing 5 or more basins, and Guinea sharing 14 basins (Sadoff et al. 2003). The challenge of Africa's international waters is great, as it is not easy to reach agreement between just two countries on one shared river.

In comparison with any other region of the world, Africa has invested much less in water management, including storage, irrigation and hydropower infrastructure, despite needs that are probably greater than anywhere else. Poverty reduction in rapidly growing Africa requires considerable international support to build the water and climate expertise and information systems, to strengthen the institutions essential to cope with its exceptionally complex hydrology, and to develop and manage the water investments that are essential for water, food and energy security and poverty reduction. Without this support, livelihoods and stability across Africa will be at growing risk, potentially leading to significant population movement, with people seeking sustainable livelihoods elsewhere in Africa and in other near and distant regions of the world.

19.5   Conclusions: Priorities and Pathways for Policy-makers[1]

Achieving and sustaining water security is a complex multidimensional challenge now, and is likely to become much more complex in the deeply uncertain future we face across the world, with growing national, regional and global economic and political 'spill-overs' from water-related events and challenges. There are no 'blueprints' to be followed, only lessons from experience. There will always be trade-offs, with significant opportunity costs, and both winners and losers. But innovative solutions can and must be identified and implemented wisely.

19.5.1   Three Priorities for Investment

Experience demonstrates that a balanced and sequenced portfolio of investments is needed in three interconnected areas (the '3 Ins') to achieve and sustain water security: institutions, information and infrastructure.

The first priority is to invest in ensuring robust institutions for water security, where institutions are the social rules that govern society and organizations are special

---

1 This concluding section draws on Sadoff et al. (2015).

institutions with structure and responsibilities. A societal objective to achieve and sustain water security cuts across and must be built into multiple economic sectors (e.g. environment, agriculture, energy, industry, housing, health), needs social rules and norms, and requires whole-of-government solutions and actions at every level of society. A necessary step is to build, reinforce and sustain the capacity and knowledge to enable innovation and effective solutions to both the general and location-specific challenges of water security. Decentralization is an instrument for building local solutions to specific challenges (upland, lowland, urban, rural, etc.) and for internalizing externalities, where, for example, agriculture protects watersheds and industry pays for treating pollution.

The second priority is to invest in robust information, including the data and knowledge needed to manage water across all aspects of the economy. The starting point is to build and maintain fit-for-purpose water resources and water services monitoring networks. Recovering costs is essential for long-term sustainability, typically through data sales and water tariffs. Water management is greatly facilitated by innovative and adaptive decision tools (e.g. systems models), to assess options, impacts and costs and to support planning and operational decisions. Investing in applied research will deepen the understanding of impacts of water insecurity, and identify economically proportionate responses.

The third priority is to invest in robust infrastructure, with innovative portfolios providing the framework for development pathways that are robust to long-term uncertainty, as the benefits and costs of investments will alter with demographic, economic and climate change. Infrastructure investment is needed across many sectors, almost invariably associated with essential, parallel, investments in institutions and information.

The highest priority here must be to invest in WSS services to provide universal coverage (clearly defined in UN SDG 6), particularly in the rapidly growing towns and cities in the developing world, as the individual benefits accrue nonlinearly across society and are essential for a healthy and stable society. Investment in sustainable groundwater development at scale will have a major role to play in reliable service provision and risk reduction. Investments in mitigating rainfall and runoff variability and in flood and drought management infrastructure (social and physical) is another major priority, as extreme events in countries across the world, both rich and poor, are already immensely harmful in lives lost, economic impacts and global spillovers.

19.5.2 Pathways to Water Security

We define a 'pathway to water security' as a sequenced portfolio of investments in institutions and infrastructure, underpinned by investments in information. We use the word 'portfolio' to denote a set of different investments ideally chosen to complement one another (Elton et al. 1997).

The idea of a 'pathway' has long been used to describe achieving complex goals, ranging from poverty reduction to resilience or adaptive capacity, highlighting the potential for alternative paths to a given outcome, the importance of 'triggers' in prompting human action, and the long-lasting effects of historic decisions and technologies.

Pathways to water security are, therefore, 'path-dependent' – past choices open some options, and foreclose others. Framing the dynamics of water security in terms of adaptive pathways helps to achieve two key goals: first, understanding the historical development paths that shape water security today; and second, gaining insight that helps navigate paths to future water security.

We may measure the success of different portfolios by different baskets of metrics – but all portfolio investments are intended to achieve more collectively than they can alone (Elton et al. 1997). Developing and implementing pathways is difficult due to the technical complexity and the political considerations involved. Historically, many investments were chosen through one incremental decision after another, with varying levels of coordination across projects. Dynamic, adaptive pathways and planning approaches are increasingly being developed (Haasnoot et al. 2013) to guide decision-making under uncertainty as part of multistage planning processes – processes with identified contingency options and opportunities for learning and adjustment.

Executing long-term, multistep plans is difficult, however, within almost any dynamic, political context. Some pragmatism is therefore necessary, to meet the challenges of water security, guided by priorities described in this chapter. Learning from the past is central to this approach. Failure to learn and to act now and into the future to achieve and sustain water security across the world could have catastrophic consequences for humanity and our 'blue planet'.

Acknowledgements

The author acknowledges: the ideas of many World Bank water colleagues across the world, in particular the late Prof. John Briscoe (latterly Harvard, USA) for countless water debates, and Dr Claudia Sadoff (now IWMI Sri Lanka) for a long water security partnership; Prof. Dustin Garrick (Oxford) for support from inception of the chapter and partnership in developing ideas within it; Profs Dadson, Hall and Penning-Rowsell (Oxford) for reviewing drafts and providing advice; and Isabel Jorgensen (Trinity College, Dublin) for patiently researching and preparing references.

References

Arezki, R. and Bruckner, M. (2011). *Food Prices and Political Instability*, Working Paper. Washington, DC: International Monetary Fund.

Asian Human Rights Commission (2012). *Lack of sanitaiton facilities in schools – obstacle in girls' education*. Relief Web. https://reliefweb.int/report/pakistan/lack-sanitation-facilities-schools-obstacle-girls-education (accessed 1 January 2019).

Attenborough, D. (2017). *Blue Planet 2*. December 10, 2017 broadcast, BBC.

Ball, P. (1999). *Life's Matrix: A Biography of Water*. Oakland, CA: University of California Press.

Ball, P. (2016). *The Water Kingdom: A Secret History of China*. Chicago: University of Chicago Press.

Banze, F., Guo, J., and Xiaotao, S. (2018). Variability and trends of rainfall, precipitation, and discharges over the Zambezi River basin, Southern Africa: a review. *International Journal of Hydrology* 2: 132–135.

Beck, M.B. and Villaroel-Walker, R. (2013). On water security, sustainability, and the water-food-energy-climate nexus. *Frontiers in Environmental Science and Engineering* 7: 626–639.

Burrough, S.L. (2016). Late quaternary environmental change and human occupation of the Southern African interior. In: *Africa from MIS 6–2: Population Dynamics and Paleoenvironments* (eds. S. Jones and B. Steward), 161–174. Dordrecht: Springer.

Checchi, F. and Robinson, W. (2013). *Mortality Among Populations of Southern and Central Somalia Affected by Severe Food Insecurity and Famine During 2010–2012*. Somalia: Food Security and Nutrition Analysis Unit. FAO/FSNAU, FEWS-NET.

Chongvilaiyan, A. (2012). *Thailand's 2011 Flooding: Its Impact on Direct Exports and Global Supply Chains*, ARTNet Working Paper Series. Bangkok: ARTNeT.

Collins, R.O. (2002). *The Nile*. New Haven, CT: Yale University Press.

Collins, M., Knutti, R., Arblaster, J. et al. (2013). Long-term Climate Change: Projections, Commitments and Irreversibility. In: *Climate Change 2013: The Physical Science Basis. Contribution of Working Group I to the Fifth Assessment Report of the Intergovernmental Panel on Climate Change*. Cambridge, UK and New York, USA: Intergovernmental Panel on Climate Change.

Cook, C. and Bakker, K. (2012). Water security: debating an emerging paradigm. *Global Environmental Change* 22: 94–102.

Cullen, H.M., Demenocal, P.B., Hemming, S. et al. (2000). Climate change and the collapse of the Akkadian empire: evidence from the deep sea. *Geology* 28: 379–382.

Dixit, A. (2009). Kosi embankment breach in Nepal: need for a paradigm shift in responding to flood. *Economic and Political Weekly* 44: 70–78.

Elton, E.J., Gruber, M.J., and Blake, C.R. (1997). *Modern Portfolio Theory, 1950 to Date*. NYU Working Paper No. FIN-97-003. https://ssrn.com/abstract=1295211.

FAO Aquastat (2016). *Water Withdrawal by Sector*. UN FAO. http://www.fao.org/nr/water/aquastat/data/query/results.html?regionQuery=true&yearGrouping=SURVEY&showCodes=false&yearRange.fromYear=1958&yearRange.toYear=2017&varGrpIds=4250%2C4251%2C4252%2C4253%2C4257&cntIds=&regIds=9805%2C9806%2C9807%2C9808%2C9809&edit=0&save=0&query_type=WUpage&lowBandwidth=1&newestOnly=true&_newestOnly=on&showValueYears=true&_showValueYears=on&categoryIds=-1&_categoryIds=1&XAxis=VARIABLE&showSymbols=true&_showSymbols=on&_hideEmptyRowsColoumns=on&lang=en (accessed 1 October 2018).

FAO, IFAD, UNICEF, WFP and WHO (2018). *The State of Food Security and Nutrition in the World*. Rome: United Nations FAO.

Finlayson, C. (2014). *The Improbable Primate: How Water Shaped Human Evolution*. Oxford: Oxford University Press.

Global Water Partnership (2017). *Glossary*. https://www.gwp.org/en/Website-Information/Glossary/?az=i.

Grey, D. and Cooke, H.J. (1977). Some problems in the Quaternary evolution of the landforms of Northern Botswana. *CATENA* 4: 123–133.

Grey, D., Garrick, D., Blackmore, D. et al. (2013). Water security in one blue planet: twenty-first century policy challenges for science. *Philosophical Transactions of the Royal Society of London. Series A* 371 https://doi.org/10.1098/rsta.2012.0406.

Grey, D. and Sadoff, C. (2007). Sink or swim? Water security for growth and development. *Water Policy* 9: 545–571.

Griffiths, J.G. (1966). Hecataeus and Herodotus on 'A gift of the river'. *Journal of Near Eastern Studies* 25: 57–61.

Gulhati, N.D. (1973). *The Indus Waters Treaty: An Exercise in International Mediation*. New Delhi: Allied Publishers.

Haasnoot, M., Kwakkel, J.H., Walker, W.E., and Ter Maat, J. (2013). Dynamic adaptive policy pathways: a method for crafting robust decisions for a deeply uncertain world. *Global Environmental Change* 23: 485–498.

Haines, D. (2017). *Indus Divided. India, Pakistan and the River Basin Dispute*. Penguin Random House India.

Hall, J.W., Grey, D., Garrick, D. et al. (2014). Coping with the curse of freshwater availability. *Science* 346: 429–430.

He, D., Wu, R., Feng, Y. et al. (2014). China's transboundary waters: new paradigms for water and ecological security through applied ecology. *Journal of Applied Ecology* 51 (5): 1159–1168.

Jansen, M. (1989). Water supply and sewage disposal at Mohenjo-Daro. *World Archaeology* 21: 177–192.

Kaniewski, D., Guiot, J., and Van Campo, E. (2015). Drought and societal collapse 3200 years ago in the Eastern Mediterranean: a review. *Wiley Interdisciplinary Reviews: Climate Change* 6: 369–382.

Kelley, C.P., Mohtadi, S., Cane, M.A. et al. (2015). Climate change in the Fertile Crescent and implications of the recent Syrian drought. *Proceedings of the National Academy of Sciences of the United States of America* 112: 3241–3246.

Khan, H.F., Morzuch, B.J., and Brown, C.M. (2017). Water and growth: an econometric analysis of climate and policy impacts. *Water Resources Research* 53: 5124–5136.

Kim, C.-R. (2011). Flood impact on Honda spreads, Toyota slowly recovers. *Reuters*. https://www.reuters.com/article/us-autos-floods/flood-impact-on-honda-spreads-toyota-slowly-recovers-idUSTRE7A945020111110.

Kimenvi, M.S. and Mbaku, M.J. (2015). *Governing the Nile Basin*. Washington, DC: Brooking Institutional Press.

Kramer, S.N. (1981). *History Begins at Sumer: Thirty-Nine Firsts in Recorded History*. Philadelphia: University of Pennsylvania Press.

Kummu, M., de Moel, H., Ward, P.J., and Varis, O. (2011). How close do we live to water? A global analysis of population distance to freshwater bodies. *PLoS One* 6 https://doi.org/10.1371/journal.pone.0020578.

Lau, W. and Kim, K. (2012). The 2010 flood and Russian heatwave: teleconnection of hydrometeorological extremes. *Journal of Hydrometeorology* 13: 392–403.

Maslin, M. (2017). *The Cradle of Humanity*. Oxford: Oxford University Press.

McKinney, D.C. (2011). *Transboundary Water Challenges: Case Studies*. Technical Report. Tradecraft Class: PE –305 Environment, Science, Technology, and Health, Foreign Service Institute. Washington, DC: US Department of State.

Mithen, S. (2012). *Thirst: For Water and Power in the Ancient World*. Cambridge, MA: Harvard University Press.

Mukhtarov, F. and Gerlak, A.K. (2015). 'Ways of knowing' water: integrated water resources management and water security as complementary discourses. *International Environmental Agreements: Politics, Law and Economics* 15: 257–272.

Odusola, A. and Abidoye, B. (2015). Effects of Temperature and Rainfall Shocks on Economic Growth in Africa. *29th Triennial Conference of the International Association of Agricultural Economists*, Milan.

Ragab, R., Evans, J.G., Battilani, A., and Solimando, D. (2017). The cosmic-ray soil moisture observation system (cosmos) for estimating the crop water requirement: new approach. *Irrigation and Drainage* 66: 456–468.

Rahi, K.A. and Halihan, T. (2018). Salinity evolution of the Tigris River. *Regional Environmental Change* 18 (7): 2117–2127.

Relief Web (2011). *Pakistan Floods 2010: The DEC Real-Time Evaluation Report.* https://reliefweb.int/sites/reliefweb.int/files/resources/Full_Report_1482.pdf (accessed 3 October 2018).

Relief Web (2017). *Southern Africa: Food Security 2015–2017.* https://reliefweb.int/disaster/dr-2015-000137-mwi (accessed 3 October 2018).

Sadoff, C.W. and Grey, D. (2002). Beyond the river: the benefits of cooperation on international rivers. *Water Policy* 4: 389–403.

Sadoff, C.W. and Grey, D. (2005). Cooperation on international rivers. *Water International* 30: 420–427.

Sadoff, C.W., Hall, J.W., Grey, D. et al. (2015). *Securing Water, Sustaining Growth: Report of the GWP/OECD Task Force on Water Security and Sustinable Growth.* Oxford: University of Oxford.

Sadoff, C.W., Whittington, D., and Grey, D. (2003). Africa's international rivers: an economic perspective. In: *Directions in Development.* Washington, DC: World Bank.

Sarfraz, H. (2012). Revisiting the 1960 Indus Waters Treaty. *Water International* 38: 204–216.

Sen, A. (1999). *Development as Freedom.* Oxford: Oxford University Press.

Shultz, S. and Maslin, M. (2013). Early human speciation, brain expansion and dispersal influenced by African climate pulses. *PLoS One* 8 https://doi.org/10.1371/journal.pone.0076750.

UNDESA (2017). *World Population Prospects 2017.* https://population.un.org/wpp (accessed 1 July 2018).

UNDESA (2018). 68% of the world population projected to live in urban areas by 2050, says UN. https://www.un.org/development/desa/en/news/population/2018-revision-of-world-urbanization-prospects.html (accessed 1 July 2018).

UNECA ACPC (2011). *Working Paper 1: Climate Science, Information, and Services in Africa: Status, Gaps and Policy Implications.* Addis Ababa: UNECA.

UNICEF (2007). *Millions still affected by floods in South Asia.* Press Centre, UNICEF. https://www.unicef.org/media/media_40946.html (accessed 5 July 2018).

Vorosmarty, C.J., Green, P., Salisbury, J., and Lammers, R.B. (2000). Vulnerability from climate change and population growth. *Global Water Resources* 289: 284–288.

Wada, Y. and Heinrich, L. (2012). Assessment of transboundary aquifers of the world – vulnerability arising from human water use. *Environmental Research Letters* 8: 024003.

Waterbury, J. (2002). *The Nile Basin. National Determinants of Collective Action.* Syracuse, NY: Syracuse University Press.

Wheeler, M. (1953). *The Indus Civilization.* New York, NY: Cambridge University Press.

Wheeler, K.G., Hall, J.W., Abdo, G.M. et al. (2018). Exploring cooperative transboundary river management strategies for the Eastern Nile Basin. *Water Resources Research* 54 (11): 9224–9254.

WHO (2017). *2.1 billion people lack safe drinking water at home, more than twice as many lack safe sanitation.* https://www.who.int/news-room/detail/12-07-2017-2-1-billion-people-lack-safe-drinking-water-at-home-more-than-twice-as-many-lack-safe-sanitation (accessed 12 July 2018).

Wittfogel, K.A. (1975). *Oriental Despotism: A Comparative Study of Total Power.* New Haven, CT: Yale University Press.

World Bank (2006). *Ethiopia – Managing water resources to maximize sustainable growth: water resources assistant strategy.* http://documents.worldbank.org/curated/en/947671468030840247/Ethiopia-Managing-water-resources-to-maximize-sustainable-growth-water-resources-assistance-strategy (accessed 15 October 2018).

World Bank (2012a). *Thai flood 2011: rapid assessment for resilient recovery and reconstruction planning: Overview* (English). World Bank.

World Bank (2012b). *Thai Flood: Overview.* http://documents.worldbank.org/curated/en/677841468335414861/pdf/698220WP0v10P106011020120Box370022B.pdf (accessed 15 October 2018).

World Bank (2018a). *Access to electricity (% of population).* World Bank. https://data.worldbank.org/indicator/EG.ELC.ACCS.ZS?locations=7E (accessed 15 October 2018).

World Bank (2018b). *People practicing open defecation (% of population).* https://data.worldbank.org/indicator/SH.STA.ODFC.ZS (accessed 15 October 2018).

World Economic Forum (2018). *The Global Risks Report 2018.* World Economic Forum.

World Hunger Education Service (2018). *World Hunger and Poverty Facts and Statistics.* https://www.worldhunger.org/world-hunger-and-poverty-facts-and-statistics/ (accessed 15 October 2018).

World Meteorological Organization (2018). *El Nino/La Nina Update – March 2018.* https://public.wmo.int/en/media/press-release/el-ni%C3%B1o-la-ni%C3%B1a-update-march-2018 (accessed 15 October 2018).

Yancheva, G., Nowaczyk, N.R., Mingram, J. et al. (2007). Influence of the intertropical convergence zone on the East Asian monsoon. *Nature* 445: 74–77.

Zeitoun, M., Lankford, B., Krueger, T. et al. (2016). Reductionist and integrative research approaches to complex water security policy challenges. *Global Environmental Change* 39: 143–154.

Zhang, P., Cheng, H., Edwards, R.L. et al. (2008). A test of climate, sun, and culture relationships from an 1810-year Chinese cave record. *Science* 322: 940–942.

20

Charting the World's Water Future?
Simon J. Dadson[1], Edmund C. Penning-Rowsell[1,2], Dustin E. Garrick[3], Rob Hope[1,3], Jim W. Hall[4], and Jocelyne Hughes[1]

[1] *School of Geography and the Environment, University of Oxford, UK*
[2] *Flood Hazard Research Centre, Middlesex University, London, UK*
[3] *Smith School of Enterprise and the Environment, University of Oxford, UK*
[4] *Environmental Change Institute, University of Oxford, UK*

20.1   Linking Water Science, Policy, and Management

The chapters in this book have offered an interdisciplinary approach to the some of the world's greatest water challenges in the twenty-first century. A distinction between science, policy and management was established, to distil the necessary components that aligned to achieve the goal. Key external drivers of change include climate change and variability, land cover and land management including economics and politics, and the emergence of new technology to solve specific water-related problems. The interplay between science and technology, and the politics, economics and governance of the water sector are inextricable, and it is the interdisciplinarity of an integrated water education that provides the resilience to address these dimensions together. Each chapter in this volume has sought to survey the challenges in its field, and offered a perspective on how the subject might develop in the coming decade. The aim of this chapter is to draw out insights and observations from the individual pieces to highlight key themes that connect across chapters, or that arise persistently across contributions.

20.2   Charting the World's Water Future: Five Key Challenges

The discussions of drivers of change, in Chapter 1, show the water environment to be dynamic and complex, requiring careful analysis using rigorous methods. They also present a number of challenges to those seeking to intervene to improve the water resource environment across the globe, and almost every one of these challenges involves issues of science, policy and management. Every researcher and commentator will have different priorities for the challenges that are presented, but there is broad

Water Science, Policy, and Management: A Global Challenge, First Edition. Edited by Simon J. Dadson, Dustin E. Garrick, Edmund C. Penning-Rowsell, Jim W. Hall, Rob Hope, and Jocelyne Hughes.
© 2020 John Wiley & Sons Ltd. Published 2020 by John Wiley & Sons Ltd.

consensus globally, regionally and locally of the unprecedented nature of the challenges and increasing urgency to develop the capacity and knowledge to address them.

The *scientific challenges* involve understanding the future states of the water environment across the globe based on a careful analysis of the existing situations. Such scientific analysis and understanding can never be perfect, but efforts over the last several decades have shown considerable promise and we anticipate further progress in the future. Understanding the links between water quality and water quantity, the vital role of ecosystem function, and their links in turn to societies affected by these dimensions, is another scientific challenge requiring an interdisciplinary approach grounded in comprehensive theoretical understandings of the water cycle and its intricacies, and ecosystem processes. A further scientific challenge is understanding the nature of risk and extremes, separating these from annual or decadal variabilities. For example, for communities in many parts of the world it is vital for us to better understand the nature of monsoon conditions, and their variability in time and space. Similarly, scientific understanding of climatic change and variability is a prerequisite to managing the extremes that can result. Our modelling of these phenomena is improving all the time, but not such as to be reliably predictive, creating challenges of how to respond to situations that are both unpredictable and serious.

The next and related challenge is one of *uncertainty*. Many aspects of the water environment are uncertain, such as the nature and extent of flooding or droughts, or the viability of various responses, for example, processes in water treatment plants or sanitation facilities. Much is known, but uncertainties remain. Climate risk in the future is uncertain, and however good our scientific understanding of climatic processes, this uncertainty will always be with us. The political environment surrounding water interventions is similarly uncertain, with governments almost routinely changing their political complexions and attitudes towards public provision and private investment. Yet we have the challenge of acting now in the context of this uncertain tomorrow. We need to make decisions on a continuous basis, yet these decisions are inevitably based on uncertain information and imperfect forecasts. We may seek no-regret solutions, but these are not always feasible. Much water management requires substantial investment over long periods, meaning that opportunity costs can be substantial; resources allocated to water cannot be allocated to education or transport, yet such investments are often desperately needed in many parts of the world. The challenge is one of making decisions under conditions of uncertainty that are robust enough under a range of different future circumstances.

This raises the challenges of *effectiveness and efficiency*. Interventions require both considerable capital investment and long-term efforts, and it is important that these are effective in delivering the enhanced water services that many communities across the globe so desperately need. Given the pressures and drivers of change discussed above, the challenge is therefore not only how to improve the effectiveness of what we do now, but also the challenge of increasing the effectiveness of interventions in the future. In terms of efficiency, the challenge is one of getting 'more for less', or at least 'more for the same' unit of resource.

Many chapters suggest interventions to improve our scientific understanding, the policy implications and the management imperatives of tackling the water problems of the future. But each of these activities is expensive, and not all will be *affordable*, especially in countries and regions where poverty is endemic. The challenge here is, how can we and others finance plans to improve water resources development, water supply, sanitation, and flood and drought risk reduction? Political issues here concern the relative merits of investment in such services vis-a-vis other private- and public-sector investment opportunities and imperatives. Also important will be the split between private finance and public sector involvement, given the probability that public sector resources will be insufficient to meet every need. This, in turn, raises the question of political will and governance; the challenge here is one of getting support from those with power for the water sector investments that are necessary when these powerful constituencies have many other competing priorities. Although we write about water, we are not short-sighted enough to see the sector standing alone, necessarily commanding top priority. Indeed, we see that certain key water resource decisions can and should be successfully managed through better education, training and public engagement. We know the problems of obtaining the necessary high levels of investment; we also know that much can be achieved without investment but by better public awareness of the issues involved, and a more concerted effort to respond to the many other issues raised by the rapid urbanization of the world's population that we see today.

Our final challenge is one of *inclusivity*. As of today, many communities across the world suffer from poverty, and very often this poverty runs alongside water insecurity and low levels of investment. In many parts of the world economic and social inequalities are growing rather than diminishing, and the challenge here is one of planning the enhancement of our water services for all, rather than for the few. We note that women often bear the brunt of these failures, with high but avoidable time, health and welfare costs, particularly in rural Africa and Asia. This is a political as well as an economic challenge, and one that should be adequately addressed. The industrialized nations of the world have spent many billions of dollars improving their water services over the last several decades, and similarly enormous sums are needed in developing countries, both today and tomorrow, for the inequalities of water security to be tackled comprehensively. How the required finance and effort is obtained is perhaps the greatest challenge of all.

20.3 A Vision for Interdisciplinary Water Education

Our vision is to cultivate a generation of water leaders whose exposure to interdisciplinary thinking primes them to evaluate, reflect, design and adapt interventions, not only to seek well-tailored solutions to the problems they face, but to evaluate them critically, reflecting on the lessons contained in history to judge for themselves the merits or otherwise of plans laid before them as well as those that they themselves are mandated to devise.

What is needed is a diverse group of expert integrators who, whilst valuing the deep domain expertise, are capable of drawing on and engaging with best practice in each of those domains in order to construct solutions that bridge intelligently between individual disciplines, cognisant of the assumptions, simplifications and limitations imposed by the range of approaches in common use today. It is to that generation of water professionals that this book is aimed: a generation of water scientists, policy analysts and managers capable of critical analysis, appraisal and implementation of solutions to the challenges with which the water sector, and those other sectors closely connected to it, is faced.

The role of education amongst such a group is paramount. Education is a transformative experience – an intergenerational exchange which should do far more than reproduce the prejudices of the past. It provides the space for new ideas to emerge and challenge orthodoxies. Each chapter in this volume constitutes a form of inter-generational conversation between the faculty researchers and student co-authors. Our ethos is not to promote a particular world-view, nor to push a particular set of normative positions within each chapter. Rather, it is to guide the reader, scholar or practitioner through the landscape, offering perspectives on each topic's origins and intellectual history, current state of the art, and future prognosis for the coming decades or so ahead.

Index

Water Science, Policy, and Management: A Global Challenge, First Edition. Edited by Simon J. Dadson,
Dustin E. Garrick, Edmund C. Penning-Rowsell, Jim W. Hall, Rob Hope, and Jocelyne Hughes.
© 2020 John Wiley & Sons Ltd. Published 2020 by John Wiley & Sons Ltd.

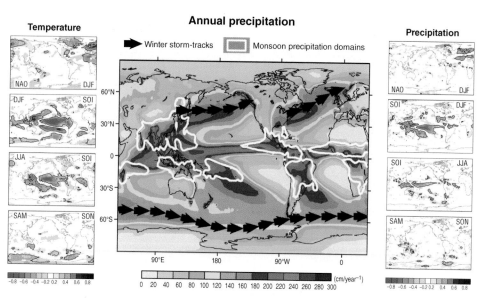

Figure 2.3 Global distribution of average annual rainfall from 1979 to 2010 with monsoon precipitation domain and winter storm-tracks shown with white contours and black arrows respectively. The left and right columns show the seasonal correlation maps of North Atlantic Oscillation (NAO), Southern Oscillation Index (SOI, the atmospheric component of El Niño – Southern Oscillation [ENSO]) and Southern Annular Mode (SAM) mode indexes compared with monthly temperature (precipitation) anomalies in boreal winter (December–February [DJF]), austral winter (June–August [JJA]) and austral spring (September–November [SON]). *Source*: Box 14.1, Figure 1 from Christensen et al (2013). Reproduced with permission of IPCC.

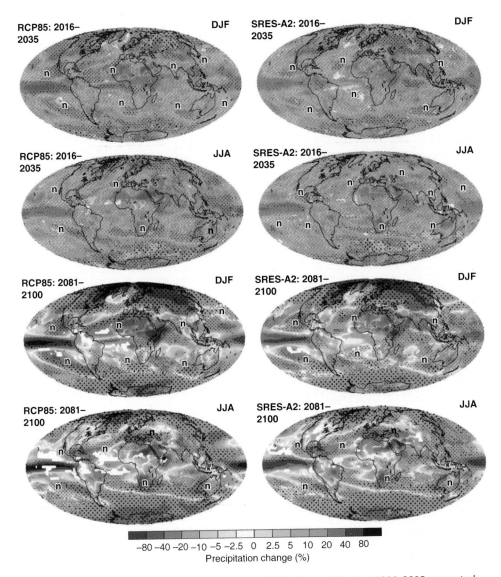

Figure 2.4 Spatial patterns of projected change in precipitation, relative to a 1986–2005 computed baseline, for the two seasons December–February (DJF) and June–August (JJA). Results from the CMIP3 experiments are shown on the left; results from the later CMIP5 experiments are on the right. Stippling indicates areas of model agreement, while hatching shows areas where projected change is within the range of natural variability. Areas with negative changes are labelled 'n' to distinguish them in monochrome. Source: Knutti and Sedláček (2013). Reproduced with permission of SpringerNature.

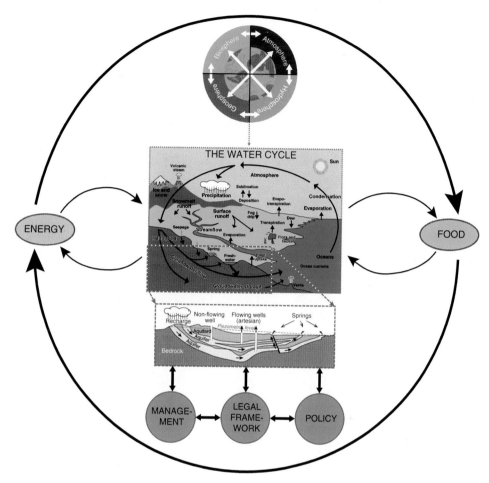

Figure 3.1 Conceptual diagram of groundwater within the water cycle, the Earth's spheres and the wider 'web of interdependencies'. *Source:* Abi Stone.

(a)

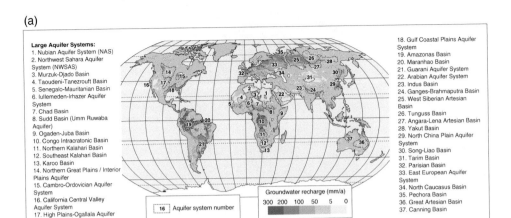

Large Aquifer Systems:
1. Nubian Aquifer System (NAS)
2. Northwest Sahara Aquifer System (NWSAS)
3. Murzuk-Djado Basin
4. Taoudeni-Tanezrouft Basin
5. Senegalo-Mauritanian Basin
6. Iullemeden-Irhazer Aquifer System
7. Chad Basin
8. Sudd Basin (Umm Ruwaba Aquifer)
9. Ogaden-Juba Basin
10. Congo Intracratonic Basin
11. Northern Kalahari Basin
12. Southeast Kalahari Basin
13. Karoo Basin
14. Northern Great Plains / Interior Plains Aquifer
15. Cambro-Ordovician Aquifer System
16. California Central Valley Aquifer System
17. High Plains-Ogallala Aquifer

18. Gulf Coastal Plains Aquifer System
19. Amazonas Basin
20. Maranhao Basin
21. Guarani Aquifer System
22. Arabian Aquifer System
23. Indus Basin
24. Ganges-Brahmaputra Basin
25. West Siberian Artesian Basin
26. Tunguss Basin
27. Angara-Lena Artesian Basin
28. Yakut Basin
29. North China Plain Aquifer System
30. Song-Liao Basin
31. Tarim Basin
32. Parisian Basin
33. East European Aquifer System
34. North Caucasus Basin
35. Pechora Basin
36. Great Artesian Basin
37. Canning Basin

Groundwater recharge (mm/a)
300 200 100 50 5 0

16 | Aquifer system number

(b)

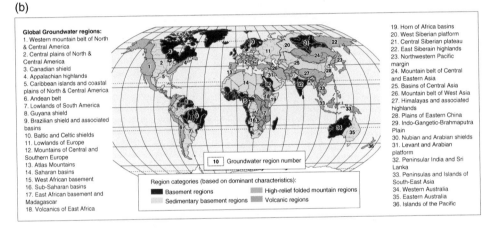

Global Groundwater regions:
1. Western mountain belt of North & Central America
2. Central plains of North & Central America
3. Canadian shield
4. Appalachian highlands
5. Caribbean islands and coastal plains of North & Central America
6. Andean belt
7. Lowlands of South America
8. Guyana shield
9. Brazilian shield and associated basins
10. Baltic and Celtic shields
11. Lowlands of Europe
12. Mountains of Central and Southern Europe
13. Atlas Mountains
14. Saharan basins
15. West African basement
16. Sub-Saharan basins
17. East African basement and Madagascar
18. Volcanics of East Africa

19. Horn of Africa basins
20. West Siberian platform
21. Central Siberian plateau
22. East Siberain highlands
23. Northwestern Pacific margin
24. Mountain belt of Central and Eastern Asia
25. Basins of Central Asia
26. Mountain belt of West Asia
27. Himalayas and associated highlands
28. Plains of Eastern China
29. Indo-Gangetic-Brahmaputra Plain
30. Nubian and Arabian shields
31. Levant and Arabian platform
32. Peninsular India and Sri Lanka
33. Peninsulas and Islands of South-East Asia
34. Western Australia
35. Eastern Australia
36. Islands of the Pacific

10 | Groundwater region number

Region categories (based on dominant characteristics):
■ Basement regions   ▨ High-relief folded mountain regions
▨ Sedimentary basement regions   ▨ Volcanic regions

(c)

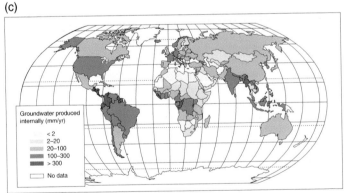

Groundwater produced internally (mm/yr)
< 2
2–20
20–100
100–300
> 300
No data

Figure 3.2 Global assessments of groundwater volumes, including (a) WHYMAP (2008) depicting the outline of the 37 'mega aquifers' and grey shading for recharge rates in the major groundwater basins. *Source:* see WHYMAP (2008) for a full version of the figure, depicting in three colours additional information about the recharge rate in two other groundwater types (areas with complex hydrogeological structure and areas with local and shallow aquifers) (reproduced in modified form with permission from WHYMAP); (b) the 36 global GW regions (IGRAC 2009), which are different from the 37 'mega aquifers'. *Source:* Margat and van der Gun (2013). Reproduced with permission from IGRAC; and (c) using country borders. *Source:* Authors using AQUASTAT data, FAO (2016).

(a)

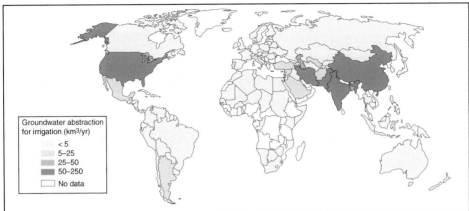

(b)

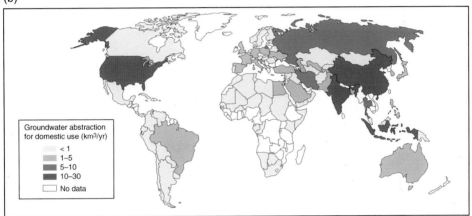

(c)

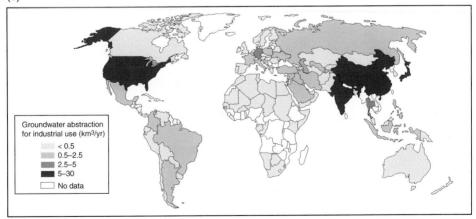

Figure 3.3 Estimates of current groundwater withdrawal as rates and as share of total freshwater withdrawal by country: withdrawal rates for (a) irrigation, (b) domestic supply, (c) industrial use; and share of groundwater in total freshwater in each country, (d) overall, and for (e) irrigation, (f) domestic supply and (g) industry. *Source:* Adapted from Margat and van der Gun © (2013). Reproduced by permission of Taylor and Francis Books UK.

(d)

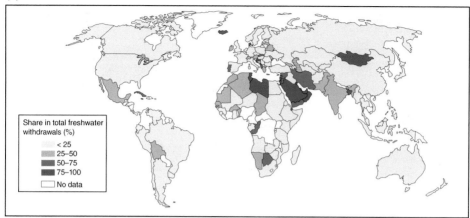

(e)

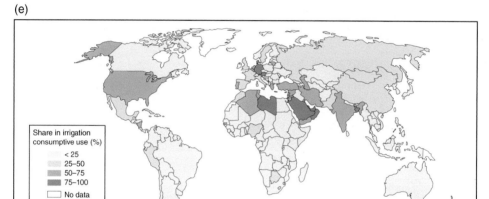

(f)

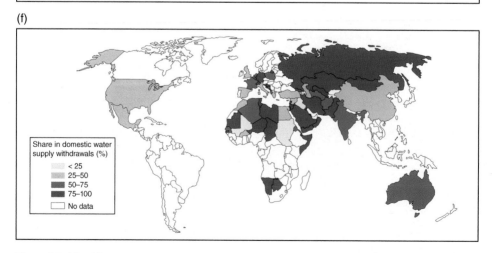

Figure 3.3 (*Cont'd*)

(g)

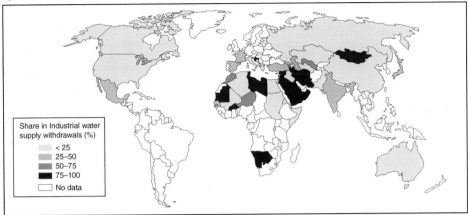

Share in Industrial water
supply withdrawals (%)

    < 25
    25–50
    50–75
    75–100
    No data

Figure 3.3 (*Cont'd*)

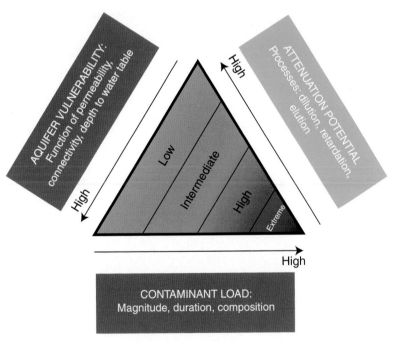

AQUIFER VULNERABILITY:
Function of permeability,
connectivity, depth to water table

High

ATTENUATION POTENTIAL
Processes: dilution, retardation,
elution

High

Low

Intermediate

High

Extreme

High

CONTAMINANT LOAD:
Magnitude, duration, composition

Figure 3.5 Schematic overview of pollution risk in groundwater. *Source:* Adapted from Morris et al. (2003).

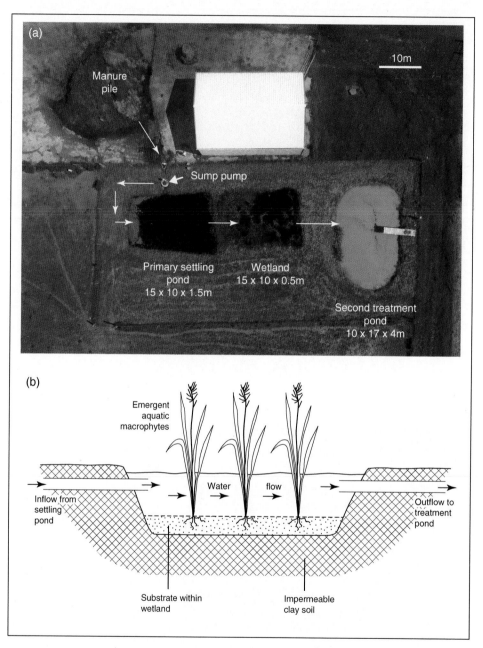

Figure 5.2 (a) Aerial view of the free water surface constructed wetland (FWSCW) system on Pemdale farm, Ontario, Canada, showing the manure pile, the settling pond, wetland, and treatment pond comprising the FWSCW; and direction of main runoff and flow pathways (white arrows). *Source:* courtesy of Larry Bond. (b) Schematic cross-section of the wetland within the FWSCW system showing the flow pathways (arrows), substrate and plants (not drawn to scale). *Source:* chapter authors.

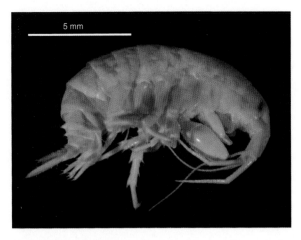

Figure 5.3 *Dikerogammarus haemobaphes* is native to the Ponto-Caspian region and has been recorded in the Thames catchment since 2012. This specimen measures 8 mm. *Source:* courtesy of J. Keble-Williams and J. Hughes.

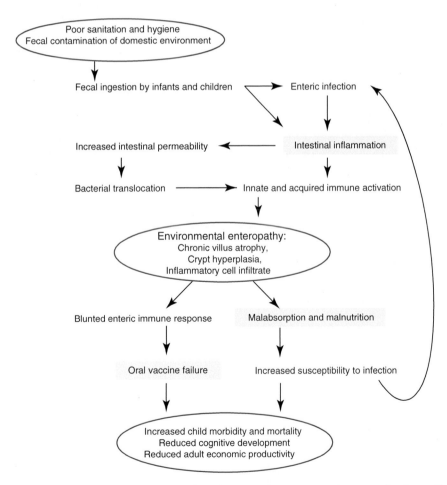

Figure 6.1 Model of the mechanism of development of environmental enteropathy. *Source:* Korpe and Petri (2012). Reproduced with permission of Elsevier Ltd.

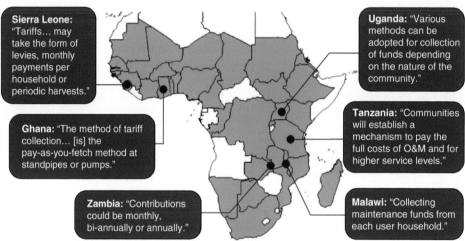

Country with rural water cost recovery policy or financing plan assuming O&M costs covered by household contributions

Sierra Leone: "Tariffs... may take the form of levies, monthly payments per household or periodic harvests."

Uganda: "Various methods can be adopted for collection of funds depending on the nature of the community."

Ghana: "The method of tariff collection... [is] the pay-as-you-fetch method at standpipes or pumps."

Tanzania: "Communities will establish a mechanism to pay the full costs of O&M and for higher service levels."

Zambia: "Contributions could be monthly, bi-annually or annually."

Malawi: "Collecting maintenance funds from each user household."

Figure 9.1 The legacy of community-based management in Africa. Based on information presented in Banerjee and Morella (2011) and WHO and UN-Water (2014). Banerjee and Morella (2011) listed countries with a rural water cost recovery strategy. WHO and UN-Water (2014) listed countries with a 'financing plan [which] defines if operating and basic maintenance is to be covered by tariffs or household contributions'. Quotes taken from the following *sources:* Malawi Ministry of Irrigation and Water Development (2010), Tanzania Ministry of Water and Livestock Development (2002), Zambia Ministry of Local Government and Housing (2007), Uganda Ministry of Water and Environment (2011), Sierra Leone Ministry of Water Resources (2013), Ghana Community Water & Sanitation Agency (2011).

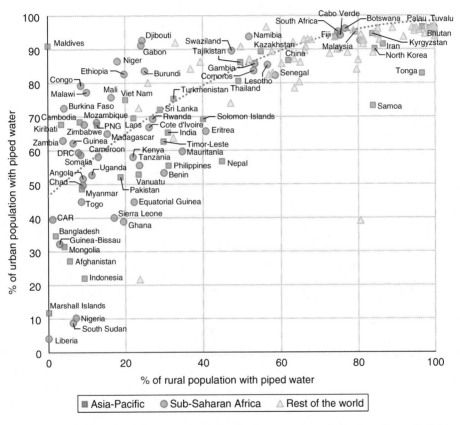

Figure 9.3 Piped water coverage as main drinking water source in rural vs. urban areas: sub-Saharan Africa, Asia, and the rest of the world in 2015. *Source:* Authors' analysis of data from UNICEF/WHO (2018).

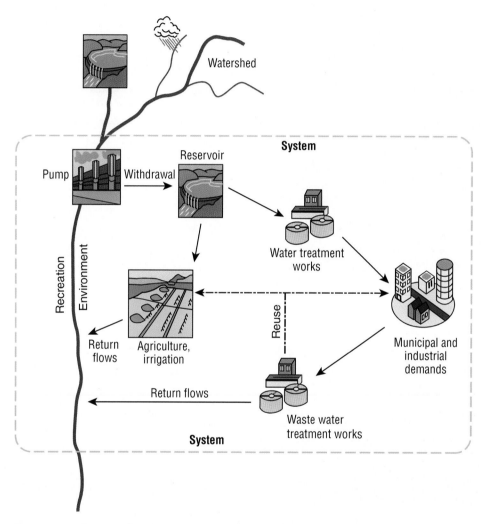

Figure 14.1 Schematic of the water resource system, including diversions for human uses and return flows. *Source:* Adapted from National University of Singapore.

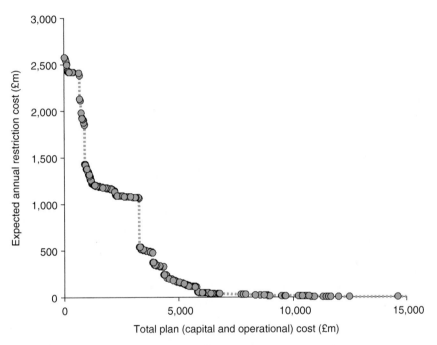

Figure 14.2 Example of presenting the trade-off between different objectives for a water resource system. This example shows the trade-off between the economic risk of water shortages (*y*-axis) and the cost of providing additional water supplies (*x*-axis). The economic risk of shortage is a quantification of the aggregate economic impact on water users due to restrictions on supply. Each dot on the graph shows a possible water resource management plan for this system. There are many other possible plans to the right of the curve (i.e. more costly plans that also incur higher risk), but the ones along the curve illustrate the inevitable trade-off between these two objectives. *Source:* Adapted from Borgomeo et al. (2018b). Reproduced with permission of John Wiley & Sons.

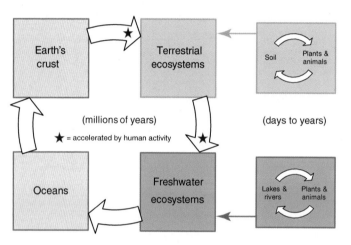

Figure 16.1 Global phosphorus cycle. *Source:* Authors.